Pro/ENGINEER Wildfire 4.0

IN SIMPLE STEPS

Pro/ENGINEER Wildfire 4.0

IN SIMPLE STEPS

Authored by:

Kogent Solutions Inc.

Published by:

dreamtech PRESS

ISBN: 978-81-7722-437-5

Edition: 2009

Printed at: Printed and Bound at Anand Book Binding House, Delhi – 31

CONTENTS

Chapter 1

Introducing Pro/ENGINEER Wildfire 4.0

In this Section

- Features of Pro/ENGINEER Wildfire 4.0
- Frequently-used Terms in Pro/ENGINEER Wildfire 4.0
- Installing Pro/ENGINEER Wildfire 4.0

Pro/ENGINEER Wildfire 4.0 is a 3 Dimensional (3D) Computer Aided Design (CAD) software package created by Parametric Technology Corporation (PTC). It is also called as Computer Aided Manufacturing (CAM) or Computer Aided Engineering (CAE) software package. It is a feature-based, parametric, and associative solid modeling software package, which helps a user to create 3D designs for mechanical engineering. It is used to create simple and complex designs faster with improved assembly performance, analyze the design, and automate the transformation of engineering designs into manufacturing processes. You can use it to design a vast range of products, such as soap cases, mobiles, trains. The product model, during the designing phase, passes through the following modes:

- **Sketch Mode:** Used to sketch geometry for a model that you want to design.
- **Part Mode:** Used to create features of a model.
- **Assembly Mode:** Used to create an assembly design. An assembly design is a design that lets you assemble various parts/features of a model.
- **Drawing Mode:** Used to generate the drawing of the parts and assemblies created in the part and assembly mode.

A model designed in Pro/ENGINEER Wildfire 4.0 incorporates the following modeling approaches:

- **Feature-based:** A model is made up of a combination of various features, such as extrusions, holes, cuts, rounds, sweeps, and slots. A feature is the smallest building block of a model.
- **Parametric:** A model consists of various features that are interrelated to each other and every feature consists of certain attributes. The parametric approach ensures that modification of an attribute of a feature affects the complete model. For example, modifying the diameter of a cylinder in an engine automatically modifies the diameter of the piston.
- **Associative:** A model design in `Pro/ENGINEER Wildfire 4.0` passes through various modes, such as **Sketch**, **Part**, and **Assembly**. The associative approach of `Pro/ENGINEER Wildfire 4.0` ensures that modifications made in one mode of a model should be reflected in other modes of the same model. For example, if you modify the length of a shaft in the drawing mode then modifications should also be reflected in other corresponding modes.
- **Solid Modeling:** The solid modeling approach ensures that the model must have properties of a solid object, such as volume and surface area. Solid modeling allows you to create models of physically tangible objects.

Let's now discuss some of the important features of Pro/ENGINEER Wildfire 4.0.

Features of Pro/ENGINEER Wildfire 4.0

In `Pro/ENGINEER Wildfire 4.0`, lots of changes and additions have been made to make it more user-friendly than its previous versions. Following are the features of `Pro/ENGINEER Wildfire 4.0` that make it popular among other `CAD/CAM` software, such as AutoCAD, GCAM

- **User-Interface:** This feature is divided into many parts, such as message area, navigation pane, browser window, drawing area, top tool chest, and right tool chest. Chapter 2, *Exploring the User Interface*, discusses user-interface in detail.
- **Auto Round:** This feature automatically rounds the selected edges of a part or an assembly. It saves the time spent in rounding edges manually.
- **Remove Surface:** This feature makes it easy to remove unwanted surfaces. This helps in modifying and simplifying the imported geometries.
- **Shell:** This feature helps you to create shells of various thicknesses for different edges of a part.
- **Dimension Attributes:** This feature appears automatically and allows you to modify dimension attributes of a 3D drawing.

- **View Manager:** This feature helps you to create extract assemblies in Pro/ENGINEER Wildfire 4.0. An extract assembly contains components selected by the user from the parent assembly. The changes made in the extract assembly are automatically reflected back in the original assembly. An extract assembly is used for concurrent cable designing. Apart from this, the **View Manager** also helps to view a 3D model in different orientations such as, RIGHT, LEFT.
- **Surface Manipulation:** This feature makes it easy to modify surfaces just by picking points on the surface. The surfaces can also be modified by just mouse dragging. This feature also allows you to merge surfaces easily by selecting multiple surfaces at a single point of time.
- **Rendering:** This feature lets you view the model in different lightning effects. For example, the **Skylight** feature, which is added newly in Pro/ENGINEER Wildfire 4.0, lets you visualize a model in sky light environment.
- **Electro-mechanical Components:** This feature lets you design electro-mechanical components. It also allows you to design ribbon cables.
- **Import Data Doctor:** This feature enables faster import and repair of geometry.
- **Feature Recognition Toolkit:** This feature enables you to import geometries of models. Apart from that, the `Feature Recognition Toolkit` automatically converts the imported geometry into Pro/ENGINEER Wildfire 4.0 features, such as holes, rounds, and chamfer.

Now, let's discuss some important terms that you must know while working in Pro/ENGINEER Wildfire 4.0.

Frequently-used Terms in Pro/ENGINEER Wildfire 4.0

Some of the commonly used terms of Pro/ENGINEER Wildfire 4.0 are as follows:

- **Entity:** An entity is defined as the basic element of a feature. An entity can be a line, an arc, a spline, or a circle.
- **Dimension:** A dimension in `Pro/ENGINEER Wildfire 4.0` is defined as the measurement of entities.

Note

> The measurement unit in Pro/ENGINEER Wildfire 4.0 is mm.

- **Parameter:** A parameter is defined as a numeric value that defines a feature, such as thickness. A parameter can be modified any time.
- **Relation:** A relation in `Pro/ENGINEER Wildfire 4.0` can be defined as an equation used to relate two entities.
- **Constraints:** In `Pro/ENGINEER Wildfire 4.0`, constraints are defined as logical operations that are performed on selected entities to make them accurate with respect to other entities. For example, the equal constraint in Pro/ENGINEER Wildfire 4.0 is used to make selected entities equal in size.
- **Weak Dimensions and Weak Constraints:** In `Pro/ENGINEER Wildfire 4.0`, weak dimensions and weak constraints appear automatically on entities in a sketch. They are considered as temporary dimensions and constraints. They appear in gray color.
- **Strong Dimensions and Strong Constraints:** The dimensions and constraints that are added manually to a sketch are called as strong dimensions and strong constraints. They appear in white color and are not deleted automatically.

Now, let's study the process of installing Pro/ENGINEER Wildfire 4.0 on Windows operating system.

Installing Pro/ENGINEER Wildfire 4.0

To design models with `Pro/ENGINEER Wildfire 4.0` software, it must be installed on your computer. Table 1.1 lists the basic system requirements for the installation of Pro/ENGINEER Wildfire 4.0:

Table 1.1: System Requirements to install Pro/ENGINEER Wildfire 4.0

Type of Requirement	Value
Operating System	Windows Vista Business Edition, Windows Vista Ultimate Edition, Windows XP Professional Edition, Windows XP Home Edition, or UNIX
Hard Disk	Minimum: 2.0 GB Recommended: 3.0 GB or higher
CPU Speed	Minimum: 500 MHZ Recommended: 2.4 GHz or higher
RAM	Minimum: 256 MB Recommended: 1024 MB or higher
SWAP Space	Minimum: 500 MB Recommended: 2048 MB or higher
Browser Support	Microsoft Internet Explorer 6.0 or higher
Monitor Resolution	1024 x 768 screen resolution with 24-bit color support; however, higher resolution with higher color support is recommended
Mouse	Microsoft approved 3-button mouse
Network	Microsoft approved Ethernet Network Adapter or Network Card

You can install the Pro/ENGINEER Wildfire 4.0 software on `License Server Machine` or on `License Client Machine` or on both. However, to install Pro/ENGINEER Wildfire 4.0 software either on `License Server Machine` or on `License Client Machine`, you have to install `FLEXnet Publisher` first. `FLEXnet Publisher` is a software license manager, which is a product of `Acresso Software`. The FLEXnet Publisher is used in corporate environment to provide floating licenses to users attached with `License Server Machine`. If you install the Pro/ENGINEER Wildfire 4.0 software only on the License Client Machine, the License Client Machine works as both server and client.

Note

A floating license allows multiple users to share an application installed on server machine.

In this section, we see the process to install the Pro/ENGINEER Wildfire 4.0 software on License Client Machine on Windows Vista Ultimate Edition. Let's now follow these steps to install Pro/ENGINEER Wildfire 4.0:

1. *Insert* the DVD containing the `Pro/ENGINEER Wildfire 4.0` software into the DVD ROM of your computer. The **AutoPlay** window appears, as shown in Fig.Pro/E-1.1.

Note

If AutoPlay window does not appear, then access it from your computer's DVD drive.

2. Now, *click* the **Open** folder to view files folder icon in **General options** section to explore the Pro/ENGINEER Wildfire 4.0 software component, as shown in Fig.Pro/E-1.1:

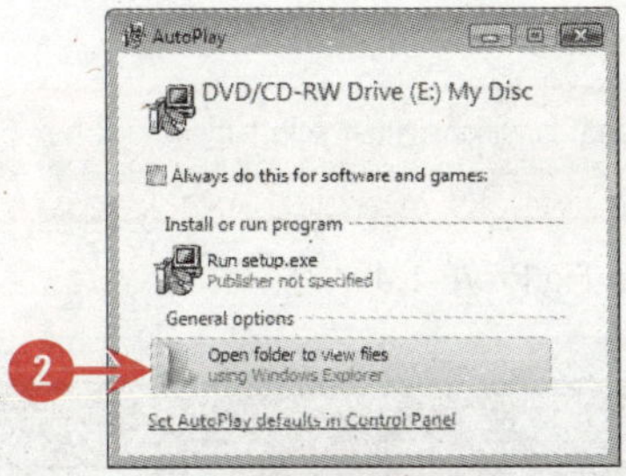

Fig.Pro/E-1.1

The Pro/ENGINEER Wildfire 4.0 software component appears.

3. *Copy* the **license.dat** file anywhere in your computer. The file is required to be copied because later on, if you update the Pro/ENGINEER Wildfire 4.0 software with the latest version then you will also have to update the **license.dat** file.
4. Now, *double-click* the **PTC.Setup** installation utility to start the process of installing the `Pro/ENGINEER Wildfire 4.0` software. The PTC welcome screen appears, as shown in Fig.Pro/E-1.2:

Fig.Pro/E-1.2

The welcome screen disappears within few seconds and the Pro/ENGINEER Wildfire 4.0 set up window appears, as shown in Fig.Pro/E-1.3.

5. *Click* the **Next** button, as shown in Fig.Pro/E-1.3:

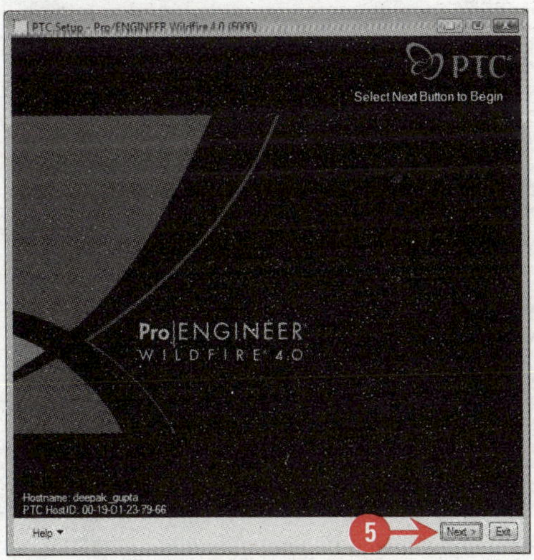

Fig.Pro/E-1.3

The license agreement window appears (Fig.Pro/E-1.4).

6. *Select* the check box beside `I Accept the Agreement Terms and Conditions` label to accept the software agreement terms and conditions.

Note

> If you do not agree with the software terms and conditions, then select the check box beside I Decline the Agreement Terms and Conditions to quit installation.

7. *Click* the **Next** button, as shown in Fig.Pro/E-1.4:

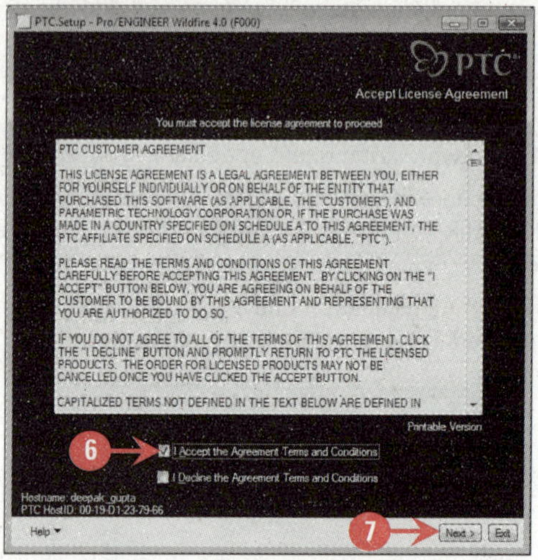

Fig.Pro/E-1.4

The **Select Product to be installed** window appears (Fig.Pro/E-1.5).

8. *Click* the **Pro/ENGINEER & Pro/ENGINEER Mechanica** link, as shown in Fig.Pro/E-1.5:

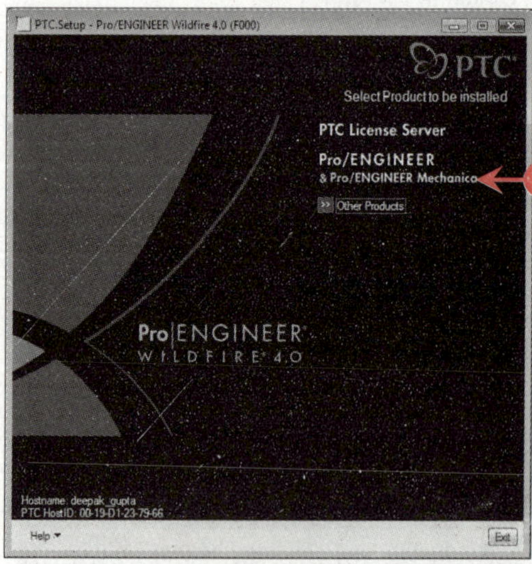

Fig.Pro/E-1.5

The **Define Installation Components** window appears, as shown in Fig.Pro/E-1.6:

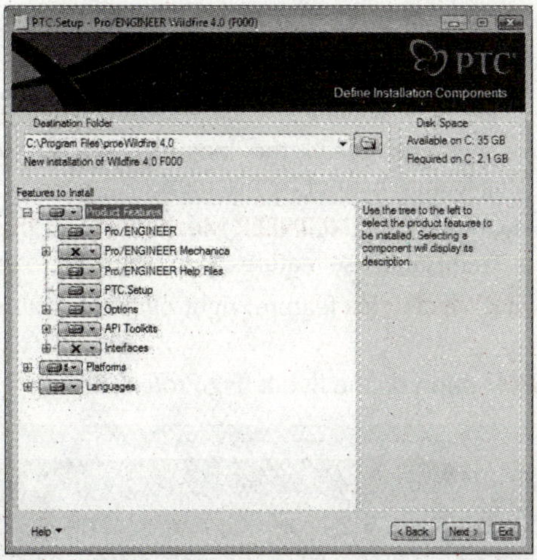

Fig.Pro/E-1.6

On this window, you see the following options:

- **Destination Folder:** It is the folder where the `Pro/ENGINEER Wildfire 4.0` software is installed. By default, the destination folder is created in the **Program Files** folder. However, you can change the location of the destination folder by clicking the folder icon shown in Fig.Pro/E-1.6. If you change the destination folder, then you must take care that the folder in which the software is to be installed must be empty or newly created.
- **Disk Space:** This option informs you about the total space available on the hard disk and the required space to install the software.
- **Features to Install:** This option lets you select the features and sub-features of the software that you want to install. It contains following main components:
  - **Product Features:** Contains the following features that can be installed:
    - **Pro/ENGINEER:** Installs the files required to run the Pro/ENGINEER software.
    - **Pro/ENGINEER Mechanica:** Helps an engineer to understand the product performance in a better way. It also helps in producing good quality products at low cost.

Note

> By default, this feature is disabled. The process to enable this feature is discussed ahead in this section.

- **Pro/ENGINEER Help Files:** Installs the Pro/ENGINEER help files in the destination folder.
- **PTC Setup:** Used to reconfigure the software after installation without executing the **PTC.Setup** utility.

Note

> To reconfigure the software configuration after installation, go to the Destination folder→bin and execute the ptcsetup file. The ptcsetup is a batch file that needs administrator privileges to execute.

- **Options:** Installs **ModelCHECK** configurator, which is used to validate the completeness of 3D drawings, and **Mold Component Catalog**, which is used for mold tooling.
- **API Toolkits:** Installs the files required to run the **Application Program Interface** toolkit, such as **Visual Basic API**, **Pro/TOOLKIT**.

- **Interfaces:** Installs interface tools for third party software, such as **Pro/INTERFACE for CATIA** (Computer Aided Three Dimensional Interactive Application), **Pro/INTERFACE for JT**.

- **Platforms:** Selects your system's default platform to install Pro/ENGINEER Wildfire 4.0 software.
- **Languages:** Lets you select and install the language for the software. By default, the English language is installed. This feature cannot be disabled.

The **Red Cross** mark beside the features, **Pro/ENGINEER Mechanica** and **Interfaces** (Fig.Pro/E-1.6) implies that they are disabled to be installed. To install these features, you first need to enable them.

9. To enable the **Pro/ENGINEER Mechanica** feature, right click the feature. A flyout appears, as shown in Fig.Pro/E-1.7.
10. *Select* **Install all sub-features** option on the flyout (Fig.Pro/E-1.7):

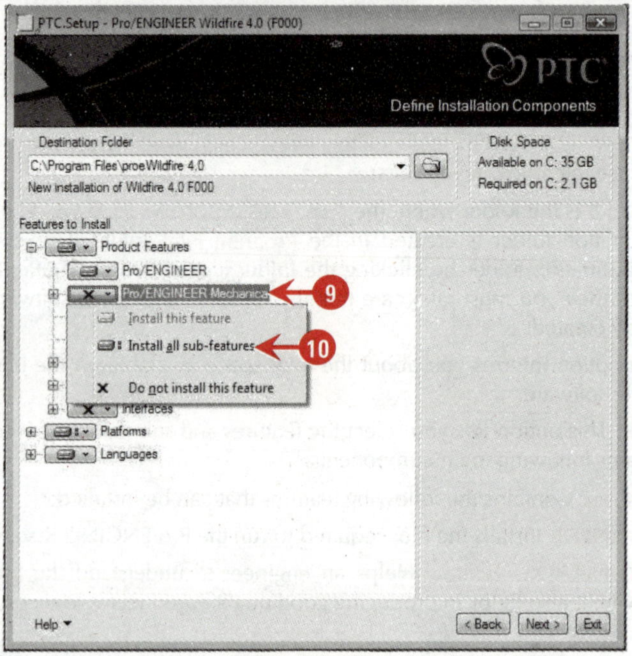

Fig.Pro/E-1.7

11. *Repeat* steps 9 and 10 to enable the **Interfaces** feature. Both the features along with their sub-features are now enabled (Fig.Pro/E-1.8).
12. *Click* the **Next** button (Fig.Pro/E-1.8) to continue the process, as shown in Fig.Pro/E-1.8:

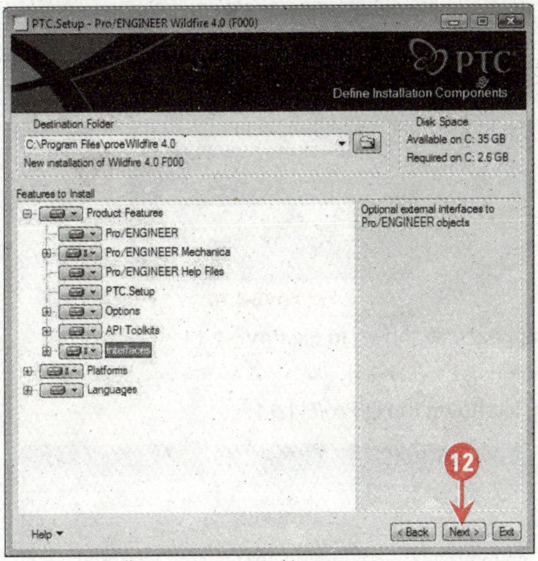

Fig.Pro/E-1.8

The FLEXnet license servers window appears, as shown in Fig.Pro/E-1.9:

13. Now, to add the license.dat file, *click* the Add button in the FLEXnet license servers window (Fig.Pro/E-1.9):

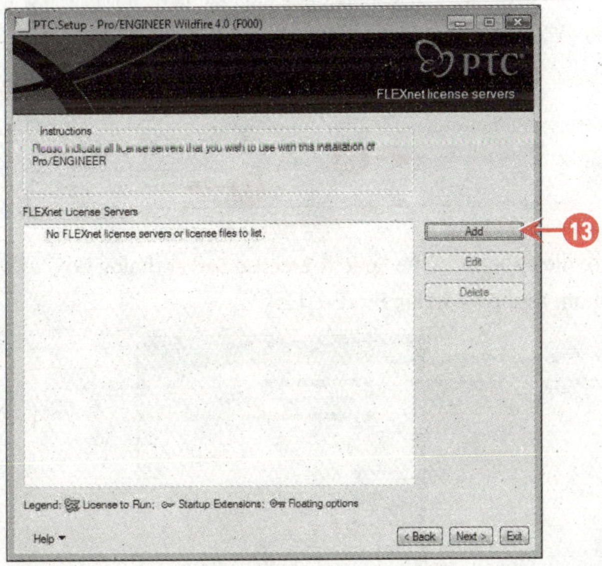

Fig.Pro/E-1.9

The Specify License Server dialog box appears, as shown in Fig.Pro/E-1.10.

14. *Select* the radio button beside the Locked license file (no server running) label (Fig.Pro/E-1.10).

15. *Click* the yellow colored folder icon, as shown in Fig.Pro/E-1.10:

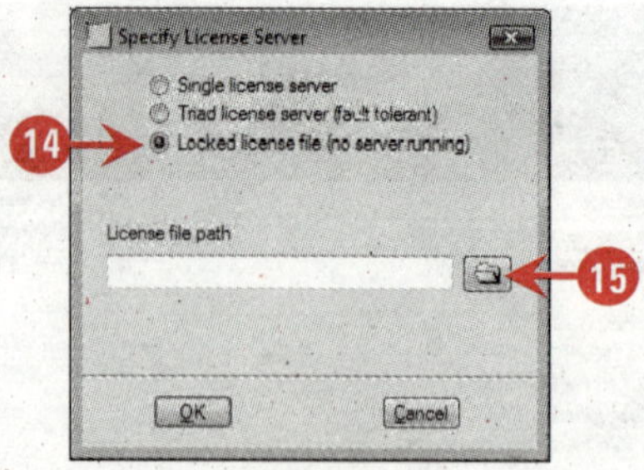

Fig.Pro/E-1.10

The **Select File** dialog box appears, as shown in Fig.Pro/E-1.11.

16. *Select* the **license.dat** file (Fig.Pro/E-1.11).
17. *Click* the **Open** button, as shown in Fig.Pro/E-1.11:

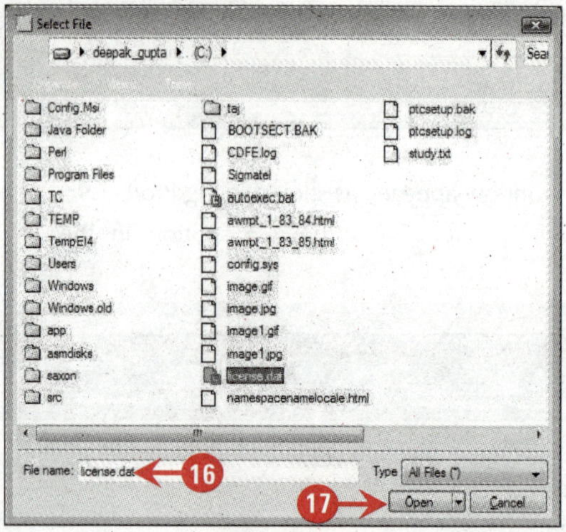

Fig.Pro/E-1.11

The path of the license file appears in the **Specify License Server** dialog box, as shown in Fig.Pro/E-1.12.

18. *Click* the **OK** button, as shown in Fig.Pro/E-1.12:

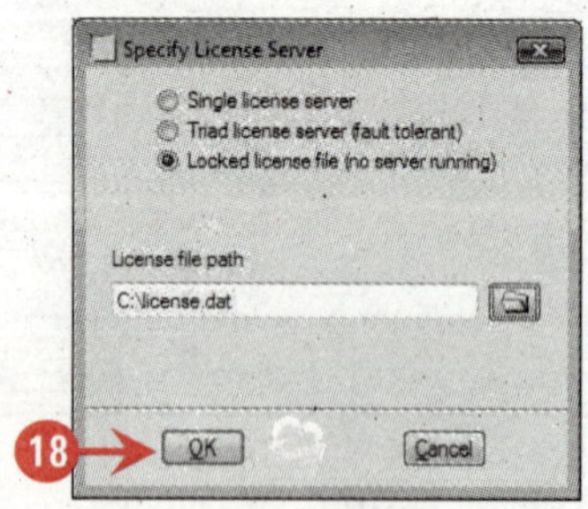

Fig.Pro/E-1.12

The **license.dat** file is added in the **FLEXnet server** window, as shown in Fig.Pro/E-1.13.

19. *Click* the **Next** button, as shown in Fig.Pro/E-1.13:

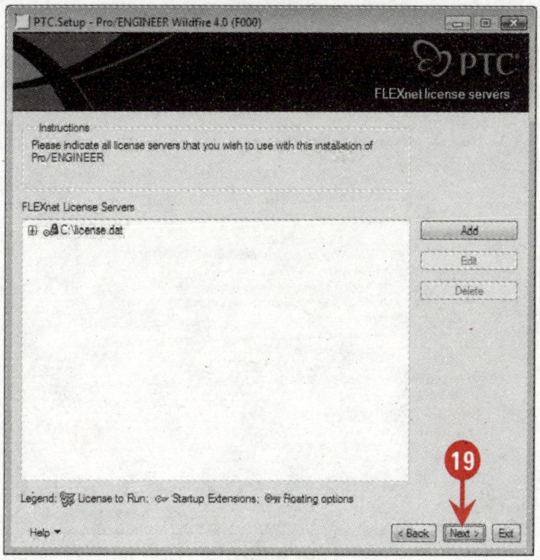

Fig.Pro/E-1.13

The Pro/ENGINEER Wildfire 4.0 window, which helps you create shortcut, appears as shown in Fig.Pro/E-1.14.

By default, the check box beside the **Program folder** label is selected. If you want to create shortcut icons in the start menu and on desktop, select the check boxes beside the **Desktop** and **Start menu** labels. In our case, we are continuing with the default settings.

20. *Click* the **Next** button, as shown in Fig.Pro/E-1.14:

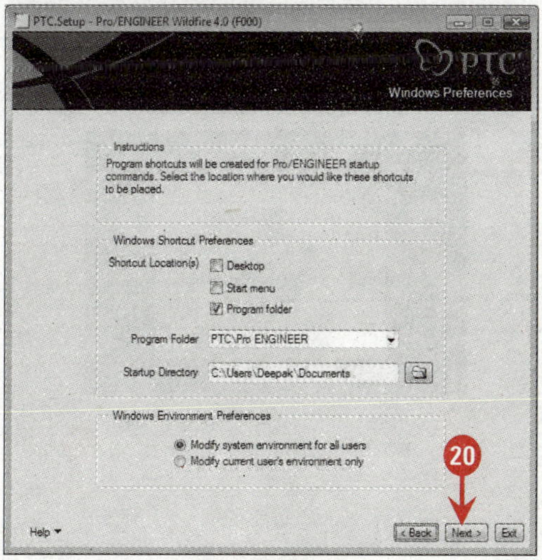

Fig.Pro/E-1.14

The Pro/ENGINEER Wildfire 4.0 window to choose additional configuration options appears, as shown in Fig.Pro/E-1.15.

21. *Click* the **Install** button (Fig.Pro/E-1.15) to continue the installation process:

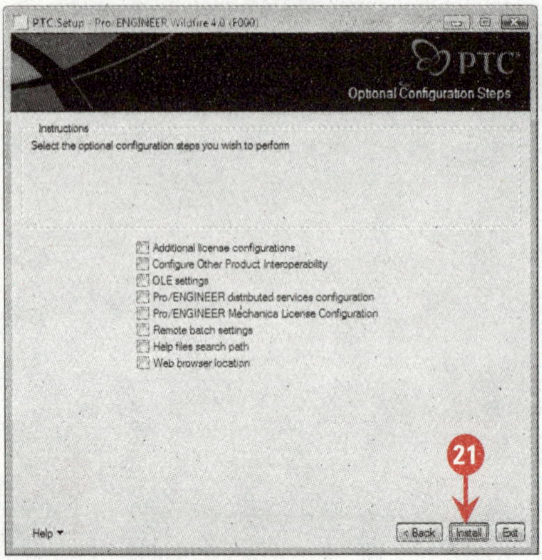

Fig.Pro/E-1.15

In order to configure any additional option, select the check box beside that option. In our case, we are not selecting additional options.

The Installation Progress window appears, as shown in Fig.Pro/E-1.16.

Buttons on Fig.Pro/E-1.16 are disabled at this moment. These buttons get enabled when the installation progress completes.

22. *Click* the **Next** button, as shown in Fig.Pro/E-1.16:

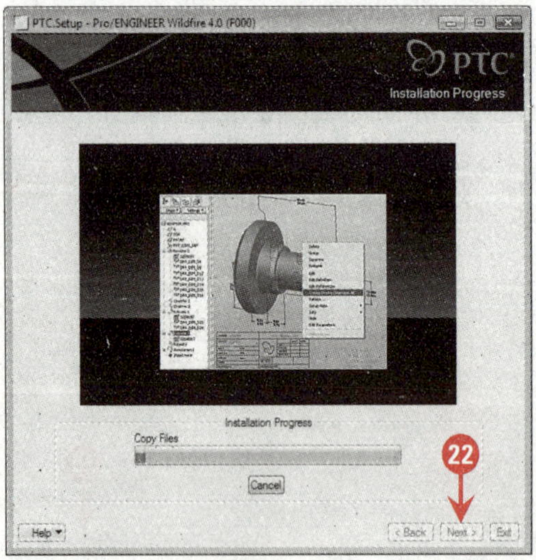

Fig.Pro/E-1.16

The Pro/ENGINEER Wildfire 4.0 is now installed on your system. The Pro/ENGINEER Wildfire 4.0 exit window appears, as shown in Fig.Pro/E-1.17.

23. *Click* the **Exit** button (Fig.Pro/E-1.17) to exit from the installation wizard:

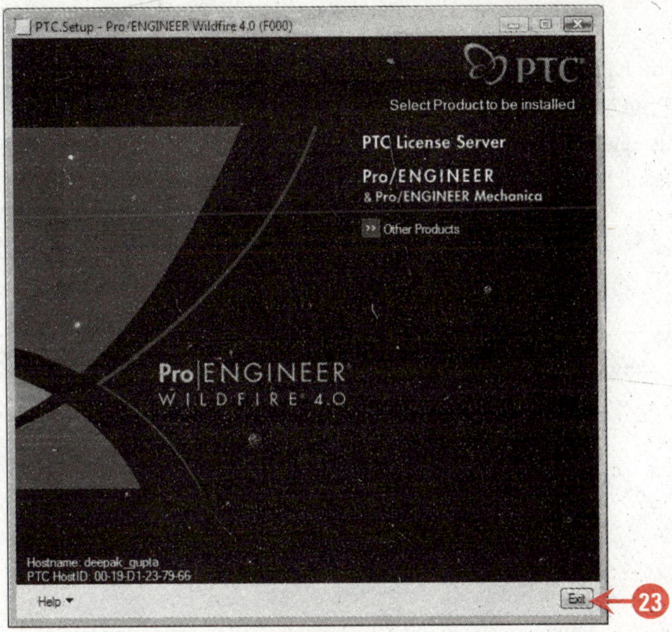

Fig.Pro/E-1.17

With this, the process to install Pro/ENGINEER Wildfire 4.0 is completed.

24. *Click* the Start→All Programs→PTC→Pro ENGINEER→Pro ENGINEER option. The Pro/ENGINEER Wildfire 4.0 initial screen appears, as shown in Fig.Pro/E-1.18:

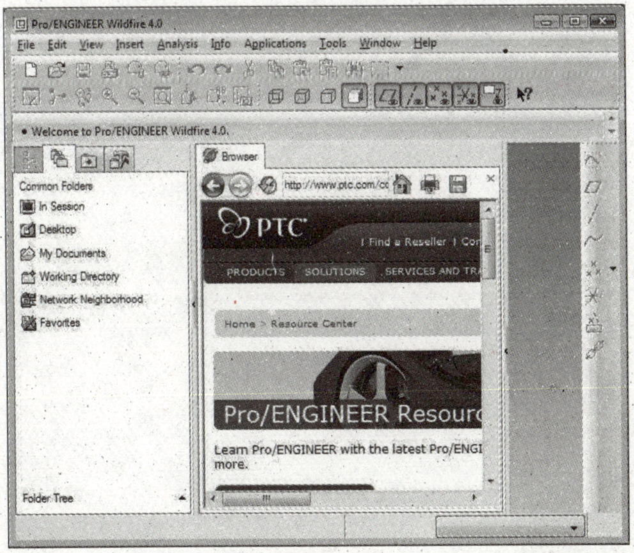

Fig.Pro/E-1.18

Now, you can start working with Pro/ENGINEER Wildfire 4.0. Let's now summarize the main points covered in this chapter.

Summary

In this chapter, you have learned about:

- Features of Pro/ENGINEER Wildfire 4.0
- Frequently used terms in Pro/ENGINEER Wildfire 4.0
- Process to install Pro/ENGINEER Wildfire 4.0 on client machine

Chapter 2

Exploring the User Interface

In this Section

- Starting Pro/ENGINEER Wildfire 4.0
- Performing File Management

Interface is a medium that allows interaction between a program and a user. To understand an interface, consider a scenario where you need to operate a television set.

To operate a television set, you need a remote control to perform various functions, such as changing the channels, controlling the volume, changing the display settings, and so on. In this case, remote control acts as an interface between you and the television set. Similarly, in computing, an interface is required for the interaction between the user and the software that is commonly known as User Interface (UI). The UI of the Pro/ENGINEER Wildfire 4.0 also acts as a medium between users and the software to help users in designing the real-world products such as cars, cell-phones, automobile parts. In order to work with a software, it is important to understand the fundamentals of its UI. Therefore, let's start by discussing the UI of Pro/ENGINEER Wildfire 4.0 in this chapter.

This chapter discusses the use of each menu on the menu bar of Pro/ENGINEER Wildfire 4.0 and proceeds with discussing about various toolbars. The chapter also discusses about managing files in Pro/ENGINEER Wildfire 4.0. Let's now see how to start Pro/ENGINEER Wildfire 4.0.

Starting Pro/ENGINEER Wildfire 4.0

Once you have installed Pro/ENGINEER Wildfire 4.0 successfully on your system, open it by *clicking* Start→All Programs→PTC→Pro ENGINEER→Pro ENGINEER, as shown in Fig.Pro/E-2.1:

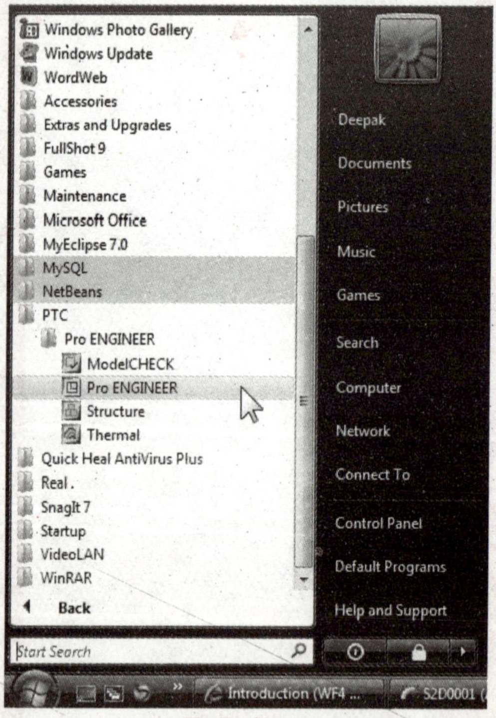

Fig.Pro/E-2.1

Note

> The process of installing the Pro/ENGINEER Wildfire 4.0 has already been discussed in Chapter 1, Introducing Pro/ENGINEER Wildfire 4.0.

The Pro/ENGINEER Wildfire 4.0 on start up displays a screen, as shown in Fig.Pro/E-2.2:

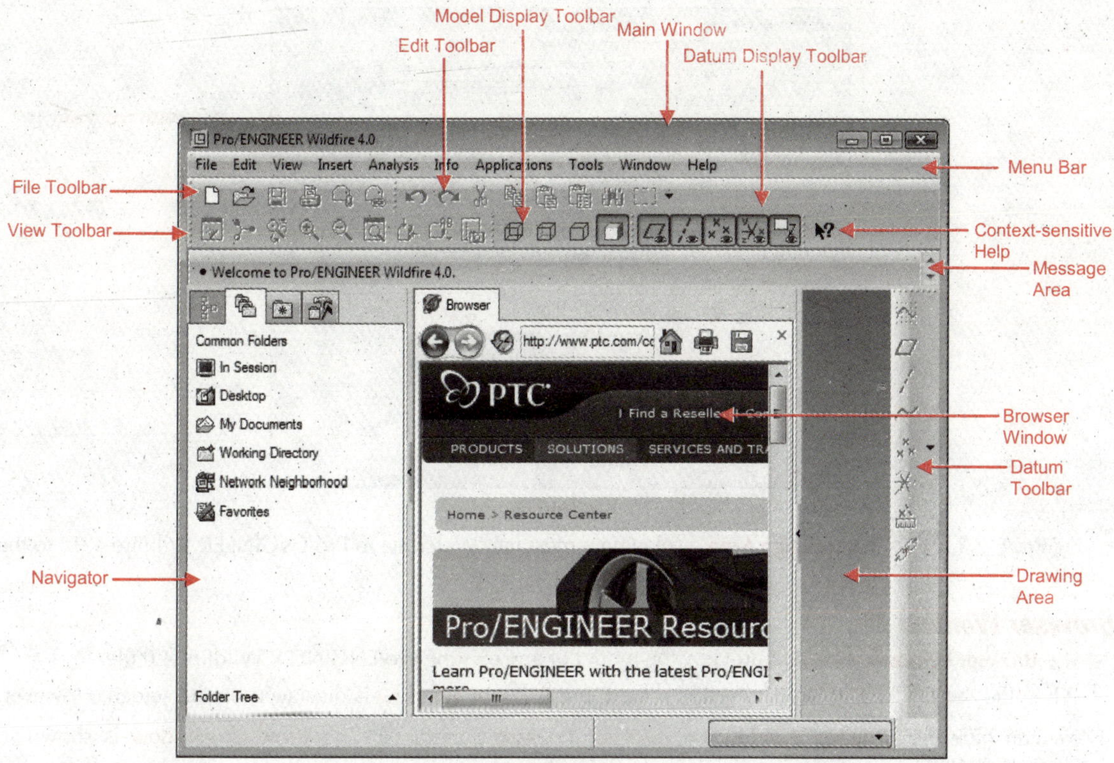

Fig.Pro/E-2.2

Fig.Pro/E-2.2 shows various windows, menu bar, and toolbars in the start-up screen of the Pro/ENGINEER Wildfire 4.0. When you start the Pro/ENGINEER Wildfire 4.0, the window that appears is known as the Main Window. All other windows are located within the Main Window.

Following are the components that are located within the Main Window of the Pro/ENGINEER Wildfire 4.0:

- Message Area
- Browser Window
- Drawing Area
- Navigator
- Menu Bar
- Toolbars

Message Area

Message Area displays messages regarding the current status of the operation performed. The Message Area also displays information to complete an operation. The Message Area is shown in Fig.Pro/E-2.3:

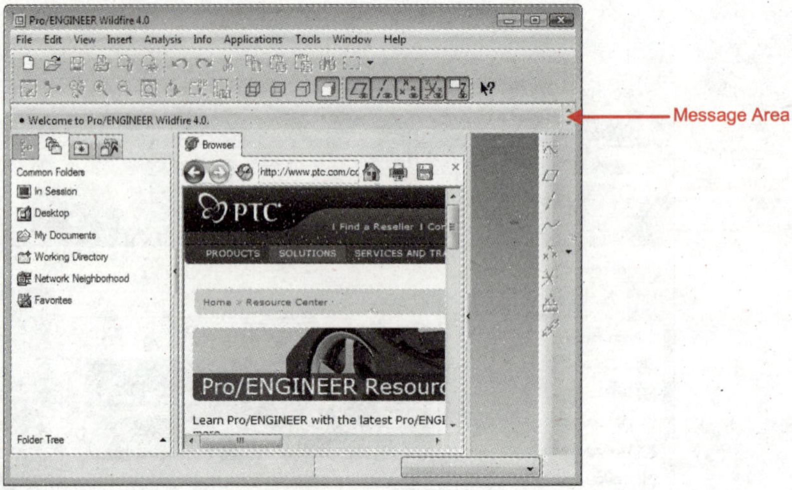

Fig.Pro/E-2.3

Fig.Pro/E-2.3 shows the Message Area displaying a message 'Welcome to Pro/ENGINEER Wildfire 4.0.' to the user.

Browser Window

The Browser Window is used to preview the model of any existing Pro/ENGINEER Wildfire 4.0 file.

When the Pro/ENGINEER Wildfire 4.0 is started, the Browser Window is displayed on the screen. However, we can hide the Browser Window by *clicking* the sash control ( ). The Browser Window is shown in Fig.Pro/E-2.4:

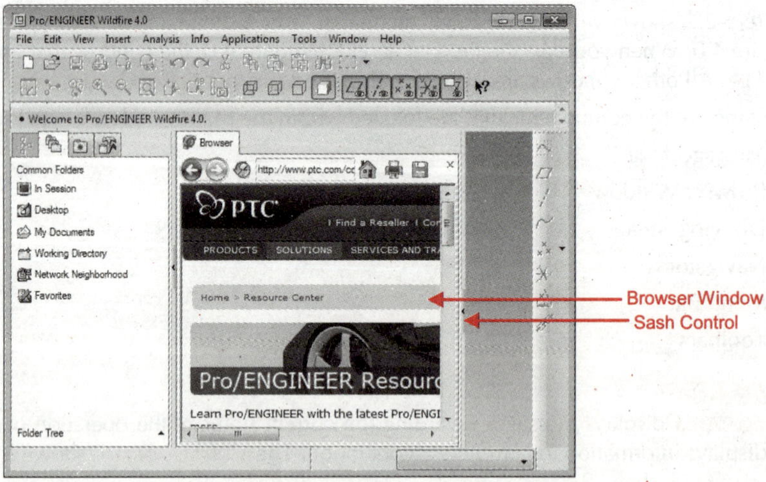

Fig.Pro/E-2.4

Drawing Area

Drawing Area is the screen where a Pro/ENGINEER Wildfire 4.0 model is created. The sketching of various entities takes place in the Drawing Area. The size of the Drawing Area can be adjusted using the sash controls located on the edges of the Navigator and the Browser Window. The Drawing Area is shown in Fig.Pro/E-2.5:

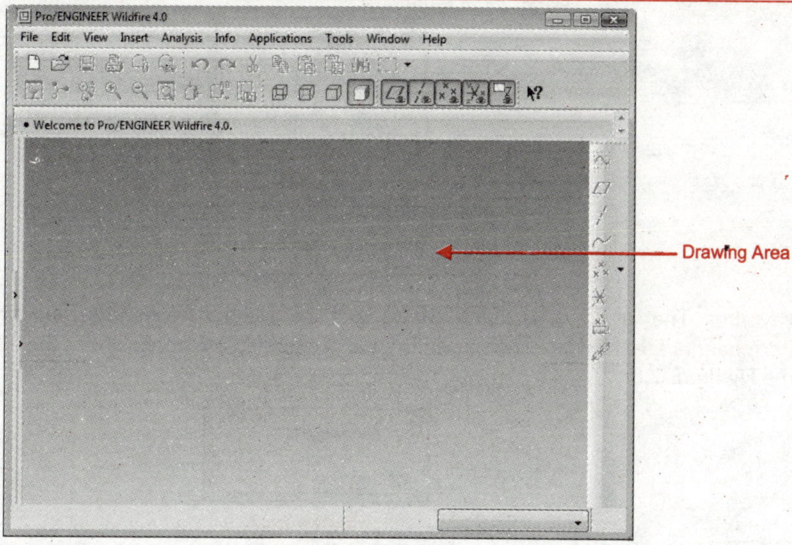

Fig.Pro/E-2.5

Navigator

Navigator is located to the left of the **Drawing Area** and contains multiple tabs for performing functions, such as browsing the Pro/ENGINEER Wildfire 4.0 files, displaying information about a specific model. Fig.Pro/E-2.6 shows the multiple tabs available in the **Navigator:**

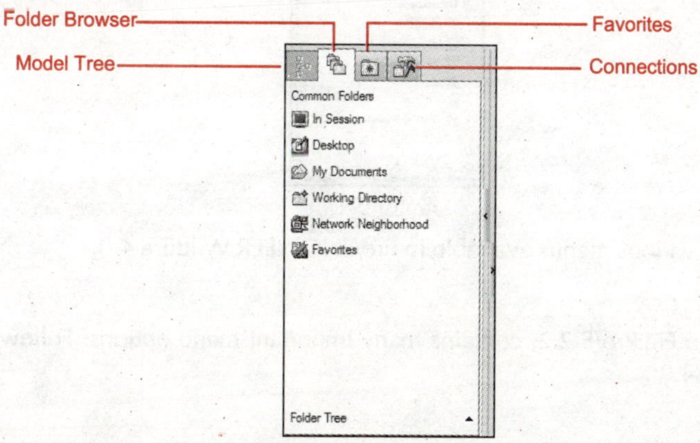

Fig.Pro/E-2.6

The tab options shown in Fig.Pro/E-2.6 are described in the following section:

- **Model Tree:** The **Model Tree** tab is used for manipulating the features in Pro/ENGINEER Wildfire 4.0 Part files. The manipulation includes deleting, suppressing, renaming, and editing the features, such as cut, hole, shell. However, the **Model Tree** also allows the user to manipulate assembly components in Assembly files. Manipulating features in Part files is discussed in Chapter 4, Exploring Pro/ENGINEER Wildfire 4.0 Part Mode.
- **Folder Browser:** The **Folder Browser** tab allows the user to browse and open folders located in the local system. In addition to these functions, the **Folder Browser** also helps in creating and setting a folder as the working directory. *Click* the **Folder Browser** tab located on the **Navigator** and a list of folders gets displayed, as shown in Fig.Pro/E-2.7:

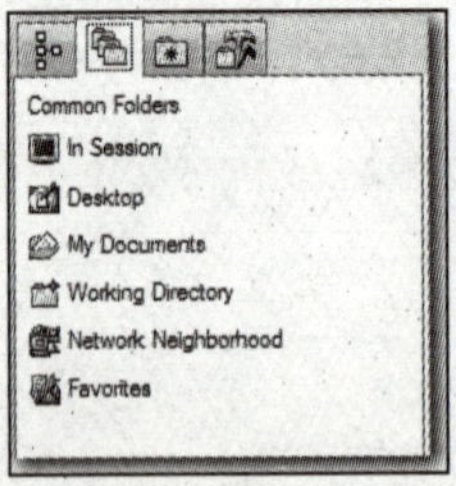

Fig.Pro/E-2.7

- Favorites: The Favorites tab allows user to save the frequently used directory folders under the Personal Favorites folder. The Personal Favorites folder is located under the Favorites tab, as shown in the Fig.Pro/E-2.8:

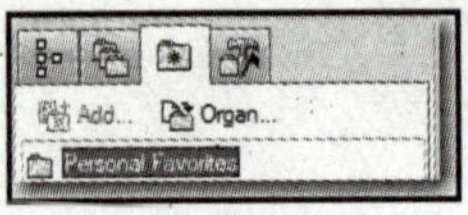

Fig.Pro/E-2.8

- Connections: The Connections tab is used to access the Parametric Technology Corporation (PTC) services and other websites within the environment of Pro/ENGINEER Wildfire 4.0. *Click* the Connections tab to select a list of options, such as Browser, 3DModelSpace, User Area, User Group, PTC.com, and Customer Support, as shown in the Fig.Pro/E-2.9:

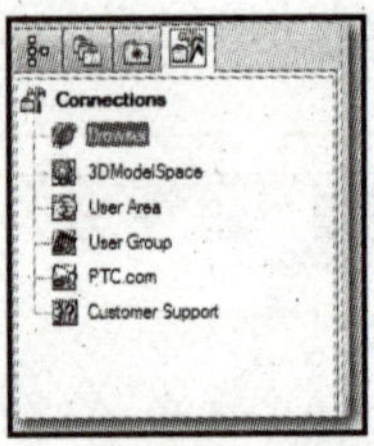

Fig.Pro/E-2.9

Let's now discuss about various menus available in Pro/ENGINEER Wildfire 4.0.

Menu Bar

The Menu Bar, shown in Fig.Pro/E-2.2, contains many important menu options. Following are the menus of Pro/ENGINEER Wildfire 4.0:

- The File menu
- The Edit menu
- The View menu
- The Insert menu
- The Analysis menu
- The Info menu
- The Applications menu
- The Tools menu
- The Window menu
- The Help menu

The File Menu

The File menu contains several options used for manipulating the files. To access options available in the File menu, *select* File from the Menu Bar and a list of options gets displayed, as shown in Fig.Pro/E-2.10:

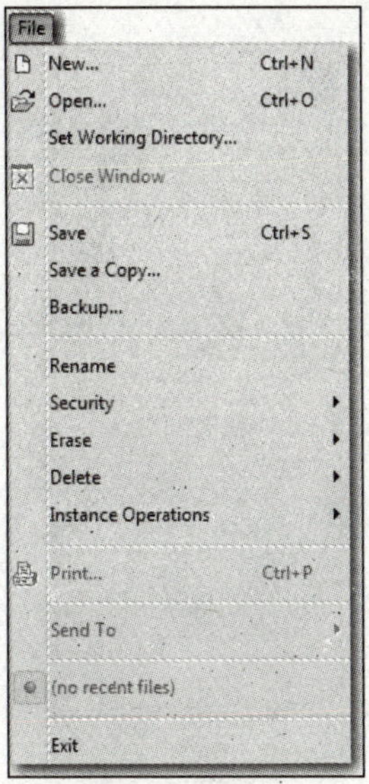

Fig.Pro/E-2.10

Some of the commonly used File menu options are as follows:

- New: Creates a new file, such as Part file, Assembly file, and Drawing file.
- Open: Opens an existing Pro/ENGINEER Wildfire 4.0 file saved in the local system.
- Set Working Directory: Selects any directory of the local system for placing the files created in Pro/ENGINEER Wildfire 4.0 during a session. In case a working directory is not selected, the files are saved in the default directory.

Note

> The default directory is selected during the installation of the Pro/ENGINEER Wildfire 4.0.

- Save: Saves a file that exists in the current session of the Pro/ENGINEER Wildfire 4.0.

Note

> In Pro/E, multiple files can exist in a particular session. A file continues to exist in a session until it is erased from the memory.

- Save a Copy: Saves a copy of a file in any location of the local system or in the current working directory. This option is mainly used for saving a file in other file formats, such as STEPS, STL.
- Backup: Creates a backup of a file existing in the current session of the Pro/ENGINEER Wildfire 4.0.
- Rename: Changes the name of a file that exists in the current session.
- Erase: Erases files from the temporary memory of the Pro/ENGINEER Wildfire 4.0. The temporary memory contains files that are not yet saved permanently onto the hard disk of the local system.
- Delete: Deletes files created in Pro/ENGINEER Wildfire 4.0 permanently from the hard disk of the local system.
- Exit: Exits the Pro/ENGINEER Wildfire 4.0 environment.

The Edit Menu

The options available under the **Edit menu** are used for modifying Pro/ENGINEER elements, such as entities. To access options available in the **Edit menu**, *select* **Edit menu** from the **Menu Bar**, as shown in Fig.Pro/E-2.11:

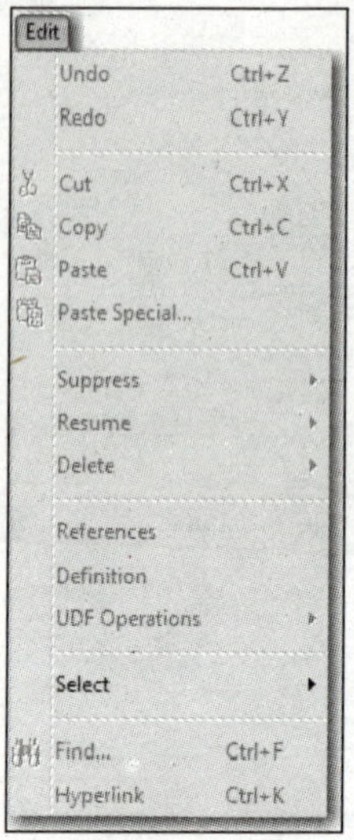

Fig.Pro/E-2.11

Some of the commonly used **Edit menu** options are as follows:

- **Undo:** Repeats the state of a file from the current state to the previous state.
- **Redo:** Reverses the action made by undo option in Pro/ENGINEER Wildfire 4.0 file
- **Cut:** Removes the selected entity, such as line, arc, circle from the drawing area
- **Copy:** Duplicates the selected entity from the **Drawing Area** to place it on the desired location
- **Paste:** Inserts the entity copied or removed from the **Drawing Area** using the Copy or **Cut** option
- **Paste Special:** Paste the original feature or a set of features on the desired location of the **Drawing Area**

The View Menu

The options available under the **View** menu are used to modify the appearance of a model and work screen of the Pro/ENGINEER Wildfire 4.0. To access options available in the **View menu**, *select* **View menu** from the **Menu Bar**, as shown in Fig.Pro/E-2.12:

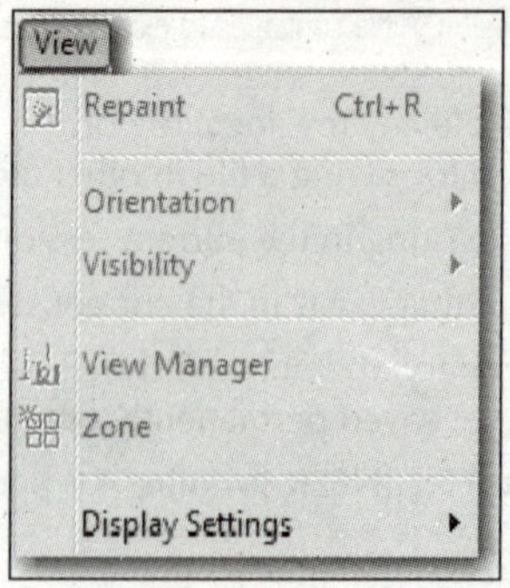

Fig.Pro/E-2.12

Some of the commonly used **View menu** options are as follows:

- **Repaint:** Removes temporary information from the drawing area.
- **Orientation:** Orients a model created in a Pro/ENGINEER Wildfire 4.0 session. Orientation of a model includes Zooming or enlarging, spinning, panning the selected model. Panning of a model refers to the horizontal or vertical movement of a model,
- **View Manager:** Creates extract assemblies in Pro/ENGINEER Wildfire 4.0. An extract assembly contains components selected by the user from the parent assembly. The changes made in the extract assembly are automatically reflected back in the original assembly. An extract assembly is used for concurrent cable designing.
- **Display Settings:** Changes the appearance and the color scheme of the model and the **Drawing Area**. It contains the following options:
    - **Model Display:** Controls the appearance of the model on the **Drawing Area**.
    - **Datum Display:** Displays datums created in a Pro/ENGINEER Wildfire 4.0 session. Datums are features that work as reference for creating a feature. Datums are discussed in detail in Chapter 4, Exploring Pro/ENGINEER Wildfire 4.0 Part Mode.
    - **System Colors:** Displays the default system colors. These colors can be allocated to various elements of a Pro/ENGINEER Wildfire 4.0 model.
    - **Performance:** Improves the display quality during dynamic orientation of a shaded model.

The Insert Menu

The **Insert menu** provides direct access to the commands for creating Pro/ENGINEER Wildfire 4.0 features, such as datum planes, holes, protrusion. However, these commands are enabled while working in the Part mode. Features are discussed in Chapter 4, Exploring Pro/ENGINEER Wildfire 4.0 Part Mode. To access options available in the **Insert menu**, select **Insert menu** from the **Menu Bar**, as shown in Fig.Pro/E-2.13:

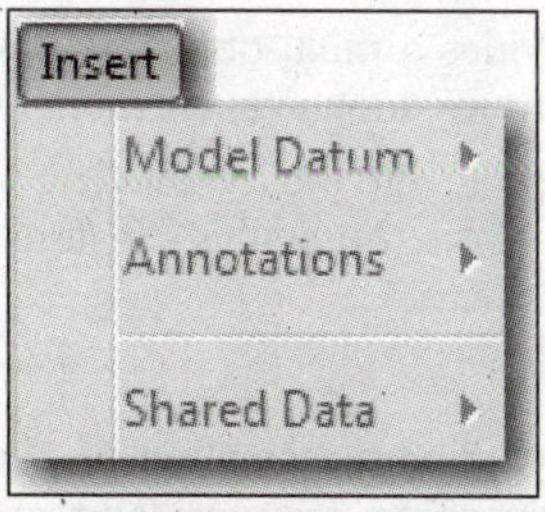

Fig.Pro/E-2.13

Some of the commonly used **Insert menu** options are as follows:

- **Model Datum:** Creates various datum features, such as datum plane, datum axis, datum curve, datum points, system coordinates.
- **Annotations:** Creates annotation elements in a model. Annotation elements in Pro/ENGINEER are data features that are used to share model information among various models.

The Analysis Menu

The options available under the **Analysis** menu are used to perform functions, such as locating properties of Assembly and Part files. To access options available in the **Analysis menu**, select **Analysis menu** from the **Menu Bar**, as shown in Fig.Pro/E-2.14:

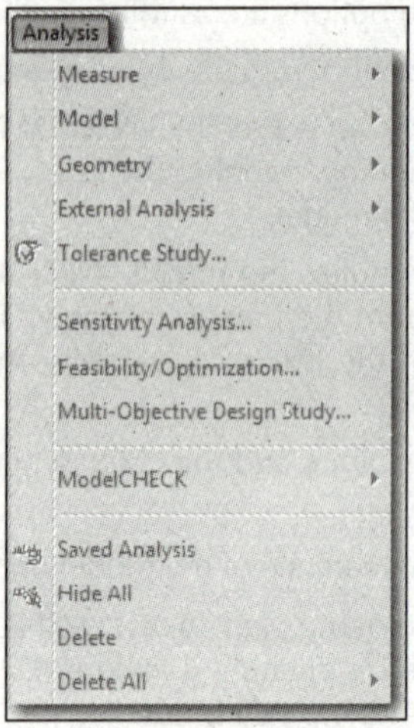

Pro/E-2.14

Some of the commonly used Analysis menu options are as follows:

- Tolerance Study: Performs tolerance analysis in a current working session. Tolerance analysis is used to predict the effects of manufacturing defects using Pro/ENGINEER Wildfire 4.0.
- Sensitivity Analysis: Performs sensitivity analysis. The sensitivity analysis allows a user to identify the effect of changing the range of certain dimensions in a model. Usually standardized fixed values are specified to the dimensions.
- Feasibility/Optimization: Provides optimized dimension values to overcome certain user defined constraints in a model.
- Multi-Objective Design Study: Provides a multi-objective design study. A multi-objective design study helps users to explore several options for designing the model in Pro/ENGINEER Wildfire 4.0.
- Saved Analysis: Saves an analysis along with the model. The Saved Analysis option is also used to create an analysis feature from a saved analysis. An analysis feature is used to automate the design process in Pro/ENGINEER Wildfire 4.0.

The Info Menu

The Info menu provides information regarding the creation of a model that is available in the session. The information includes parent-child relationships, features, and geometry of the model. To access options available in the Info menu, select Info menu from the Menu Bar, as shown in Fig.Pro/E-2.15:

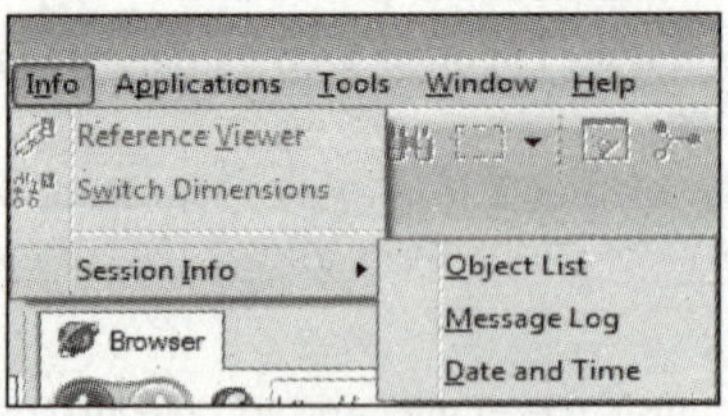

Fig.Pro/E-2.15

Some of the commonly used Info menu options are as follows:

- Reference Viewer: Allows user to graphically view a complex model to locate problem areas. The problem areas generally appear after a feature gets modified in the model.
- Switch Dimensions: Allows user to switch between numeric and symbolic dimension formats.

- **Session Info:** Allows user to manage a Pro/ENGINEER Wildfire 4.0 session. It contains the following options:
    - **Object List:** Displays a list of all the models, which are opened in a current working session. This option also allows user to perform manipulations, such as saving, editing, and closing the list containing the models.
    - **Message Log:** Displays a list of messages received in a current session. These messages include error messages resulting from regeneration failures. A regeneration failure is an error that occurs when a specified model is failed to generate in a session. The Message Log option also allows a user to perform manipulations, such as saving, editing, and closing the list containing the messages.
    - **Date and Time:** Displays the current date and time in the Message Area of the Pro/ENGINEER Wildfire 4.0.

The Applications Menu

The Applications menu contains options to access manuals provided by PTC. Users can also switch between various modes of the design process available in Pro/ENGINEER Wildfire 4.0 using the Applications menu.

Note

> The options under Applications menu are not discussed here, as these options are not available at this stage. Options under Applications menu get displayed once we start working in any of the Pro/ENGINEER Wildfire 4.0 modes.

The Tools Menu

The Tools menu contains options for customizing the UI of the Pro/ENGINEER Wildfire 4.0. The customization of the UI includes changing the appearance of the Drawing Area, creating mapkeys, and so on. To access options available in the Tools menu, select Tools menu from the Menu Bar, as shown in Fig.Pro/E-2.16:

Fig.Pro/E-2.16

Some of the commonly used Tools menu options are as follows:

- **Play Trail/Training File:** Allows user to run a trail file. A trail file is a collection of all selected menu options and keyboard entries of a specific Pro/ENGINEER working session.
- **Distributed Computing:** Allows user to enable a hardware to get accessed by multiple computers over a network.
- **Mapkeys:** Allows user to create, delete, view, and save mapkeys in the configuration file, mapkey. A mapkey is a collection of frequently used commands that are executed in sequence using a specified key on the keyboard.

- **Floating Modules:** Allows user to obtain a floating module license. A floating module license allows Pro/ENGINEER Wildfire 4.0 to run on any system on a network.
- **Auxiliary Applications:** Allows user to manage auxiliary applications in Pro/ENGINEER Wildfire 4.0. Auxiliary applications are those applications, which are created outside the environment of Pro/ENGINEER. The managing of auxiliary applications includes registering, starting, stopping, and deleting the auxiliary applications.
- **Environment:** Allows user to change the environment settings for the current working session of the Pro/ENGINEER Wildfire 4.0.

The Window Menu

The Windows menu allows user to open, close, and resize the Pro/ENGINEER windows. Pro/ENGINEER 4.0 allows you to open multiple windows of a single model or different models simultaneously; however, the user is allowed to work only in a single window at a time. Users can switch among multiple windows by selecting the desired window from the list of open windows under the Windows menu. To access options available in the Windows menu, select Windows menu from the Menu Bar, as shown in Fig.Pro/E-2.17:

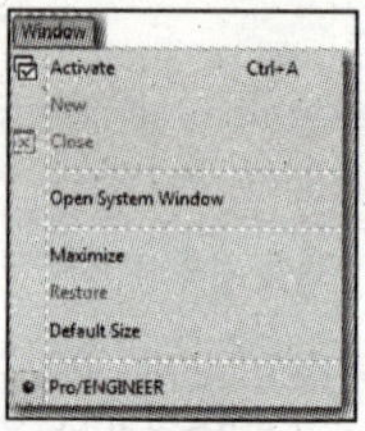

Fig.Pro/E-2.17

Some of the commonly used Windows menu options are as follows:

- **Activate:** Used to make the selected window as active. The window in which the user is currently working is known as active window. The active window enables the file in it to utilize all the features supported by Pro/ENGINEER.
- **New:** Creates a new window containing the same file which is present in the parent window. The parent window is the one from which the new window is created.
- **Close:** Closes the current window. However, if any object present in that window, it retains in the memory.
- **Maximize:** Used to increase the size of the current window to its full size.
- **Default Size:** Used to resize the current window to its default size.

The Help Menu

The Help menu allows users to access help information on using Pro/ENGINEER Wildfire 4.0. It provides context sensitive help options to locate information about Pro/ENGINEER Wildfire 4.0 menus. To access options available in the Help menu, select Help menu from the Menu Bar, as shown in the Fig.Pro/E-2.18:

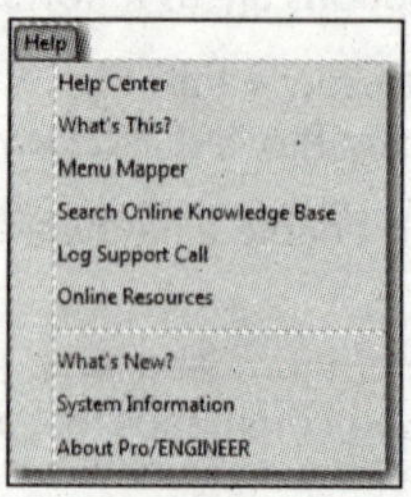

Fig.Pro/E-2.18

Some of the commonly used Help menu options are as follows:

- Help Center: Allows user to access help on any Pro/ENGINEER topic. The Pro/ENGINEER Help window opens after selecting this option.
- What's This?: Enables context-sensitive help. The context sensitive help displays information about the selected menu options.
- Menu Mapper: Opens a tool named Menu Mapper, which helps user to find the menu options available in all the previous versions of Pro/ENGINEER. The user can find menu options of Pro/ENGINEER versions, such as Pro/ENGINEER Wildfire 3.0, Pro/ENGINEER Wildfire 2.0, and Pro/ENGINEER Wildfire 2001.
- Online Resources: Allows user to access online information on various Pro/ENGINEER topics.
- What's New?: Allows user to access new functionalities introduced with Pro/ENGINEER Wildfire 4.0, such as latest tools, menus, tutorials.

Toolbars

The toolbars contain buttons that are short-cut commands to the options available under the Menu Bar. These toolbars help users in accessing the menu options with a click of a button on the toolbar.

In Fig.Pro/E-2.2, the toolbars are located at two locations on the Main window. The first location where toolbars, such as File toolbar, View toolbar, Model Display toolbar, and Datum Display toolbar are located is called the Top tool chest. The Top tool chest is present at the top of the Drawing area in the Main window. The second location is at the right side of the Drawing Area and this location is known as the Right tool chest. The toolbar located at this position is known as Datum toolbar.

The Pro/ENGINEER Wildfire 4.0 provides the following toolbars:

- File Toolbar
- Edit Toolbar
- View Toolbar
- Model Display Toolbar
- Datum Display Toolbar
- Datum Toolbar

File Toolbar

The File toolbar contains buttons, which are short-cut commands to some of the commonly used File menu options, as shown in Fig.Pro/E-2.19:

Fig.Pro/E-2.19

The buttons available in the File toolbar are as follows:

- New: The New button allows a user to create a new Pro/ENGINEER Wildfire 4.0 file
- Open: The Open button allows a user to open an existing Pro/ENGINEER Wildfire 4.0 file
- Save: The Save button allows a user to save a Pro/ENGINEER Wildfire 4.0 file created in the current working session
- Print: The Print button allows a user to print a selected model created in the current session
- Mail Recipient (as Link): The Mail Recipient (as Link) button allows a user to send a link of the current model with an e-mail

- **Mail Recipient (as Attachment):** The Mail Recipient (as Attachment) button allows a user to send the current model as an attachment with an e-mail

The Edit Toolbar

The **Edit** toolbar contains buttons that are short-cut commands to some of the commonly used **Edit** menu options. The buttons available under the **Edit** toolbar are used to modify the entities created in a Pro/ENGINEER session. Fig.Pro/E-2.20 shows the **Edit** toolbar:

Fig.Pro/E-2.20

The buttons available in the **Edit** toolbar are as follows:

- **Undo:** The **Undo** button repeat the state of a file from current to previous state.
- **Redo:** The **Redo** button reverses the action made by the undo command in Pro/ENGINEER Wildfire 4.0 file
- **Cut:** The **Cut** button removes the selected entity, such as line, arc, circle from the drawing area
- **Copy:** The **Copy** button duplicates the selected entity from the drawing area to place it on the desired location
- **Paste:** The **Paste** button inserts the entity copied or removed from the drawing area using the **Copy** or **Cut** button
- **Paste Special:** The **Paste Special** button pastes the original feature or a set of features on the desired location of the **Drawing Area**
- **Find:** The Find button allows user to search for entities from the model created in a session
- **Inside Box:** The **Inside Box** button selects drawing items, such as points and triangles that lie completely inside the rectangular region. *Click* the button on the Edit toolbar to explore the remaining buttons on the Edit toolbar. Fig.Pro/E-2.21 shows the remaining buttons available on the Edit toolbar:

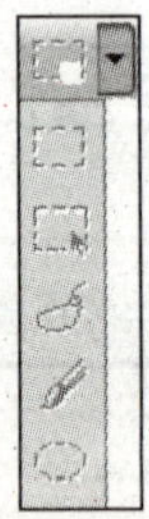

Fig.Pro/E-2.21

Following are the remaining buttons available under the Edit toolbar:

- **Across Box:** The **Across box** button selects drawing items that lie either partially or completely inside the rectangular region
- **Inside Polygon:** The **Inside polygon** button selects drawing items that lie completely inside the polygon region
- **Paintbrush:** The **paintbrush** button selects drawing items on which the stroke of a paintbrush can be applied

- Inside Ellipse: The Inside ellipse button selects drawing items inside the ellipse region

The View Toolbar

The View toolbar contains buttons that are short-cut commands to some of the commonly used View menu options, such as Repaint, View Manager. Fig.Pro/E-2.22 shows the View toolbar:

Fig.Pro/E-2.22

The buttons available in the View toolbar are as follows:

- Repaint: The Repaint button allows user to repaint the Drawing Area. Repainting is meant for removing temporary information from the Drawing Area.
- Spin Center: The Spin Center button allows user to enable or disable the Spin Center. The Spin Center is located at the center of rotation in a model.
- Orient Mode: The Orient Mode button allows user to enable or disable the visibility of the datums. The Orient Mode also allows user to provide more capability than the standard spinning, panning, and zooming techniques.
- Zoom In: The Zoom In button allows user to enlarge the selected entity on the Drawing Area. You can manually enlarge an entity using scroll key of the mouse.
- Zoom Out: The Zoom Out button allows user to reduce the view of the selected entity to its half on each click of this button.
- Refit: The Refit button allows user to adjust the size of an entire sketch such that it can be seen completely on the Drawing Area.
- Reorient: The Reorient button allows user to orient a model created in a Pro/ENGINEER Wildfire 4.0 session. The orientation of the view includes setting orientation center besides spinning, panning, and zooming the model.
- Named View List: The Named View List button allows user to display saved views. The saved views shows the changed orientation when the default orientation of the model changes.
- View Manager: The View Manager button allows user to create extract assemblies in Pro/ENGINEER Wildfire 4.0. An extract assembly contains components selected by the user from the parent assembly. The changes made in the extract assembly are automatically reflected back in the original assembly. Extract assembly is used for concurrent cable designing.

The Model Display Toolbar

The Model Display toolbar contains buttons that are short-cut commands to the options present under the submenu of the View menu option called Display Settings. These buttons are used for controlling the appearance of model on the Drawing Area. Fig.Pro/E-2.23 shows the Model Display toolbar:

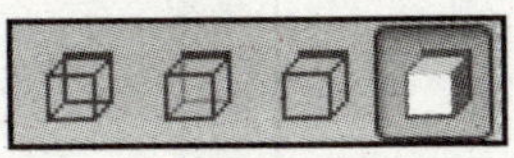

Fig.Pro/E-2.23

The buttons available in the Model Display toolbar are as follows:

- Wireframe: The Wireframe button is used to show hidden lines, such as edges on the selected model. These hidden lines are shown in white color.

- **Hidden line:** The **Hidden line** button is used to display hidden lines in grey color.
- **No hidden:** The **No hidden** button is used to remove hidden lines from the current model.
- **Shading:** The **Shading** button is used to display the selected model as shaded. The **Shading** button is applied by default to a model in a session.

The Datum Display Toolbar

The **Datum Display** toolbar also contains short-cut commands to the options under the submenu of the **View** menu option called **Display Settings**. These short-cut commands are executed using buttons available in the Datum Display toolbar. These buttons are mainly used to control the visibility of some datum features, such as datum planes, datum axes, and datum points. Fig.Pro/E-2.24 shows the **Datum Display** toolbar:

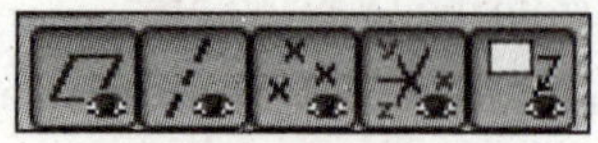

Fig.Pro/E-2.24

The buttons available in the **Datum Display** toolbar are as follows:

- **Plane Display:** The **Plane Display** button allows user to display datum planes along with their names.
- **Axis Display:** The **Axis Display** button allows user to display datum axes along with their names.
- **Point Display:** The **Point Display** button allows user to display datum points along with their names.
- **Csys Display:** The **Csys Display** button allows user to display coordinate systems along with their names. Coordinate systems are used in Pro/ENGINEER 4.0 to perform tasks, such as detailed element analysis and modeling applications.
- **Annotation Element Display:** The **Annotation Element Display** button allows user to display annotation elements along with their names. Annotation elements in Pro/ENGINEER 4.0 are data features that are used to share model information among various models.

The Datum Toolbar

The **Datum** toolbar contains buttons for creating various datum features, such as datum plane, datum axis, datum curve. The datum features created using the **Datum** toolbar act as base features to create other features in Pro/ENGINEER Wildfire 4.0. Fig.Pro/E-2.25 shows the buttons available on the **Datum** toolbar:

Fig.Pro/E-2.25

Fig.Pro/E-2.25 shows buttons available in the **Datum Display** toolbar. Let's discuss these buttons one by one:

- **Sketch:** The **Sketch** button is used to sketch datum curves. The sketched datum curves consist of one or more sketch segments.
- **Datum Plane:** The **Datum Plane** button is used to create datum planes, in addition to the three datum planes available by default in every Pro/ENGINEER session.
- **Datum Axis:** The **Datum Axis** button is used to create datum axes in a model. The datum axes are like centerline needed at the centre of orthographic models, such as holes and cylinders.
- **Datum Curve:** The **Datum Curve** button is used to create datum curves in a model. The datum curves are useful in creating advanced solid features in Pro/ENGINEER.
- **Datum Point:** The **Datum Point** button is used to create datum points in a model. The datum points are useful in creating Pipe features.
- **Coordinate System:** The **Coordinate System** button is used to create coordinate systems in a model.
- **Analysis Feature:** The **Analysis Feature** button is used to create analysis features in a model.
- **Reference:** The **Reference** button is used to create datum reference features in Pro/ENGINEER Wildfire 4.0 session. Datum reference features are user-defined datum planes, datum axes, and datum points.
- **Annotation Feature:** The **Annotation Feature** button is used to create annotation features in a model.
- **Annotation Element Propagate:** The **Annotation Element Propagate** button is used to insert an Annotation Element Propagate feature in a model.

Performing File Management

The objects that are created in a Pro/ENGINEER Wildfire 4.0 session are stored in the form of files. The user can perform manipulations, such as creating, opening, saving, and deleting on these files. Pro/ENGINEER Wildfire 4.0 provides ways for manipulating files similar to those in Windows operating system. However, there are major differences between managing files in Pro/ENGINEER Wildfire 4.0 and Windows operating system. The following is a list of differences in managing files by Pro/ENGINEER Wildfire 4.0 and Windows operating system:

- The rules for naming files in Pro/ENGINEER Wildfire 4.0 are more stringent than those for Windows. Pro/ENGINEER filenames are limited up to 31 characters and cannot contain periods, brackets, and spaces. However, files created in Windows operating system do not comply with any complex file naming conventions.
- Each time a Pro/ENGINEER Wildfire 4.0 file is saved, a new version of that file gets created. However, in Windows operating system, the original version of the file gets updated each time the file is saved.

Before exploring various options available to manipulate Pro/ENGINEER Wildfire 4.0 files, let's discuss some of the general file extensions available in Pro/ENGINEER Wildfire 4.0.

Understanding the General File Extensions

As discussed earlier, the objects created in a Pro/ENGINEER Wildfire 4.0 session are stored in the form of files. These files are saved with certain file extensions. Pro/ENGINEER Wildfire 4.0 supports various file extensions based on the working mode being selected by the user. Table 2.1 lists the file extensions in Pro/ENGINEER Wildfire 4.0:

Table 2.1: General File Extensions in Pro/ENGINEER Wildfire 4.0	
Mode	**Extension**
Sketch	***.sec**
Part	***.prt**
Assembly	***.asm**
Drawing	***.drw**

Now, let's discuss about the working directory, where all the files created in a session are stored.

Note

Various Pro/ENGINEER modes are discussed in Chapter 1, Introducing Pro/Engineer Wildfire 4.0.

Introducing the Working Directory

Pro/ENGINEER Wildfire 4.0 allows a user to select any directory on the local system as a working directory to save all the files created during a session. A working directory should be selected before starting work in Pro/ENGINEER Wildfire 4.0; else, the files will be saved in the default directory. The default directory is selected during the installation of the Pro/ENGINEER Wildfire 4.0.

Following are the two ways to select the working directory in Pro/ENGINEER Wildfire 4.0:

- Using the Navigator
- Using the Select Working Directory Dialog Box

Using the Navigator

The Navigator is a pane that is displayed on the left side of the drawing area and contains multiple tabs for performing functions, such as browsing the Pro/ENGINEER Wildfire 4.0 files, displaying information about a specific model. The Navigator can be used to select a folder as the working directory for a Pro/ENGINEER session. The detailed discussion on various tabs available under the Navigator has already been given in the section, Navigator of this chapter.

Perform the following steps to select a working directory using the Navigator:

1. *Click* the Folder Tree button on the Navigator, as shown in Fig.Pro/E-2.26:

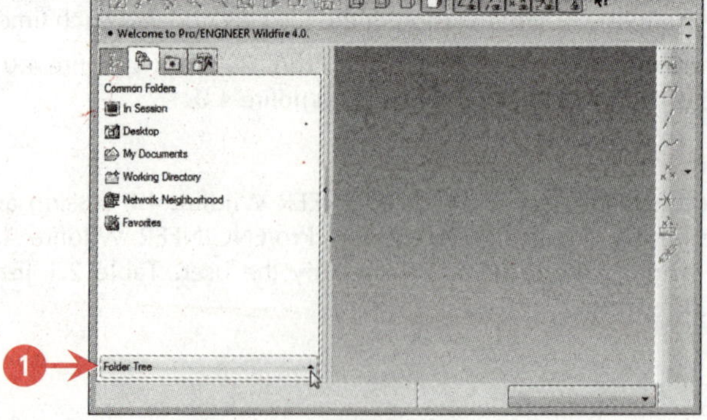

Fig.Pro/E-2.26

A list of folders and directories available under the local system gets displayed (Fig.Pro/E-2.27).

2. *Right-click* any **folder**, say **ProE**, on the local system, as shown in Fig.Pro/E-2.27.

The short-cut menu appears displaying some **File menu** options.

3. Now, *select* the **Set Working Directory** option from the short-cut menu to set the selected folder as the working directory (Fig.Pro/E-2.27):

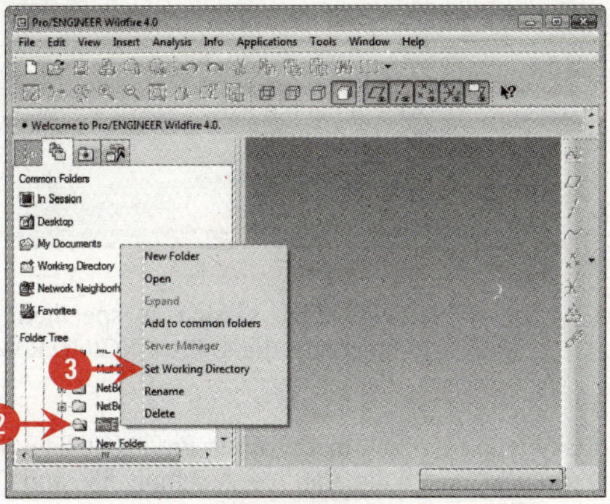

Fig.Pro/E-2.27

The ProE folder is set as a Working Directory for a Pro/ENGINEER Wildfire 4.0 session.

Using the Select Working Directory Dialog Box

Another way to set a folder as the working directory is by using the **Select Working Directory Dialog Box**.

Perform the following steps to select a working directory using the **Select Working Directory Dialog Box:**

1. *Select* **File→Set Working Directory** option from the menu bar, as shown in Fig.Pro/E-2.28:

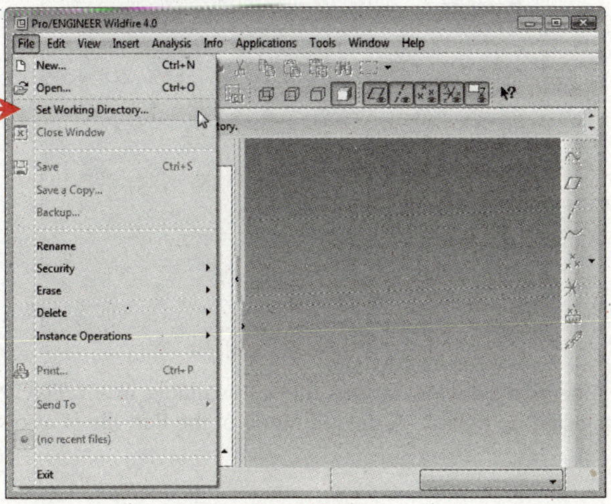

Fig.Pro/E-2.28

The Select Working Directory Dialog Box appears (Fig.Pro/E-2.29).

2. *Select* any folder, say **ProE** on the local system and *click* the **OK** button to set it as the working directory, as shown in Fig.Pro/E-2.29:

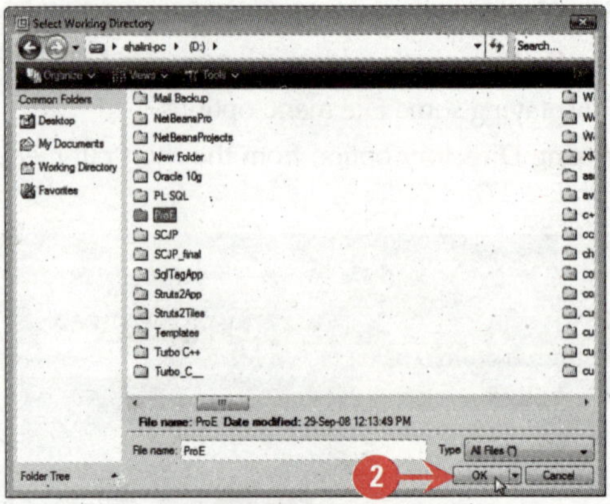

Fig.Pro/E-2.29

Once the working directory for a Pro/ENGINEER session is specified, you can start working with Pro/ENGINEER Wildfire 4.0. Let's start with creating a file in Pro/ENGINEER Wildfire 4.0.

Creating a File

The user can create files of any type from any mode of the Pro/ENGINEER Wildfire 4.0. The file created in Pro/ENGINEER Wildfire 4.0 can be a Sketch file, Part file, Assembly file, and Drawing file.

Perform the following steps to create a file in Pro/ENGINEER Wildfire 4.0:

1. *Start* the Pro/ENGINEER Wildfire 4.0. The process to start the Pro/ENGINEER Wildfire 4.0 has already been discussed in the section, Starting the Pro/ENGINEER Wildfire 4.0 of this chapter.
2. *Select* File→New option from the Menu Bar, as shown in Fig.Pro/E-2.30:

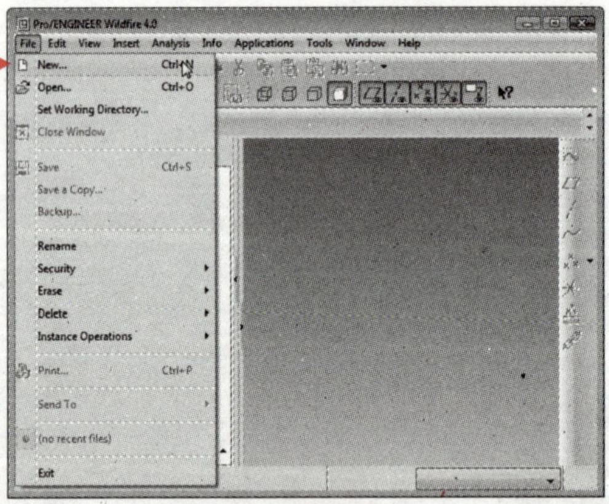

Fig.Pro/E-2.30

The New dialog box appears displaying various modes available in Pro/ENGINEER Wildfire 4.0. The Part mode radio button is selected by default. A default name of the Part file is also shown in the Name text box (Fig.Pro/E-2.31).

Note

> The Use default template check box is selected in the New dialog box. This creates a new file using a default template including three default datum planes.

3. *Click* the OK button in the New dialog box, as shown in Fig.Pro/E-2.31:

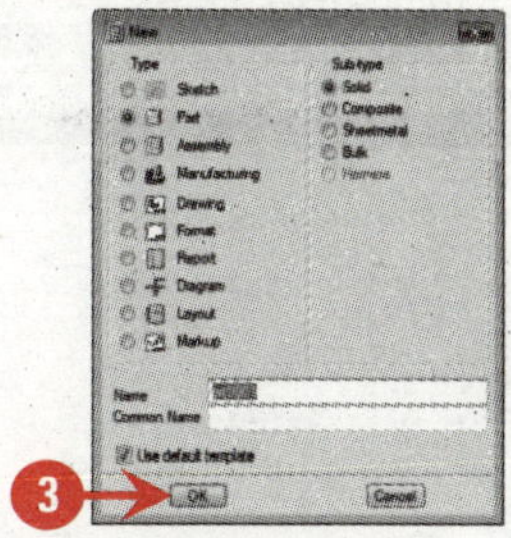

Fig.Pro/E-2.31

An initial screen containing the three default datum planes as a result of entering the Part mode is shown in Fig.Pro/E-2.32:

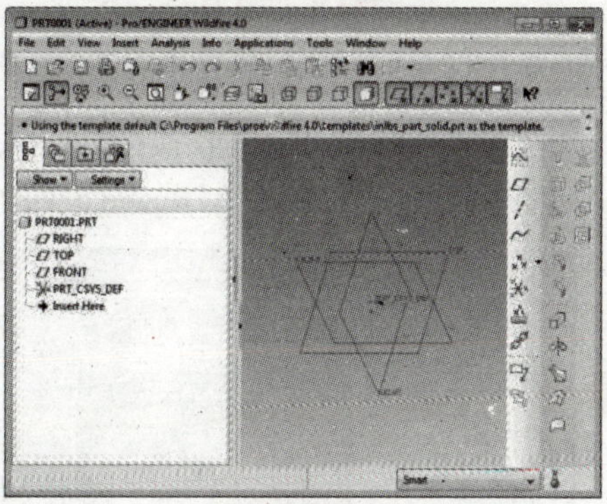

Fig.Pro/E-2.32

Saving a File

The **Save** option allows user to save files of any type from any mode of Pro/ENGINEER Wildfire 4.0.

Perform the following steps to save a file in Pro/ENGINEER Wildfire 4.0:

1. *Select* **File→Save** option from the menu bar as shown in Fig.Pro/E-2.33:

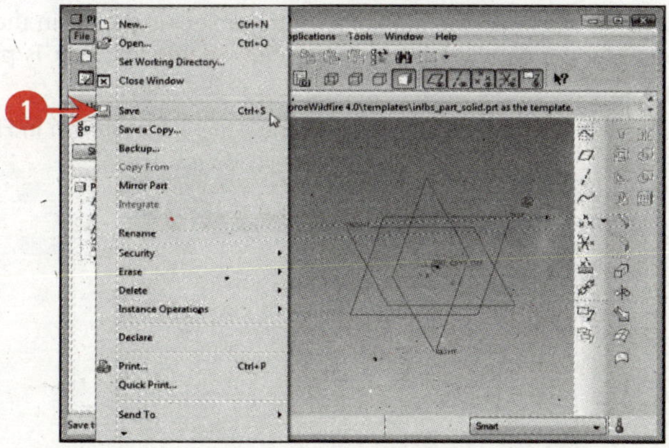

Fig.Pro/E-2.33

The **Save Object** dialog box appears and the default file name of the object is shown in the **Model Name** text box (Fig.Pro/E-2.34).

2. *Click* the **OK** button from the **Save Object** dialog box to save the file, as shown in Fig.Pro/E-2.34:

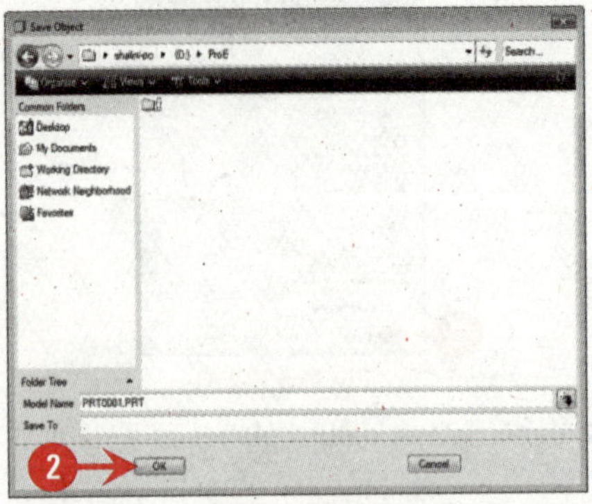

Fig.Pro/E-2.34

You can also save a copy of the current file in any location of the local system or in the current working directory using the **Save a Copy** option.

Perform the following steps to save a copy of the current file in Pro/ENGINEER Wildfire 4.0:

1. *Select* **File→Save a Copy** option from the menu bar, as shown in Fig.Pro/E-2.35:

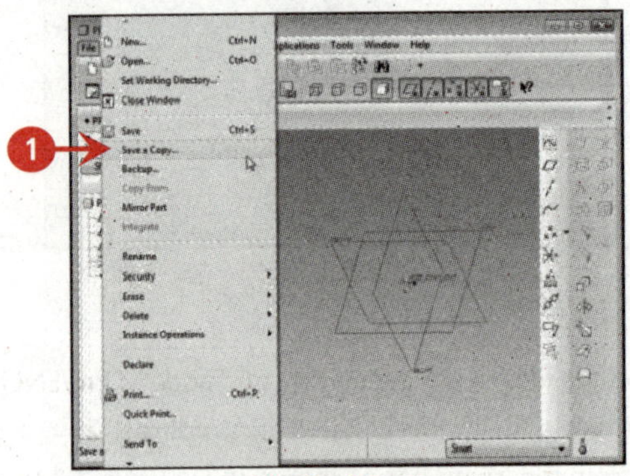

Fig.Pro/E-2.35

The **Save a Copy** dialog box appears and the file name of the object is shown in the **Model Name** box along with the target directory to place a copy of the file. The **New Name** text box is provided to specify a new name for the file to be saved.

2. *Click* the **OK** button from the **Save a Copy** dialog box to save the file, as shown in Fig.Pro/E-2.36:

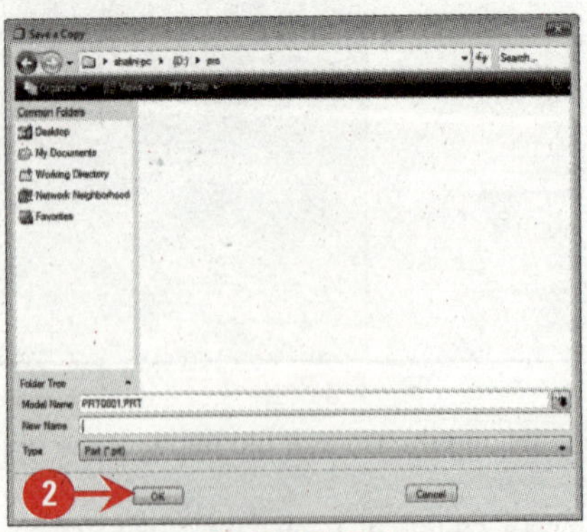

Fig.Pro/E-2.36

Fig.Pro/E-2.36 shows that a copy of the file, PRT0001.PRT, is created outside the working directory. The copy gets created in the folder named **pro**.

Backing up a File

The Backup option allows user to create a backup of a file that exists in a session.

Perform the following steps to create backup of a file in Pro/ENGINEER Wildfire 4.0:

1. *Select* **File→Backup** option from the **Menu Bar**, as shown in Fig.Pro/E-2.37:

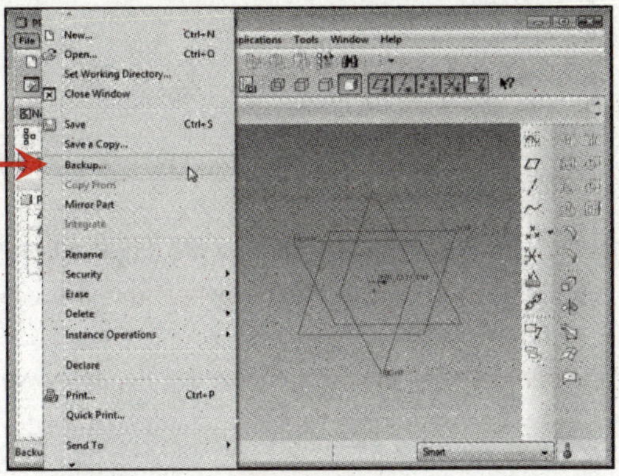

Fig.Pro/E-2.37

The **Backup** dialog box appears (Fig.Pro/E-2.38) and the file name of the object for which the backup is to be taken is shown in the **Model Name** box.

2. *Select* the target directory (backup) to place the backup of the file in the Backup dialog box. The name of the selected directory is placed automatically in the **Backup To text** box (Fig.Pro/E-2.38).
3. *Click* the **OK** button from the **Backup** dialog box to create the backup of the file, as shown in Fig.Pro/E-2.38:

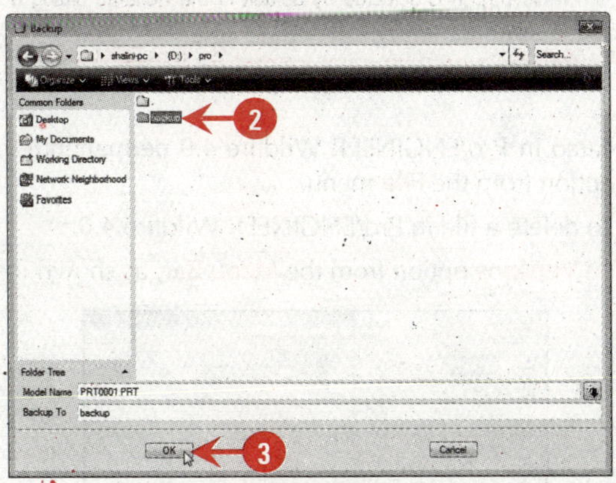

Fig.Pro/E-2.38

A backup of the file, PRT0001.PRT, is created in a folder named backup.

Renaming a File

The **Rename** option allows user to change the name of a Pro/ENGINEER file. Perform the following steps to change the name of an existing file in Pro/ENGINEER Wildfire 4.0:

1. *Select* **File→Rename** option from the **Menu Bar**, as shown in Fig.Pro/E-2.39:

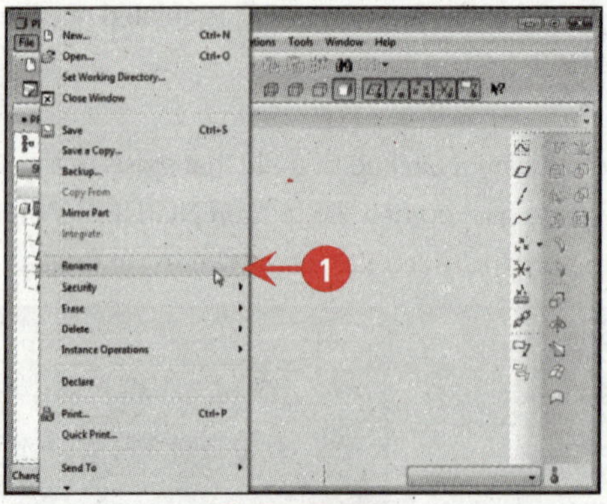

Fig.Pro/E-2.39

The **Rename** dialog box appears to provide a new name for a file (Fig.Pro/E-2.40).

2. *Specify* a new name to the file (prt0001.prt) in the **New Name** text box(Fig.Pro/E-2.40). In our case, we have specified **DemoPart** as a new name.
3. *Click* the **OK** button from the **Rename** dialog box to rename the file, as shown in Fig.Pro/E-2.40:

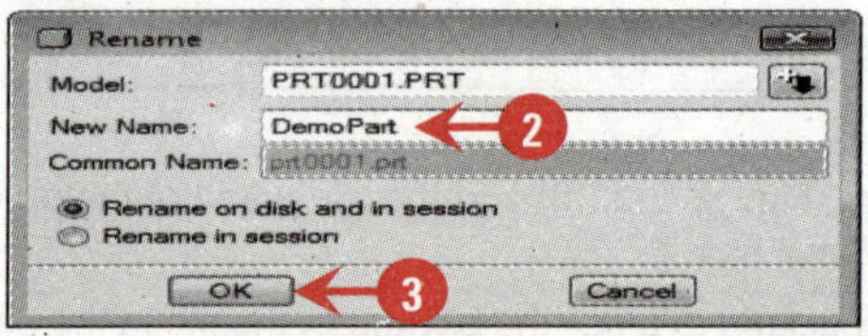

Fig.Pro/E-2.40

Note

> The Rename on disk and in session radio button is selected by default in the Rename dialog box to rename the file on the disk as well as in the current session.

Deleting a File

The user can delete files created in Pro/ENGINEER Wildfire 4.0 permanently from the hard disk of the local system by using the **Delete** option from the **File** menu.

Perform the following steps to delete a file in Pro/ENGINEER Wildfire 4.0:

1. *Select* **File→Delete→Old Versions** option from the **Menu Bar**, as shown in Fig.Pro/E-2.41:

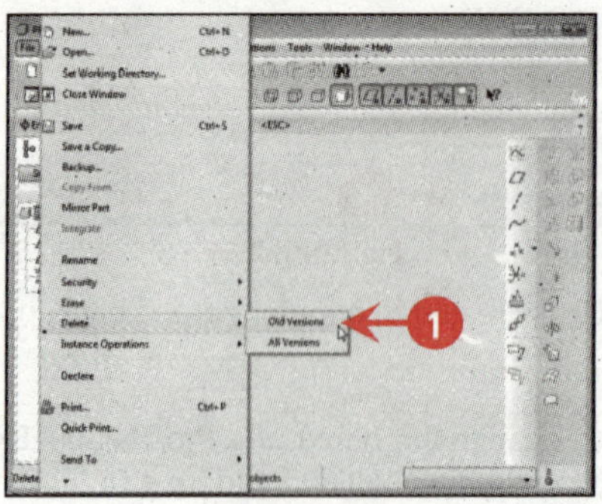

Fig.Pro/E-2.41

A Message Input Window gets displayed in the Message Area to accept the name of the file whose old versions are to be deleted (Fig.Pro/E-2.42).In the Message Input Window, specify the name of a file whose version you want to delete. In our case, the file name is PRT0001.PRT.

2. *Click* the Accept Value (☑) button on the Message Input Window to delete the old versions of the selected file, as shown in Fig.Pro/E-2.42:

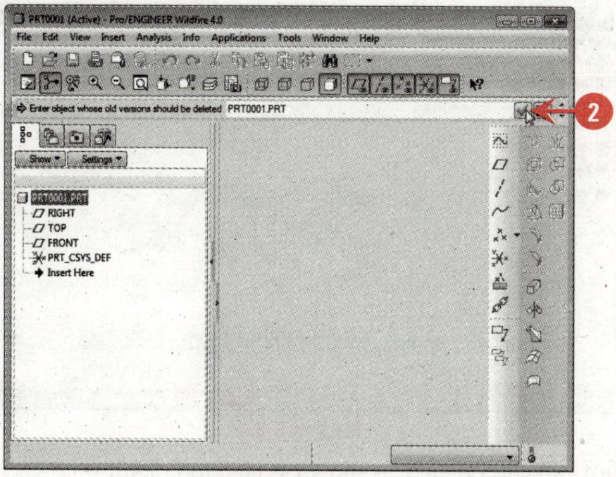

Fig.Pro/E-2.42

Note

> The File→Delete→All Versions option from the Menu Bar deletes all the versions of the selected file from the local system.

Opening an Existing File

The user can open an existing Pro/ENGINEER Wildfire 4.0 file by browsing the local system using the Open option from the File menu or Open button from the File toolbar.

Perform the following steps to open a file in Pro/ENGINEER Wildfire 4.0:

1. *Select* File→Open option from the Menu Bar as shown in Fig.Pro/E-2.43:

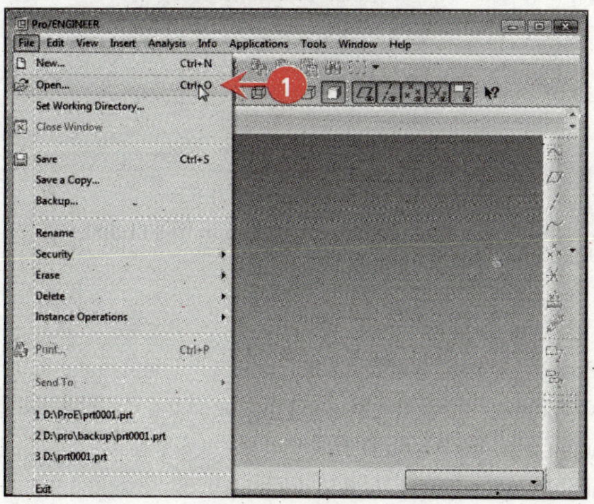

Fig.Pro/E-2.43

The **File Open** dialog box appears and displays a list of Pro/ENGINEER Wildfire 4.0 files.

2. *Select* the appropriate file and *click* the **Open** button from the **File Open** dialog box to open the file, as shown in Fig.Pro/E-2.44:

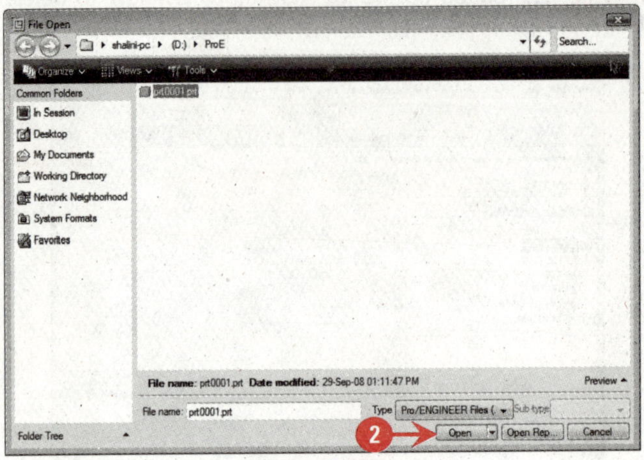

Fig.Pro/E-2.44

The selected file (prt0001.prt) gets opened, as shown in Fig.Pro/E-2.45:

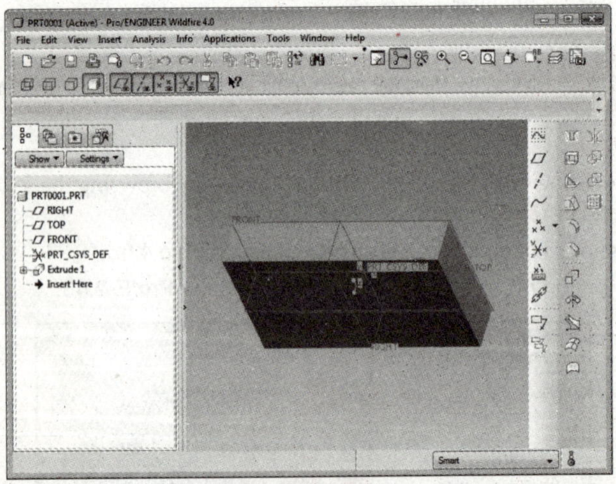

Fig.Pro/E-2.45

Let's now have a glance on the topics that we have covered in this chapter.

Summary

In this chapter, you have learned about:

- The process to start Pro/ENGINEER Wildfire 4.0
- Various components available within the Main Window of the Pro/ENGINEER Wildfire 4.0.
- Menu options available under the Menu Bar.
- Various toolbars available in Pro/ENGINEER Wildfire 4.0.
- Performing file management in Pro/ENGINEER Wildfire 4.0.

Chapter 3

Exploring the Pro/ENGINEER Wildfire 4.0 Sketch Mode

In this Section

- Starting the Sketch Mode
- Sketcher Tools Toolbar
- Working with Sketching Entities
- Working with Dimensions
- Applying Constraints
- Modifying Dimensions
- Trimming Entities
- Mirroring Entities
- Scaling and Rotating Entities
- Importing 2D Drawings
- Deleting Entities

The Sketch mode in Pro/ENGINEER Wildfire 4.0 is used to draw two-dimensional (2D) sketch of a three-dimensional (3D) feature. For example, to create a rectangular slot, you first have to draw its 2D sketch and then that sketch is converted into a 3D rectangular slot in **Part** mode. To draw a sketch in Pro/ENGINEER Wildfire 4.0, various sketching entities such as line, circle, and rectangle are used. These sketching entities are drawn by using various buttons such as line button, rectangle button available on the Sketcher Tools toolbar in the Sketch mode. Combination of various sketching entities leads into a complete sketch. However, to create a sketch of a feature, we first have to invoke the Sketch mode. Therefore, it's worthy to start this chapter by invoking or starting the Sketch mode. Then, we depict the Sketcher Tools toolbar that helps you drawing various sketching entities. Once a sketch is completed then that sketch is needed to be dimensioned and constrained. This is because a sketch will be completed only when it is fully dimensioned and constrained. You may also need to modify dimensions of an entity or to delete an entity. We discuss the process of modifying dimensions and deleting entity at the end of this chapter. Let's start this chapter with the sketch mode.

Starting the Sketch Mode

To start the Sketch mode, you first have to start the Pro/ENGINEER Wildfire 4.0. The process to start the Pro/ENGINEER Wildfire 4.0 is already discussed in chapter 2, *Exploring the User Interface* in the section, Starting Pro/ENGINEER Wildfire 4.0. Let's now follow the steps given here to start the Sketch mode:

1. *Select* the **File** menu on the menu bar in the Pro/ENGINEER Wildfire 4.0 main window, as shown in Fig.Pro/E-3.1:

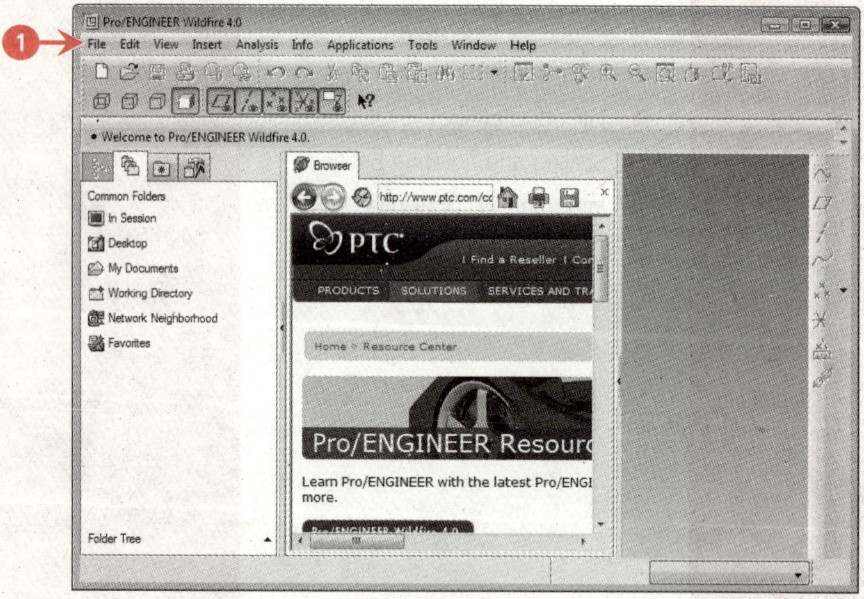

Fig.Pro/E-3.1

The **File** menu options appear (Fig.Pro/E-3.2).

2. Now, *click* the **New** option, as shown in Fig.Pro/E-3.2:

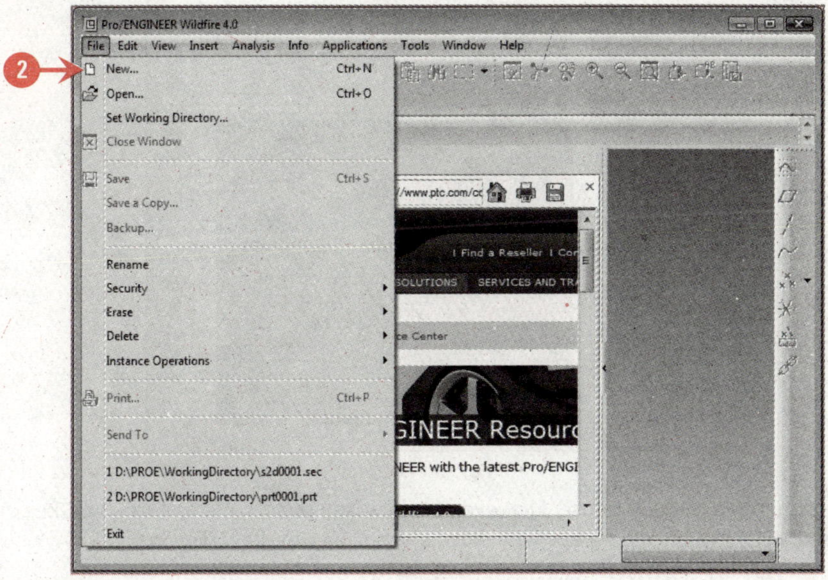

Fig.Pro/E-3.2

The **New** dialog box appears (Fig.Pro/E-3.3).

3. Now, *select* the **Sketch** radio button to enter into the `Sketch` mode, as shown in Fig.Pro/E-3.3.

Note

> By default, the Part radio button is selected.

4. Now, *specify* a name to the sketch. To assign a name to the sketch either accepts the default name specified in the text box beside the **Name** label or specifies a new name. In our case, we have accepted the default name (Fig.Pro/E-3.3):
5. Next, *click* the **OK** button, as shown in Fig.Pro/E-3.3:

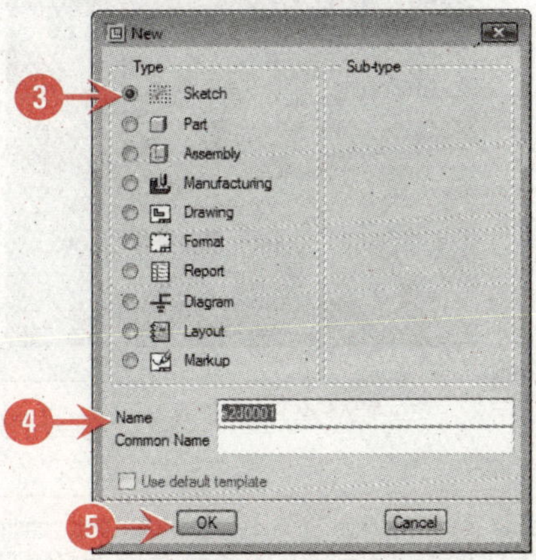

Fig.Pro/E-3.3

The `Pro/ENGINEER Wildfire 4.0` window with the `Sketch` mode appears (Fig.Pro/E-3.4):

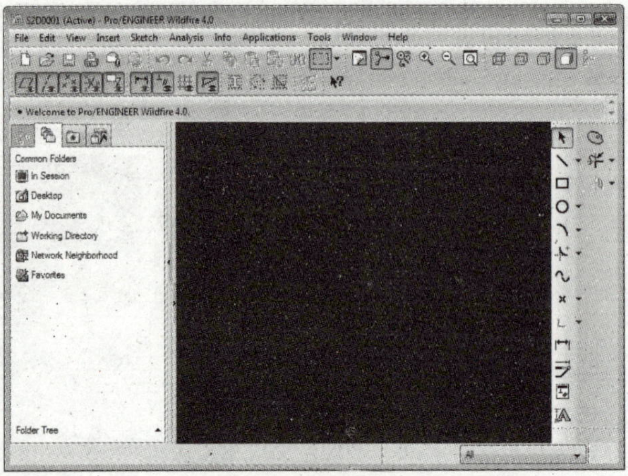

Fig.Pro/E-3.4

Now, the `Sketch` mode is ready to use. However, to draw any sketch of a feature, the `Sketcher Tools toolbar` is used. Let's explore the `Sketcher Tools toolbar` to draw sketches.

Sketcher Tools Toolbar

The `Sketcher Tools toolbar` contains various buttons that are used to draw sketching entities such as lines, circles, and rectangle. The Sketcher Tools toolbar also contains buttons to dimension and constrain the sketching entities. The sketching entities when used together create a sketch of a feature. The `Sketcher Tools toolbar` is situated at the right tool chest in the `Pro/ENGINEER Wildfire 4.0` window with the Sketch mode, as shown in Fig.Pro/E-3.5 (see encircled part):

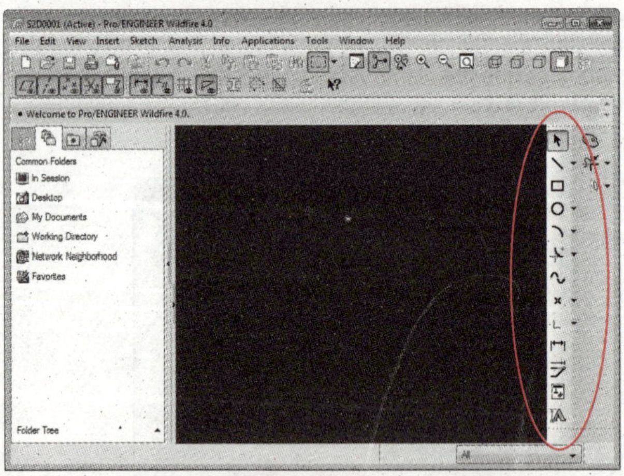

Fig.Pro/E-3.5

Note

Apart from using the Sketcher Tools toolbar to draw sketching entities, dimension and constrain the sketching entities, you can use these options such as, Rectangle, Arc, Circle from the Sketch menu in the menu bar in the Pro/ENGINEER Wildfire 4.0 window.

In this chapter, we use the `Sketcher Tools toolbar` to draw sketching entities, dimension and constrain the sketching entities. Now, let's discuss working with the sketching entities next.

Working with Sketching Entities

Sketching entities are called as the basic entities that are used to draw a 2D sketch of a 3D feature. In Pro/ENGINEER Wildfire 4.0, a sketching entity can be drawn in two ways: first, by using the buttons available on the Sketcher Tools toolbar; second, by accessing the options given in the Sketch menu. Sketching entities that are most commonly used to draw a 2D sketch are as follows:

- Points
- Line
- Rectangle
- Circle
- Arcs
- Circular Fillet
- Elliptical Fillet
- Splines
- Text

Let's discuss the process to draw sketching entities.

Points

Points are geometrical entities that are used to dimension the vertices. Following are the steps to create points:

1. *Click* the **Point** ( ) button on the Sketcher Tools toolbar (Fig.Pro/E-3.5).
2. Now, *click* anywhere on drawing area to place a point.

Repeat the step 2 to create more points.

Note

If you create only one point then dimensions will not appear. However, if more than one point is created then they are automatically dimensioned. After creating all the points click the middle mouse button to terminate using the Point button.

Line

In Pro/ENGINEER Wildfire 4.0, three different types of lines can be drawn. To draw three different types of lines, three different buttons are available on the Sketcher Tools toolbar. Following are the buttons available on the Sketcher Tools toolbar in Pro/ENGINEER Wildfire 4.0 to draw three different types of lines:

- **Line** ( ): Draws a simple line.
- **Line Tangent** ( ): Creates a tangent between two entities.
- **Centerline** ( ): Defines the axis of revolution to create a revolved feature, mirroring the sketched entities, and so on.

Note

- A revolved feature is created by revolving a section by a specified angle around a central axis/line.
- By mirroring the sketching entities, the time to create symmetrical entities is reduced.

To access these buttons, *click* the down arrow beside the Line ( ) button in the Sketcher Tools toolbar (Fig.Pro/E-3.5). A flyout appears, as shown in Fig.Pro/E-3.6:

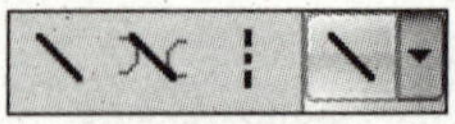

Fig.Pro/E-3.6

As you can see, the flyout contains three buttons to draw different type of lines. Let's perform the following steps to draw a line using the **Line** button:

1. *Click* the down arrow beside the Line ( ) button in the Sketcher Tools toolbar (Fig.Pro/E-3.5). A flyout appears (Fig.Pro/E-3.6).
2. *Select* the Line button ( ) (Fig.Pro/E-3.6) to create a simple line.
3. Now, *click* on any location in the drawing area (Fig.Pro/E-3.7) in the Sketcher window to specify the starting location of the line. A yellow rubber-band line appears which is attached with the cursor.

Note

> The drawing area appears as a black screen in the Sketcher window.

4. Now, move the cursor over the drawing area and *click* on the desired location to end the line in the drawing area. A line is created, as shown in Fig.Pro/E-3.7:

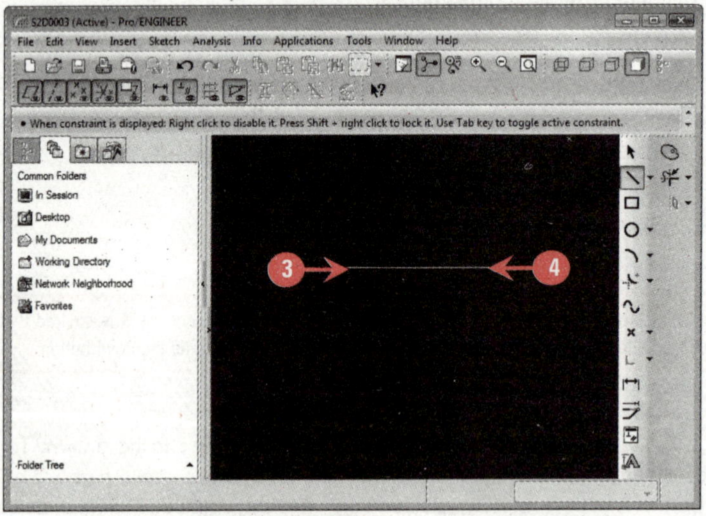

Fig.Pro/E-3.7

Another rubber-band is attached to the cursor when you click on the second location to end the line. Now, by clicking on any other location, you can create another line. However, if you do not want to create another line then click the middle mouse button to stop the functioning of the **Line** button. When you click the middle mouse button then the rubber-band line attached with the cursor disappears.

Rectangle

A rectangle in `Pro/ENGINEER Wildfire 4.0` is drawn using the **Rectangle** button on the Sketcher Tools toolbar. Following steps are used to draw a rectangle:

1. *Click* the **Rectangle** ( ) button on the **Sketcher Tools** toolbar (Fig.Pro/E-3.5).
2. Now, *click* any location in the drawing area (Fig.Pro/E-3.8) and drag the mouse to create a rectangle of desired size.
3. Next, *click* the left mouse button to end the rectangle. A rectangle is created, as shown in Fig.Pro/E-3.8:

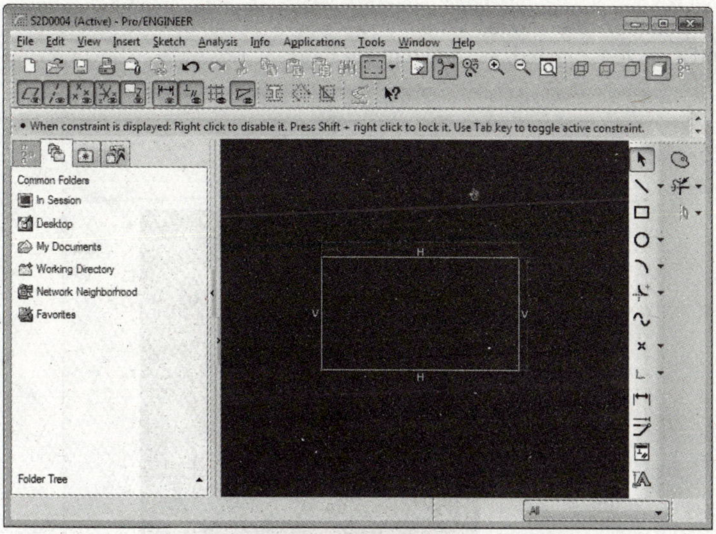

Fig.Pro/E-3.8

The V and H shown on the rectangle sides (Fig.Pro/E-3.8) are vertical and horizontal constraints respectively that are representing the horizontal and vertical lines of the rectangle. The gray lines (Fig.Pro/E-3.8) are called as weak dimensions that appear automatically and represent the horizontal and vertical dimensions of a rectangle.

Circle

In `Pro/ENGINEER Wildfire 4.0`, four buttons are available on the Sketcher Tools toolbar (Fig.Pro/E-3.5) to create a circle and one for ellipse. Following buttons are available on the Sketcher Tools toolbar in Pro/ENGINEER Wildfire 4.0 to create a circle:

- **Center and Point ( ):** Creates a circle by defining a center of the circle and a point on the circle.
- **Concentric ( ):** Creates a concentric circle.
- **3 Point ( ):** Creates a circle through three points.
- **3 Tangent ( ):** Creates a circle that is tangent to three existing entities.
- **Ellipse ( ):** Creates an ellipse.

To access these buttons, *click* the down arrow beside the Center and Point button (Fig.Pro/E-3.5) in the Sketcher Tools toolbar. A flyout appears, as shown in Fig.Pro/E-3.9:

Fig.Pro/E-3.9

Let's perform the following steps to create a circle by using the Center and Point button:

1. *Click* the down arrow beside the Center and Point ( ) button in the Sketcher Tools toolbar (Fig.Pro/E-3.5). A flyout appears, as shown in Fig.Pro/E-3.9.
2. Now, *select* the Center and Point button ( ) to draw a circle (Fig.Pro/E-3.9).
3. Next, *click* anywhere in the drawing area to specify the center of the circle. A yellow rubber-band gets attached to the cursor at the center point.
4. *Drag* the cursor to get the desired circle size.

5. Now, *click* a point on the circle. When you click a point on the circle the yellow rubber-band disappears and a circle is created, as shown in Fig.Pro/E-3.10:

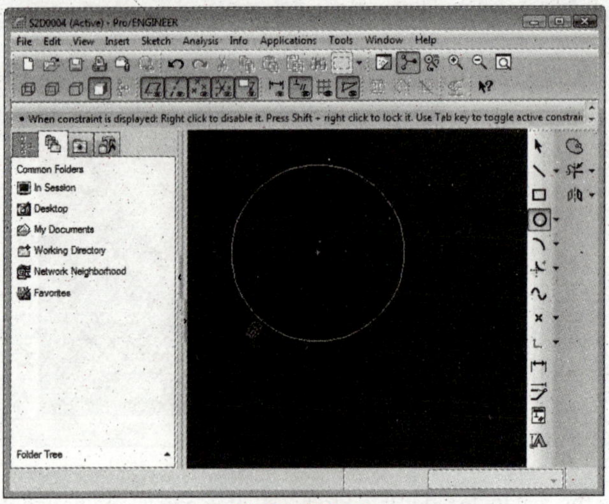

Fig.Pro/E-3.10

Continue the process to create more circles. However, if you do not want to use the **Center and Point** button further, then click the middle mouse button anywhere in the drawing area.

Arc

In `Pro/ENGINEER Wildfire 4.0`, five buttons are available on the `Sketcher Tools toolbar` (Fig.Pro/E-3.5) to draw an arc. Following are the buttons available on the `Sketcher Tools toolbar` in `Pro/ENGINEER Wildfire 4.0` to draw an arc:

- **3 Point/Tangent End ( ):** Draws an arc either by specifying three points on the drawing area or by drawing a tangent from the end of an existing entity.
- **Concentric ( ):** Draws an arc, which is concentric to an already existing entity. The existing entity must be an arc or a circle.
- **Center and Ends ( ):** Draws an arc by specifying a center point and an end point on the drawing area.
- **3 Tangent ( ):** Draws an arc, which is a tangent to three selected entities.
- **Conic ( ):** Draws a conic arc along with the selected entity.

To access these buttons, *click* the down arrow beside 3 Point/Tangent ( ) button in the Sketcher Tools toolbar. A flyout appears, as shown in Fig.Pro/E-3.11:

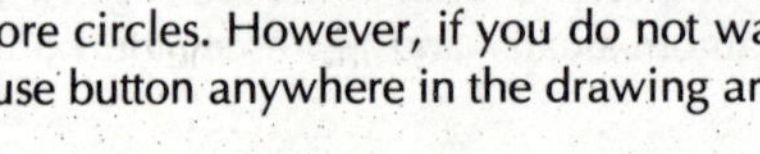

Fig.Pro/E-3.11

Let's perform the following steps to create an arc using **3 Point/Tangent** button:

1. *Click* the down arrow beside the **3 Point/Tangent** ( ) button in the Sketcher Tools toolbar (Fig.Pro/E-3.5). A flyout appears, as shown in Fig.Pro/E-3.11.
2. Now, *select* the **3 Point/Tangent** button ( ) (Fig.Pro/E-3.11) to draw an arc.
3. Next, *click* anywhere on the drawing area to specify the first point.

4. *Move* the cursor to size the arc.
5. Next, *click* on the drawing area to specify the second point.
6. Now, *click* anywhere on the arc to specify the third point to complete the arc creation. A sketched arc is shown in Fig.Pro/E-3.12:

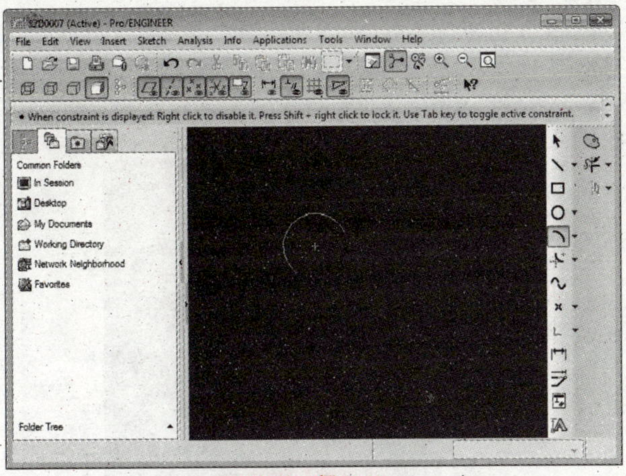

Fig.Pro/E-3.12

This is how to create an arc by specifying three points on the drawing screen.

Let's perform the following steps to create an arc, which is tangent to an already existing entity. In our case, a line already exists; therefore, we create an arc, which is tangent to that line:

1. *Click* the down arrow beside the **3 Point/Tangent** button in the Sketcher Tools toolbar (Fig.Pro/E-3.5). A flyout appears, as shown in Fig.Pro/E-3.11.
2. Now, *select* the **3 Point/Tangent** button (Fig.Pro/E-3.11) to create an arc.
3. Next, *click* at the end of the line. A green circle, which is known as the **Target** symbol, appears, as shown in Fig.Pro/E-3.13:

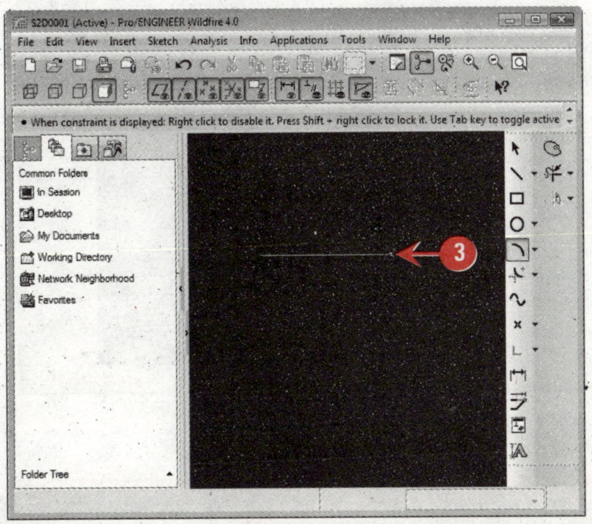

Fig.Pro/E-3.13

4. *Move* the cursor along the tangent direction to draw an arc. A yellow rubber-band gets attached to the cursor.
5. *Click* anywhere in the tangent direction to complete the arc. As soon as you click, the **Target** symbol disappears. Repeat steps 3 to 5 to create another arc at the other end of the line; a figure is created, as shown in Fig.Pro/E-3.14:

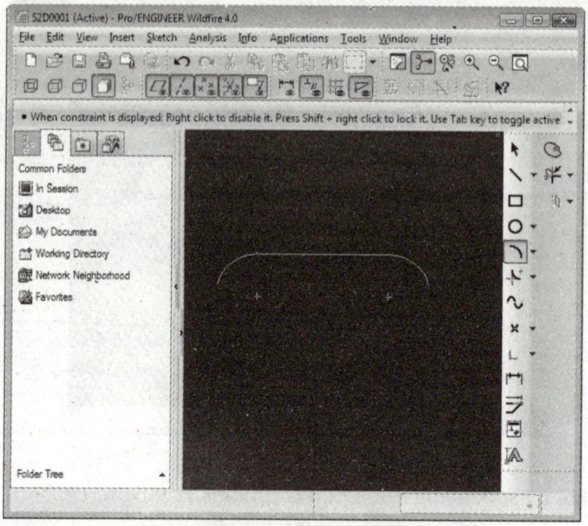

Fig.Pro/E-3.14

6. *Click* the middle mouse button to stop using the **3 Point and Tangent** button. As soon as, you click the middle mouse button, weak dimensions appear to let you know the dimensions of the arc you have created, as shown in Fig.Pro/E-3.15:

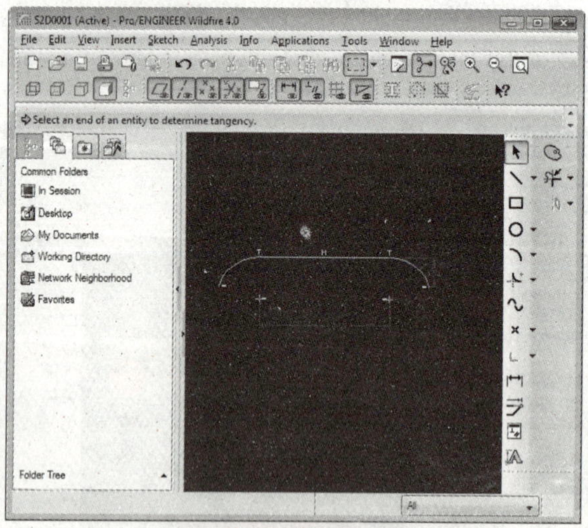

Fig.Pro/E-3.15

Note

If you do not want to create an arc, which is tangent to an entity then move the cursor perpendicular to the end point of an entity.

Circular Fillet

A circular fillet is an arc, which is used to join two entities such as two lines, two arcs, or one line-one arc. Following are the steps to create a circular fillet:

1. *Click* the down arrow beside the **Circular** ( ) button in the Sketcher Tools toolbar (Fig.Pro/E-3.5). A flyout appears, as shown in Fig.pro/E-3.16.

2. Now, *select* the **Circular** button ( ) (Fig.Pro/E-3.16) to create a circular fillet, as shown in Fig.Pro/E-3.16:

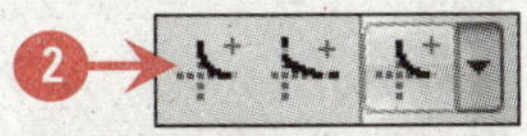

Fig.Pro/E-3.16

In our case, we create a circular fillet to join two lines, shown in Fig.Pro/E-3.17:

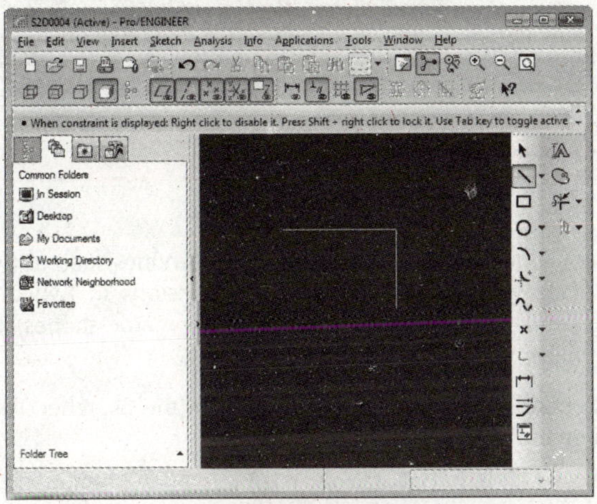

Fig.Pro/E-3.17

Note

> The creation of the circular fillet depends on the positions selected on the entities. For example, if you click anywhere on two entities then a circular fillet is created between the clicked points. On the other hand, if you select the two corners of two entities then a circular fillet joins the two entities from their selected corners.

3. As soon as, you *click* the **Circular** button, a **Select** dialog box appears which prompts you to select an entity, as shown in Fig.Pro/E-3.18:

Fig.Pro/E-3.18

Now, *click* anywhere on the first entity. In our case, we clicked near about right end on the horizontal line (Fig.Pro/E-3.17).

4. Now, you are prompted again to select the second entity; therefore, *click* anywhere on the second entity. In our case, we clicked near about upper end on the vertical line (Fig.Pro/E-3.17). As soon as, you *click* on the second entity, a circular fillet is automatically created, as shown in Fig.Pro/E-3.19:

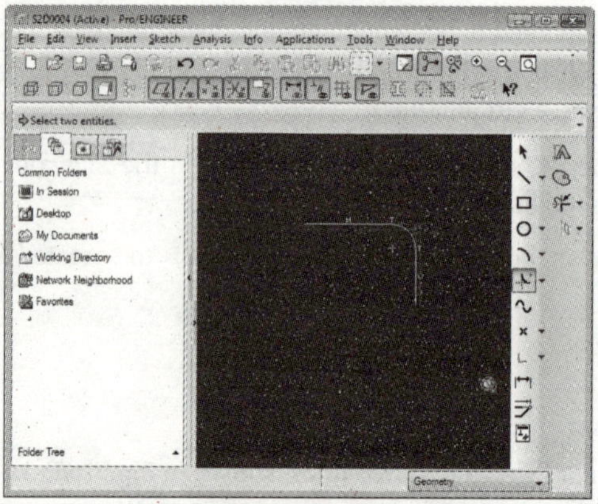

Fig.Pro/E-3.19

5. Next, *click* the **OK** button on the **Select** dialog box (Fig.Pro/E-3.18) to complete the circular fillet.

Note

> A circular fillet cannot be created between two parallel lines.

Elliptical Fillet

An elliptical fillet is an elliptical arc, which is used to join two entities such as two lines, one line-one arc, or two arcs. The geometry of an ellipse is controlled by dimensions in two directions: x and y direction. Therefore, when an elliptical fillet is modified, more curved geometric shapes are created. Following are the steps to create an elliptical fillet:

1. *Click* the down arrow beside the **Circular** ( ) button in the Sketcher Tools toolbar (Fig.Pro/E-3.5). A flyout appears, as shown in Fig.pro/E-3.16.
2. Now, *select* the **Elliptical** ( ) button (Fig.Pro/E-3.16) to create an elliptical fillet. In our case, we create an elliptical fillet to join two lines, shown in Fig.Pro/E-3.20:

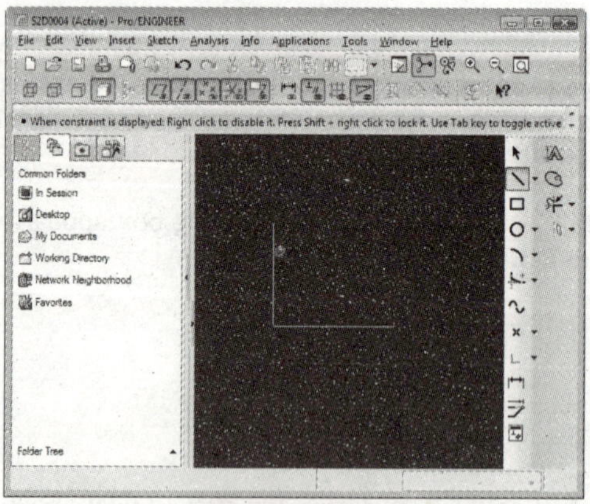

Fig.Pro/E-3.20

Note

> The creation of an elliptical fillet depends on the positions selected on the entities. For example, if you click anywhere on two entities then an elliptical fillet is created between the clicked points.

3. As soon as you click the **Elliptical** button, a **Select** dialog box (Fig.Pro/E-3.18) appears, which prompts you to select an entity. Therefore, *click* anywhere on the first entity. In our case, we clicked near about lower end on the vertical line (Fig.Pro/E-3.20).
4. Now, you are prompted again to select the second entity; therefore, *click* anywhere on the second entity. In our case, we clicked near about left end on the horizontal line (Fig.Pro/E-3.20). As soon as you *click* on the second entity, an elliptical fillet is automatically created, as shown in Fig.Pro/E-3.21:

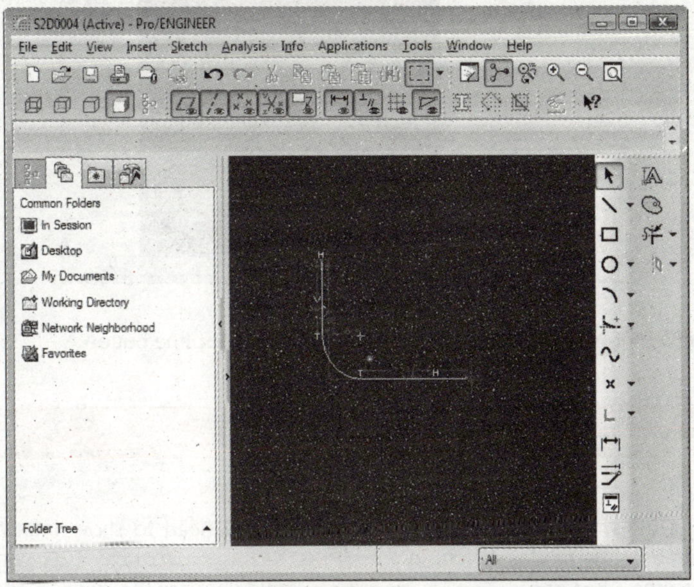

Fig.Pro/E-3.21

5. Next, *click* the **OK** button on the *Select* dialog box to complete creating an elliptical fillet, as shown in Fig.Pro/E-3.22:

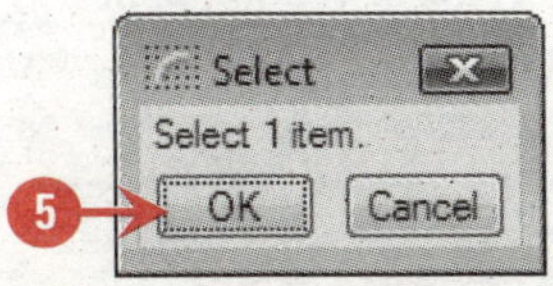

Fig.Pro/E-3.22

This is how an elliptical fillet is created. Let's move ahead by depicting another sketching entity, splines.

Splines

`Splines` are known as the curves that can smoothly pass through many numbers of intermediate points. Splines are very flexible; therefore, they are generally used to sketch surface of a model such as, a car, bus, and aero plane. Following are the steps to sketch a spline:

1. *Click* the **Spline** (∿) button on the Sketcher Tools toolbar (Fig.Pro/E-3.5).
2. Now, *click* anywhere in the drawing area to select the starting point of a spline.
3. Next, *click* again on drawing area to select the next point of the spline.
4. Repeat the step 3 to select multiple points of the spline.
5. To complete the process of creating the spline *double-click* at any point in the drawing area. A spline as shown in Fig.Pro/E-3.23 is created:

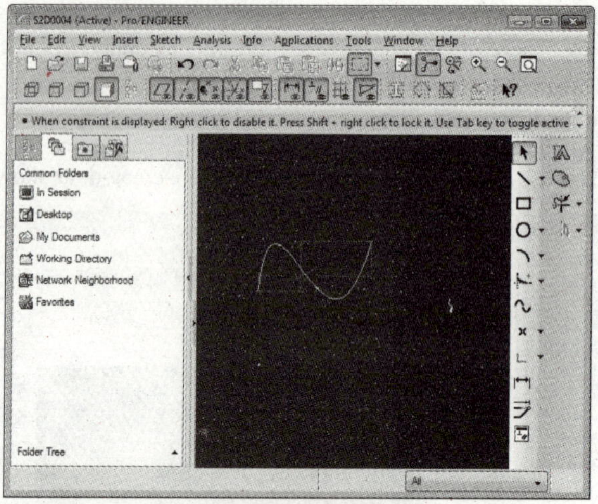

Fig.Pro/E-3.23

6. *Click* the middle mouse button to stop the functioning of the Spline button.

Note

The points from which a spline passes are known as interpolation points.

Text

When you create a 2D model of a 3D feature then you may need to show information such as model number, model name on the model sketch. In `Pro/ENGINEER Wildfire 4.0`, you can do this by using the `Text` button on the `Sketcher Tools toolbar`. Following are the steps to add the text on a model sketch:

1. *Click* the **Text** ( ) button on the Sketcher Tools toolbar (Fig.Pro/E-3.5). A message appears in the message area that prompts you to select the start point of line to determine text height and orientation, as shown in Fig.Pro/E-3.24:

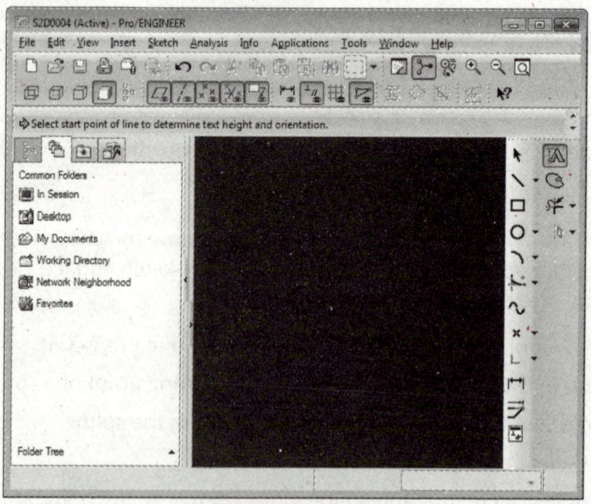

Fig.Pro/E-3.24

2. *Click* anywhere in the drawing area to select the start point. As soon as you click in the drawing area to select a start point; another message appears in the message area. This new message prompts you to select second point of line to determine text height and orientation, as shown in Fig.Pro/E-3.25:

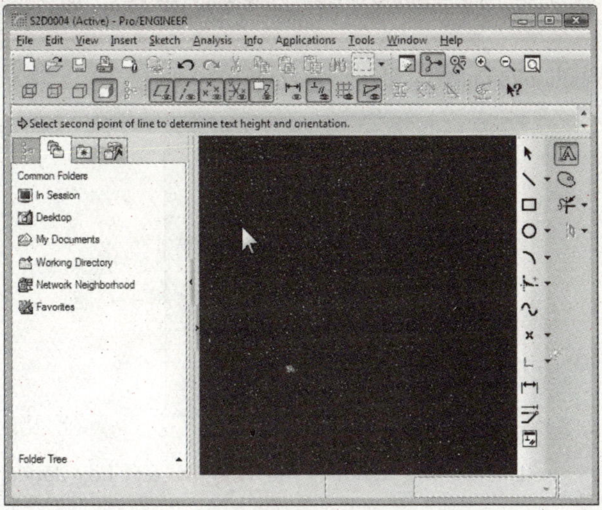

Fig.Pro/E-3.25

When you move the cursor to select the second point, a construction line appears between the start point and the end point. The length of the construction line determines the height of the text and the angle of the text determines the orientation of the text

3. *Click* either in upward or downward direction to the start point to select the second point. The **Text** dialog box appears (Fig.Pro/E-3.26).

Note

> If you select the second point on the upward direction from the start point then the text is written from left to right direction and in upright state. However, if you want to write text in downside and right to left state then select the second point in the downward direction from the start point.

There are various options available in the **Font** section on **Text** dialog box to modify the appearance of the text entered in the **Text line** text box. The options available in the **Font** section on the **Text** dialog box are as follows:

- **Font:** By default, the font for the text entered in the **Text Line** text box is **font3D**. To change the font type, click the down arrow beside the **Font** label and select the font as per your requirement.
- **Position:** The text is positioned in horizontal and vertical directions. By default, the Horizontal position of the text is Left while the Vertical position is Bottom. To change the horizontal position, click the down arrow beside the **Horizontal** label in the **Position** option and select any other available position. The available horizontal text positions are: **Left**, **Center**, and **Right**.
- To change the vertical position, click the down arrow beside the **Vertical** label in the **Position** option and select any other available position. The available horizontal text positions are: **Bottom**, **Middle**, and **Top**.
- **Aspect ratio:** The aspect ratio is modified by moving the slider on left or right direction.
- **Slant angle:** The slant angle can be modified by moving the slider on left or right direction.

4. *Enter* some text in the **Text line** text box. By default, the **Text Line** text box is empty. In our case, we have entered "Text Button" in this text box(Fig.Pro/E-3.26).

Note

> You can enter the text in the Text line text box up to the length of 79 characters. To add any symbol in this text box, click the Text Symbol button on the Text dialog box. The Text Symbol dialog box appears, select the symbol to insert in the text box, and click the Close button to close the Text Symbol dialog box.

5. Now, *click* the **Ok** button to finish, as shown in Fig.Pro/E-3.26:

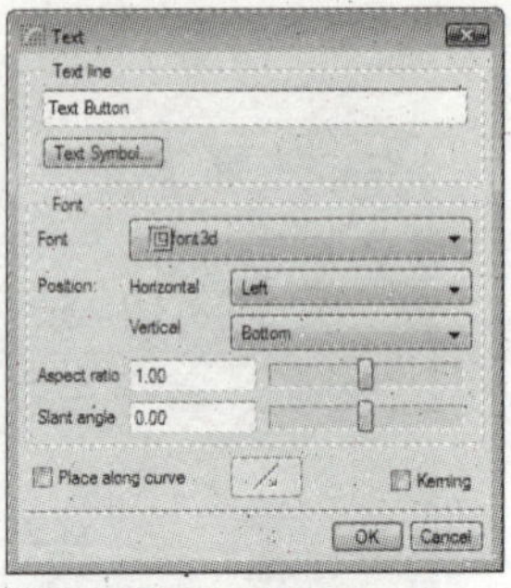

Fig.Pro/E-3.26

The **Text** dialog box is closed. The text entered in the **Text line** text box appears on the drawing area, as shown in Fig.Pro/E-3.27:

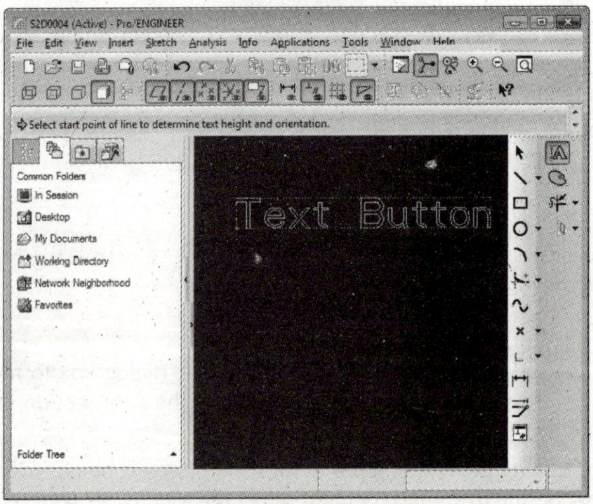

Fig.Pro/E-3.27

Note

> You can click the Cancel button in the Text dialog box to exit from it without doing any action.

This is all about drawing the sketch entities. Now, the next step in the sketch creation is dimensioning the sketch entities.

Working with Dimensions

Whenever a sketching entity is created, some dimensions are automatically applied on that entity. These dimensions are known as `weak dimensions` and are `gray` in color. These dimensions are called as weak dimensions because they are generated and removed automatically without any confirmation.

Apart from the weak dimensions, you can also apply your own dimensions on the sketched entities. Dimensions that you apply on sketched entities are known as `strong dimensions` that are mainly used to control their size. When you apply strong dimensions on any sketched entity then weak dimensions are automatically deleted.

Dimensions can be added on sketched entities in two ways: first, by selecting the **Dimensions** option in the **Sketch** menu in Pro/ENGINEER Wildfire 4.0 window; second, by using the **Normal** button available on the Sketcher Tools toolbar in the Pro/ENGINEER Wildfire 4.0 window. In this section, we see dimensioning the sketched entities using the **Normal** button. Let's now follow the steps given here to dimension the sketched entities using the **Normal** button:

1. *Click* the **Normal** (↦) button on the **Sketcher Tools toolbar** (Fig.Pro/E-3.5). A **Select** dialog box, as shown in Fig.Pro/E-3.28 appears that prompts you to select an entity:

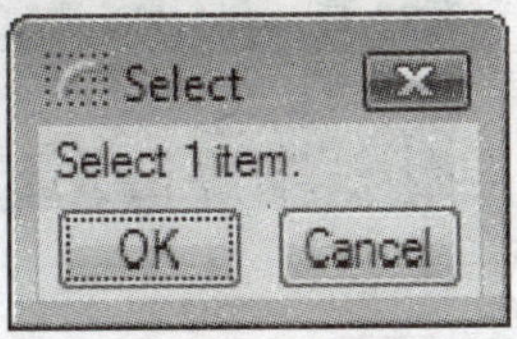

Fig.Pro/E-3.28

In our case, we have sketched a rectangle.

2. Now, *select* one of the horizontal lines of the rectangle to apply dimensions on it. As soon as you select the horizontal line, the color of the line is changed to red from yellow (Fig.Pro/E-3.29).
3. Next, move the cursor on the selected line and *click* the middle mouse button to apply dimensions on it (Fig.Pro/E-3.29):

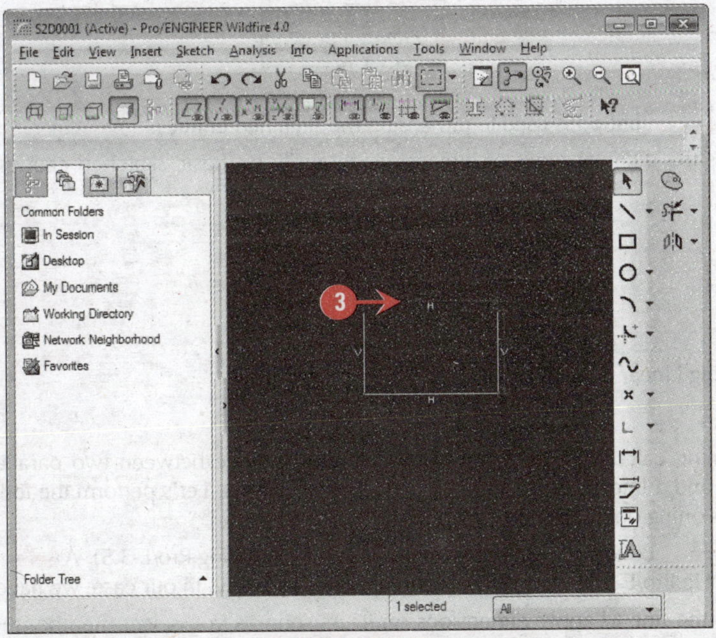

Fig.Pro/E-3.29

As soon as, you click the middle mouse button, dimension is applied on the selected line of the rectangle, as shown in Fig.Pro/E-3.30:

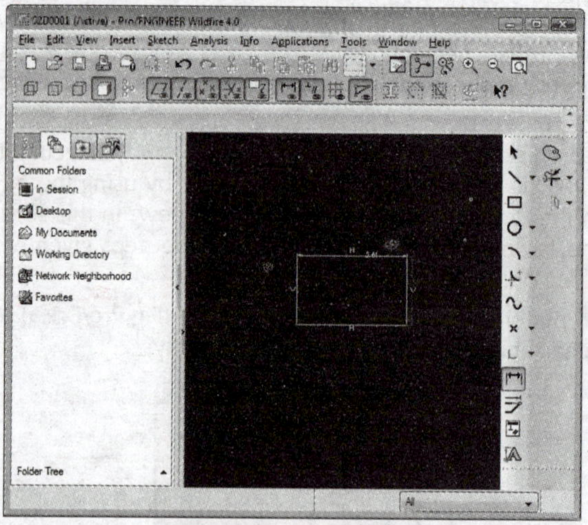

Fig.Pro/E-3.30

Note

The color of the horizontal line is changed to yellow once again.

4. Now, *click* the OK button to stop using the Normal button, as shown in Fig.Pro/E-3.31:

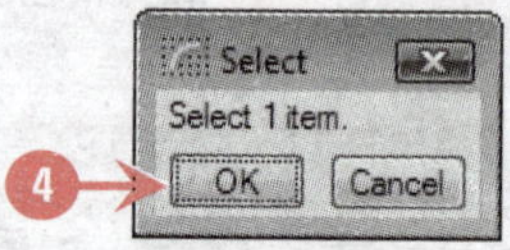

Fig.Pro/E-3.31

This is how the Normal button is used to dimension the sketched entities.

Now, we discuss the different types of dimensions to apply on basic sketched entities under the following subheadings:

- Linear
- Radial
- Angular
- Diameter

Let's start depicting Linear dimension.

Linear

Linear dimensioning calculates the length of a line, the distance between two parallel lines, the distance between a point and a line, and the distance between two points. Let's perform the following steps to apply the linear dimensioning on a line:

1. *Click* the Normal button (|↔|) on the Sketcher Tools toolbar (Fig.Pro/E-3.5). A Select dialog box appears, as shown in Fig.Pro/E-3.31 that prompts you to select an entity. In our case, we have sketched a line.
2. Now, *select* the line to apply dimensions on it. As soon as you select the line, the color of the line is changed to red from yellow (Fig.Pro/E-3.32).
3. Next, *move* the cursor on the selected line and *click* the middle mouse button to apply dimensions on it (Fig.Pro/E-3.32):

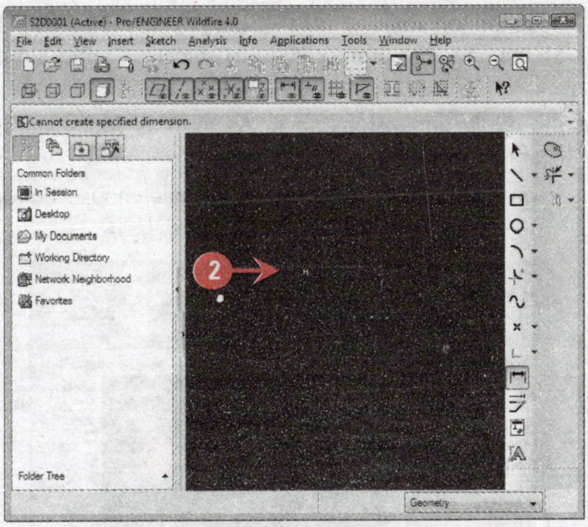

Fig.Pro/E-3.32

As soon as, you click the middle mouse button, dimension is applied on the line, as shown in Fig.Pro/E-3.33:

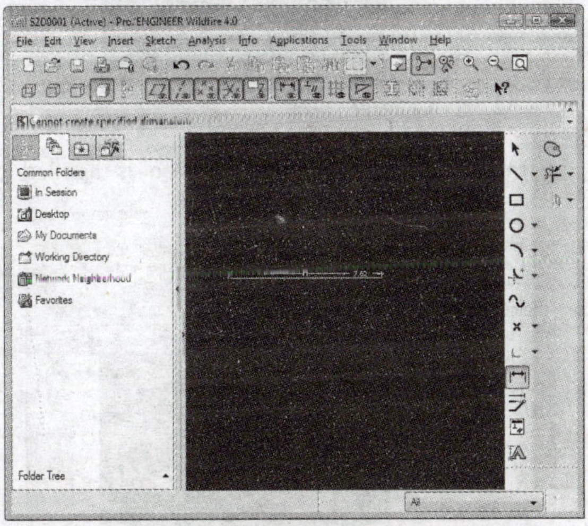

Fig.Pro/E-3.33

The color of the line gets changed to yellow once again.

Note

> - The location on the line where the dimension will appear depends on the point where you click the middle mouse button.
> - A center line cannot be dimensioned because it is infinite.

4. Now, *click* the **OK** button to stop using the **Normal** button (Fig.Pro/E-3.31).

Radial

`Radial` dimensioning is used to calculate the radius of an arc or a circle. In this section, we see the process to apply radial dimension on a circle. Following are the steps to apply radial dimension on a circle:

1. *Click* the **Normal** (↦) button on the **Sketcher Tools toolbar** (Fig.Pro/E-3.5). A **Select** dialog box, as shown in Fig.Pro/E-3.31 appears that prompts you to select an entity. In our case, we use circle as an entity.
2. Now, *select* the circle to apply dimensions on it. As soon as you select the circle, weak dimension is applied on the circle (Fig.Pro/E-3.34).
3. Next, *move* the cursor on the circle and *click* the middle mouse button to apply dimensions on it (Fig.Pro/E-3.34):

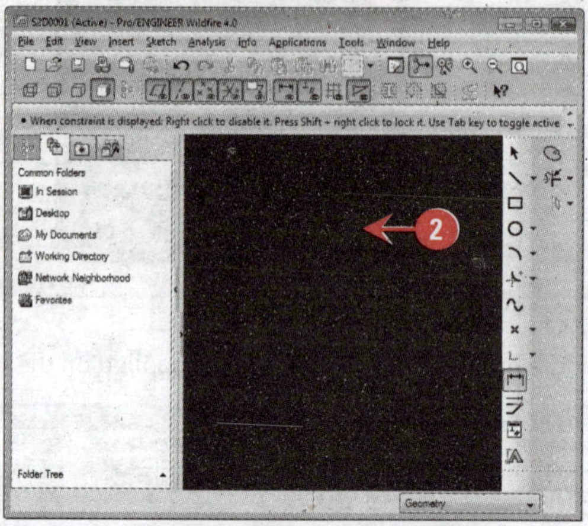

Fig.Pro/E-3.34

As soon as, you click the middle mouse button, dimension is applied on the circle, as shown in Fig.Pro/E-3.35:

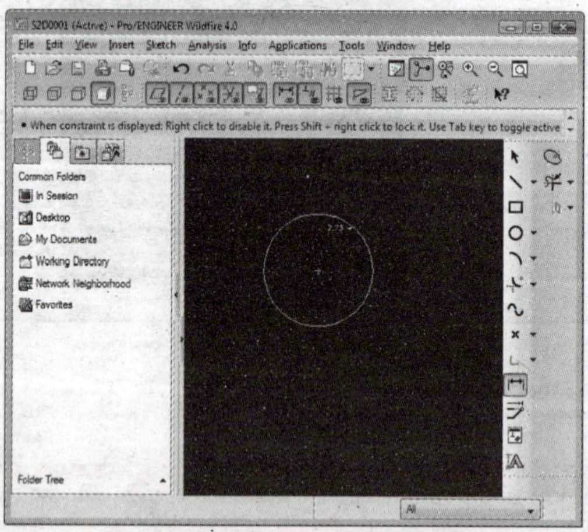

Fig.Pro/E-3.35

Radius of the circle appears in white, as shown in Fig.Pro/E-3.35. In our case, radius of the circle is 2.75.

Note

> The location on the circle where the radius will appear depends on the point where you click the middle mouse button.

4. Now, *click* the **OK** button to stop using the **Normal** button (Fig.Pro/E-3.31).

Angular

`Angular` dimensioning is used to calculate the angle of an arc or the angle between two lines. In this section, we see the process to calculate angle between two lines. Following are the steps to apply angular dimension on a circle:

1. *Click* the **Normal** (⊢⊣) button on the **Sketcher Tools toolbar** (Fig.Pro/E-3.5). A **Select** dialog box appears, as shown in Fig.Pro/E-3.31 that prompts you to select an entity. In our case, we use two lines, shown in Fig.Pro/E-3.36.
2. Now, *select* the horizontal line (Fig.Pro/E-3.36). As soon as you select the horizontal line, the **Select** dialog box appears again that prompts you to select a line. The prompt select a line here means to select the second line (Fig.Pro/E-3.31).
3. Next, *select* the second line, as shown in Fig.Pro/E-3.36:

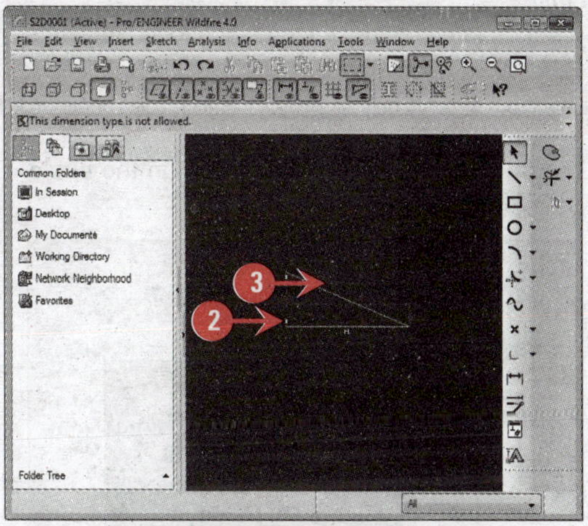

Fig.Pro/E-3.36

4. Now, *click* the middle mouse button on any of the line. Angular dimension is applied on the lines, as shown in Fig.Pro/E-3.37:

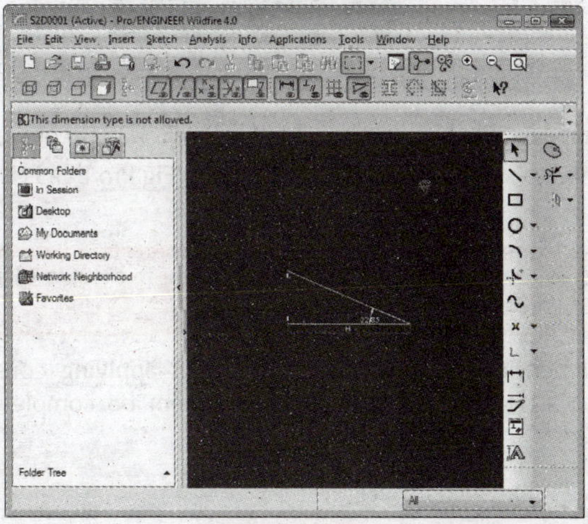

Fig.Pro/E-3.37

Angle between two lines appears in white, as shown in Fig.Pro/E-3.37. In our case, angle is 22.85 °.

Note

> Depending on the location of the dimension, the angle can be acute or obtuse.

5. Finally, *click* the **OK** button to stop using the **Normal** button (Fig.Pro/E-3.31).

Diameter

`Diameter` dimensioning is used to calculate the diameter of a circle or on an arc. In this section, we discuss the process to apply diameter dimensioning on a circle. Follow the steps given here to apply diameter dimensioning on a circle:

1. *Click* the **Normal** (⊢⊣) button on the **Sketcher Tools toolbar** (Fig.Pro/E-3.5). A **Select** dialog box as shown in Fig.Pro/E-3.31 appears that prompts you to select an entity. In our case, we use a circle as an entity.
2. Now, *double-click* the circle.

Note

> By double clicking the diameter is applied but if you do only single click then the radius of the circle is applied.

3. Next, *click* the middle mouse button to apply dimension on the circle. Diameter dimension is applied on the circle, as shown in Fig.Pro/E-3.38:

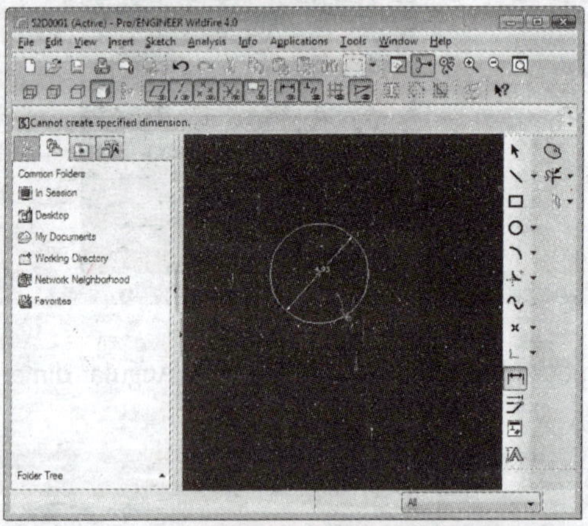

Fig.Pro/E-3.38

The diameter of the circle appears in white, as shown in Fig.Pro/E-3.38. In our case, the diameter is 4.95.

4. Now, *click* the **OK** button to stop using the **Normal** button (Fig.Pro/E-3.31).

Note

> Apart from dimensioning line, arc, and circle, you can also dimensioned splines. To dimensioned splines, use the Coordinate System button available on the Sketcher Tools toolbar.

This is how entities are dimensioned. Now, we study about applying constraints on sketching entities because an entity in `Pro/ENGINEER Wildfire 4.0` cannot be completed until it is not completely dimensioned and constrained.

Applying Constraints

Constraints are applied on entities in a 2D sketch to make them accurate in respect to the position of other entities. Constraints are applied on an entity by using the buttons available in the **Constraints** dialog box.

Note

> The process to access Constraints dialog box is discussed ahead in this section.

Let's now see Fig.Pro/E-3.39 to view the **Constraints** dialog box:

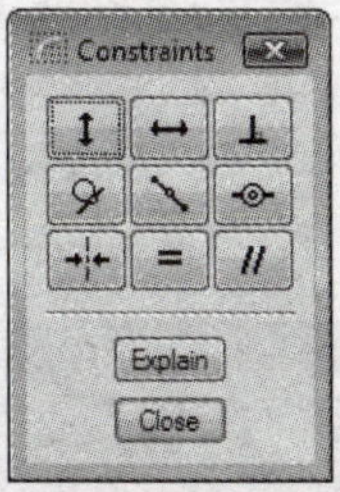

Fig.Pro/E-3.39

Table 3.1 lists the buttons available in the **Constraints** dialog box:

Table 3.1: Constraints dialog box

Button	Function
	Causes a selected line or two vertices to be vertical.
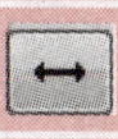	Causes a selected line or two vertices to be horizontal.
	Causes the two entities to be perpendicular.
	Causes the two entities to be tangent.
	Places a selected point or vertex on the middle of a line.
	Collinear the two selected entities. This button also makes two points coincident.
	Causes two points or vertices to be symmetric about a centerline.
	Makes the two selected entities such as lines, arcs, circles, or ellipses become equal in dimensions. If you apply this constraint on two lines then those lines become equal in their length. However, if you apply this constraint on arcs, circles, or ellipses then these entities become equal with respect to their radii.
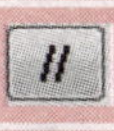	Causes two lines to be parallel.
Explain	Provides information about the constraints applied to a sketch.
Close	Closes the Constraints dialog box.

Now, you are acquainted with the buttons available on the **Constraints** dialog box; therefore, now we can study the process to create constraints on a sketch. In this section, we discuss the use of (=) button. Follow the steps given here to apply constraints on a sketch:

1. *Click* the **Constrain** ( ) button on the **Sketcher Tools toolbar** (Fig.Pro/E-3.5). The **Constraints** dialog box appears (Fig.Pro/E-3.39).

2. Now, *click* (=) button in the **Constraints** dialog box. A **Select** dialog box (Fig.Pro/E-3.31) appears that prompts you to select an entity. In this section, we apply equal length constraints on two lines of unqual length(Fig.Pro/E-3.40).
3. Next, *select* the left vertical line (Fig.Pro/E-3.40).
4. Now, *select* the right vertical line, as shown in Fig.Pro/E-3.40:

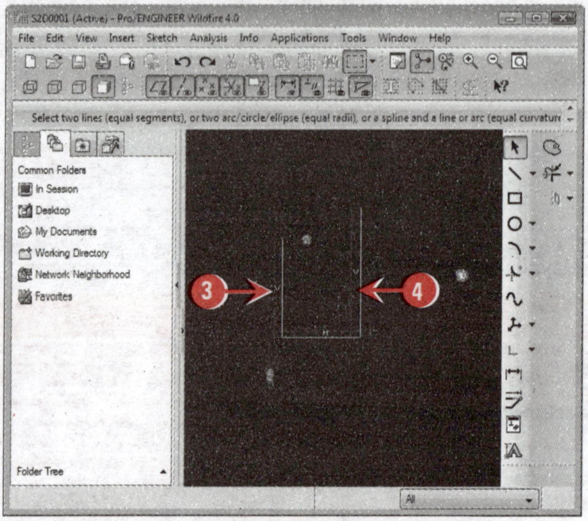

Fig.Pro/E-3.40

As soon as, you select the right vertical line both of the lines become equal, as shown in Fig.Pro/E-3.41:

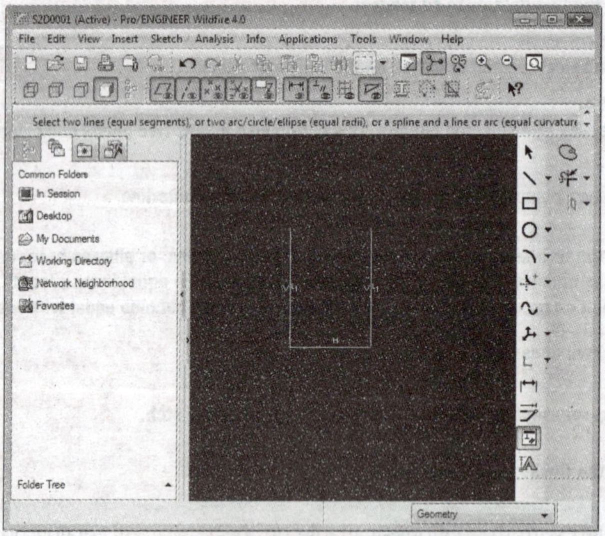

Fig.Pro/E-3.41

5. Next, *click* the **OK** button to close the **Select** dialog box (Fig.Pro/E-3.31).
6. Finally, *click* the **Close** button on the **Constraints** dialog box (Fig.Pro/E-3.39) to close it.

This is how; constraints are applied on a sketch. Now, we discuss about modifying dimensions.

Modifying Dimensions

Once you have completed a sketch, you may need to modify the dimensions of one or more entities in that sketch to make those entities accurate with other entities. To modify the dimensions of the entities in a sketch, following four options are available in Pro/ENGINEER Wildfire 4.0:

- Modify Button
- Modify Option in the Edit Menu
- Double-clicking a Dimension
- Modifying Dimension Dynamically

Let's start with modify button.

Modify Button

The **Modify** (⇁) button is available on the Sketcher Tools toolbar (Fig.Pro/E-3.5) and is the most commonly used option to modify dimensions. Following are the steps that let you modify dimensions using the **Modify** button:

1. *Click* a dimension in a sketch that you want to modify. When you click a dimension then its color becomes red. In our case, we are modifying the dimensions of a line, as shown in Fig.Pro/E-3.42:

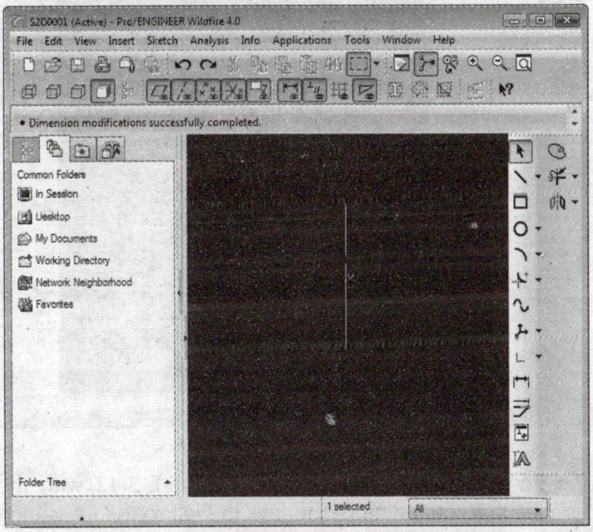

Fig.Pro/E-3.42

Note

Hold CTRL key, if you want to select more than one dimension to modify.

2. Now, *click* the **Modify** (⇁) button on the Sketcher Tools toolbar (Fig.Pro/E-3.5). **Modify Dimensions** dialog box appears (Fig.Pro/E-3.43).
3. *Modify* the dimension value appearing in the text box (Fig.Pro/E-3.43) by entering the new dimension value. The value appearing in the text box (Fig.Pro/E-3.43) is the current dimension of the entity, you want to modify. In our case, we entered 2.78872346793 as new value.
4. *Click* the **Build** (✔) button to reflect the changes in the drawing area, as shown in Fig.Pro/E-3.43:

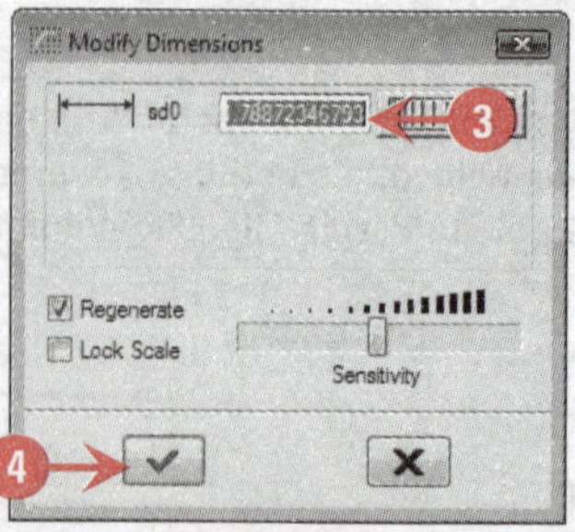

Fig.Pro/E-3.43

Dimension of the selected line is changed, as shown in Fig.Pro/E-3.44:

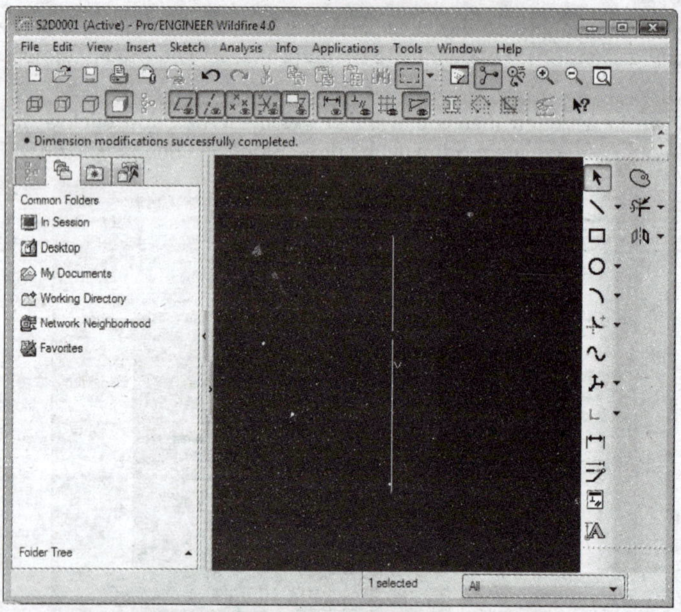

Fig.Pro/E-3.44

Now, if you compare the line dimension appearing in Fig.Pro/E-3.42 with the dimension appearing in Fig.Pro/E-3.44, then you notice that the length of the line in Fig.Pro/E-3.44 is increased. This is because the new dimension value, 2.78872346973 is more than the older value, 1.79.

Modify Option in the Edit Menu

Dimensions in a sketch can also be modified using the Modify option in the Edit menu. To modify dimensions using the Modify options in the Edit menu, follow the steps given here:

1. *Click* Edit→Modify... in the menu bar. A check mark appears on the left of the Modify... option in the Edit menu, which means the Modify... option can be used, now.
2. Now, *select* a dimension in a sketch to modify. As soon as, you select a dimension, the Modify Dimensions dialog box appears (Fig.Pro/E-3.43).
3. *Enter* new dimensions to modify the current dimensions in the text box on the Modify Dimensions dialog box.
4. *Click* the Build () button to reflect the changes in the drawing area.

Double-clicking a Dimension

Another way to modify a dimension is to double-click a dimension. Follow the steps given here to modify a dimension by double-clicking:

1. *Double-click* a dimension in a sketch. A text box appears showing you the current dimension.
2. *Edit* the current dimension.
3. Now, either press the ENTER key or click the middle mouse button on the drawing area. Changes are reflected in the drawing area.

Modifying Dimension Dynamically

Dimension of an entity can also be modified by dragging the dimension. Following are the steps that help you modifying a dimension by dragging it:

1. *Move* the cursor on the entity whose dimension you have to modify. When you move the cursor on an entity, its color becomes cyan.
2. Now, place the cursor on the end of an entity and drag it. When you drag the entity then along with the entity its dimension is also changed. The changed dimension appears as weak dimension along with the entity.

This is all about the modifying dimensions. Now, we move ahead by depicting the process to trim entities.

Trimming Entities

Whenever a sketch is created, some unwanted areas may also be created in various entities in that sketch. To remove those areas, you need to trim the entities. In Pro/ENGINEER Wildfire 4.0, following three buttons that are available in the Sketcher Tools toolbar can be used to trim entities:

- Delete Segment Button
- Corner Button
- Divide Button

Let's discuss the process of trimming entities using these buttons.

Delete Segment Button

The Delete Segment ( ) button deletes the selected entity. Follow the steps given here to use the Delete Segment button to trim an entity in a sketch:

1. *Click* the down arrow beside the Delete Segment ( ) button in the Sketcher Tools toolbar (Fig.Pro/E-3.5). A flyout appears (Fig.Pro/E-3.45).
2. Now, *select* the Delete Segment ( ) button, as shown in Fig.Pro/E-3.45:

Fig.Pro/E-3.45

3. Next, move the cursor over the entity you want to trim in the sketch. In our case, the sketch is shown in Fig.Pro/E-3.46:

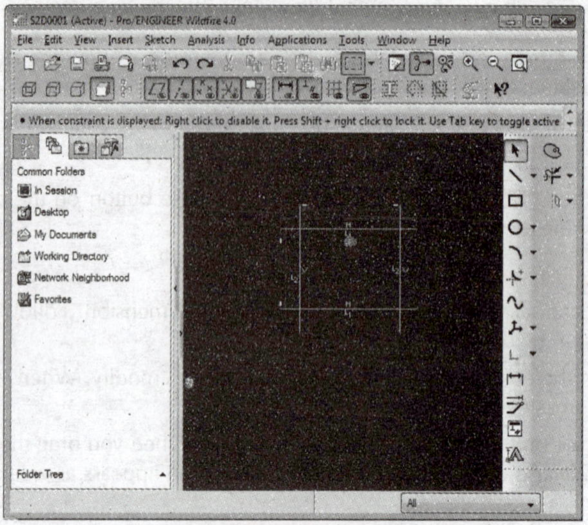

Fig.Pro/E-3.46

When you move the cursor on an entity, the color of that entity becomes cyan.

4. *Click* the left mouse button. The entity is deleted. In our case, the sketch after trimming an entity is shown in Fig.Pro/E-3.47:

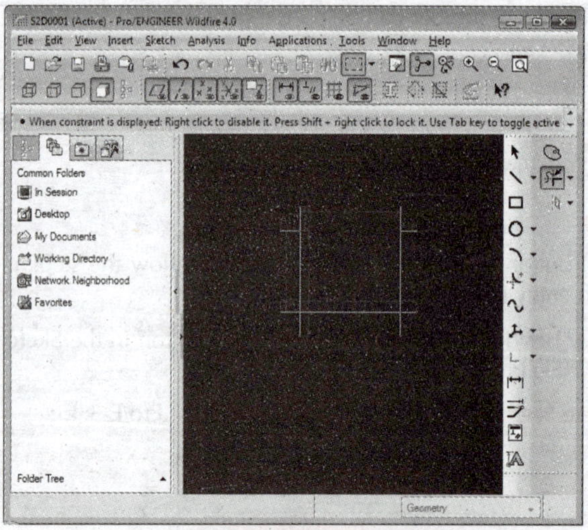

Fig.Pro/E-3.47

Corner Button

The **Corner** button in `Pro/ENGINEER Wildfire 4.0` trims two entities at their corner. When using the **Corner** button to trim entities, you must remember that only the selected portion of an entity is retained and the other is trimmed. Follow the steps given here to trim entities using the **Corner** button:

1. *Click* the down arrow beside the **Delete Segment** ( ) button in the **Sketcher Tools toolbar** (Fig.Pro/E-3.5). A flyout appears, as shown in Fig.Pro/E-3.45.

2. Now, *select* the **Corner** ( ) button (Fig.Pro/E-3.45). The **Select** dialog box appears (Fig.Pro/E-3.31).

3. *Select* the first entity; you want to trim in the sketch (Fig.Pro/E-3.46). In our case, we select the left vertical line in the sketch (Fig.Pro/E-3.46). Now, you prompt again to select the second entity.
4. *Select* the second entity in the sketch (Fig.Pro/E-3.46). In our case, we selected the upper horizontal entity in the sketch (Fig.Pro/E-3.46).

As soon as, you select the second entity, both the entities are trimmed, as shown in Fig.Pro/E-3.48:

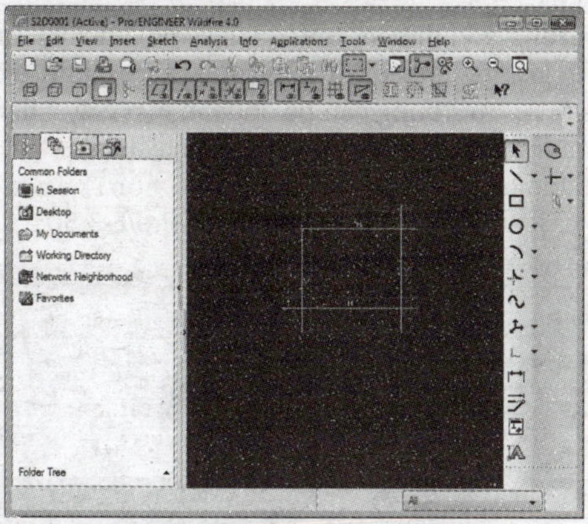

Fig.Pro/E-3.48

5. Now, *click* the **OK** button (Fig.Pro/E-3.31) on the **Select** dialog box to stop using the **Corner** button.

Divide Button

The **Divide** button divides an entity in many parts depending on the points specified in that entity. Follow the steps given here to use the **Divide** button:

1. *Click* the down arrow beside the **Delete Segment** ( ) button in the **Sketcher Tools toolbar** (Fig.Pro/E-3.5). A flyout appears, as shown in Fig.Pro/E-3.45.
2. Now, *select* the **Divide** ( ) button (Fig.Pro/E-3.45).
3. *Move* the cursor over the entity (Fig.Pro/E-3.46).
4. Now, *click* on the point over the entity where you want to divide. In our case, we *click* in the middle of the upper horizontal line (Fig.Pro/E-3.46). The selected entity is divided in two entities, as shown in Fig.Pro/E-3.49:

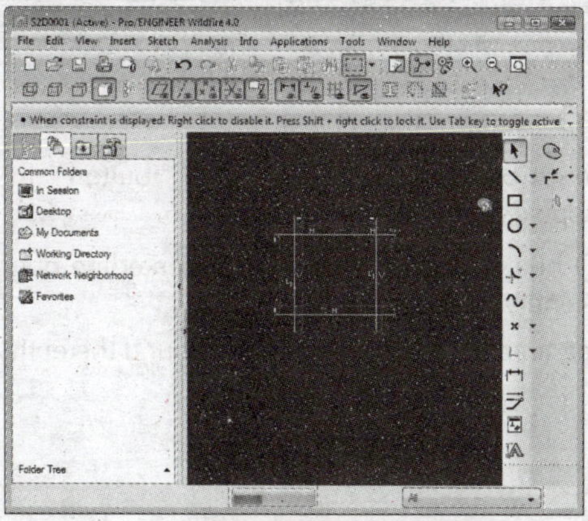

Fig.Pro/E-3.49

You can see a very small blue dot in the middle of the upper horizontal line, as shown in Fig.Pro/E-3.49, which means the line is divided in two parts. Now, this line can be treated as two separate entities. Apart from dividing a line, the **Divide** button can also be used to divide entities such as an arc, a circle.

Now, we discuss the process of mirroring entities.

Mirroring Entities

Mirroring is the process of creating and dimensioning the symmetrical entities. Mirroring saves your time by creating symmetrical entities in just few steps. How mirroring saves time is discussed ahead in this section. In `Pro/Engineer Wildfire 4.0`, mirroring can be done in two ways: first, through using the **Mirror** button on the `Sketcher Tools toolbar`; second, by accessing the **Mirror** option in the **Edit** menu. In this chapter, we use the **Mirror** button on the `Sketcher Tools toolbar` to mirror entities. To mirror entities, you need a sketch and in our case, the sketch is shown in Fig.Pro/E-3.50:

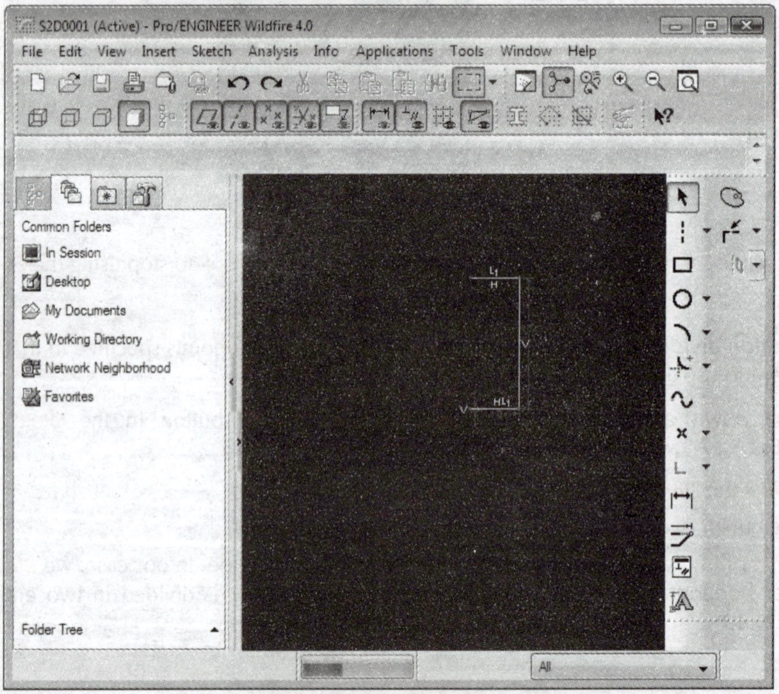

Fig.Pro/E-3.50

Let's perform the following steps to mirror the entities:

1. *Click* the down arrow beside the **Line** ( ) button in the **Sketcher Tools toolbar** (Fig.Pro/E-3.5). A flyout appears, as shown in Fig.Pro/E-3.6.
2. *Select* the **Centerline** ( ) button (Fig.Pro/E-3.6). The Centerline button is selected to create an axis of revolution along with the entity will be mirrored.
3. Now, *click* on the drawing area to create a centerline around the entity, you want to mirror, as shown in Fig.Pro/E-3.51:

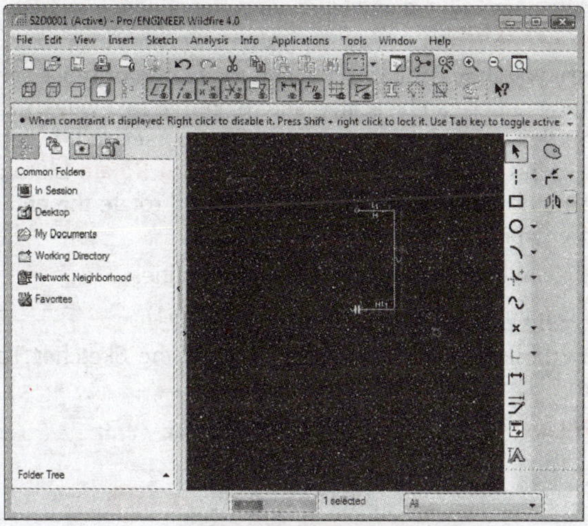

Fig.Pro/E-3.51

The infinite red colored line (Fig.Pro/E-3.51) is called as the centerline.

4. Next, *select* an entity to mirror. If you want to select multiple entities, hold the **CTRL** key and select entities. In our case, we select all the entities appearing in Fig.Pro/E-3.51.

When you select an entity then the **Mirror** ( ) button is enabled.

Note

By default, the Mirror button is disabled.

5. Now, *click* the **Mirror** button on the `Sketcher Tools toolbar`. As soon as, you click the **Mirror** button, the **Select** dialog box (Fig.Pro/E-3.31) along with a message on the **Message** area appears that prompts you to select the centerline.
6. Next, *click* the centerline. The entities, you have selected are mirrored, as shown in Fig.Pro/E-3.52:

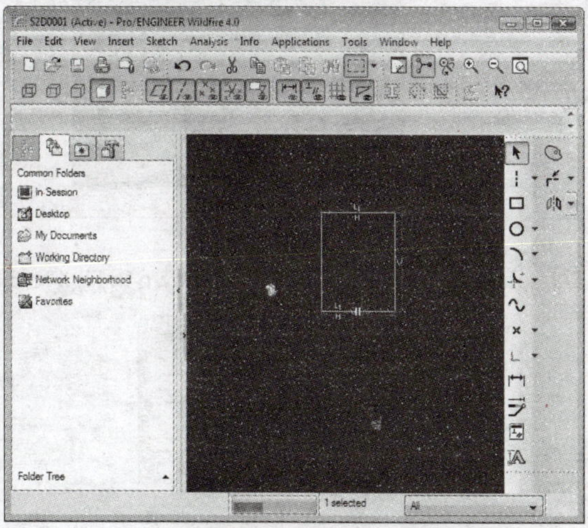

Fig.Pro/E-3.52

In this way, entities are mirrored. Let's now discuss the process to scale and rotate the entities.

Scaling and Rotating Entities

While drawing a 2D sketch, you may need to scale and rotate one or more entities to get the desired sketch. In Pro/Engineer Wildfire 4.0, entities can be scaled and rotated in two ways: first, through using the **Scale and Rotate** button available on the `Sketcher Tools toolbar`; second, by accessing the **Scale and Rotate** option in the **Edit** menu. In this chapter, we use the **Scale and Rotate** button available on the `Sketcher Tools toolbar` to scale and rotate the entities. To scale and rotate the entities, you need a sketch and in our case, the sketch is shown in Fig.Pro/E-3.51.

Let's now follow the steps given here to scale and rotate the entities:

1. *Select* the entity, you want to scale and rotate (Fig.Pro/E-3.51).
2. *Click* the down arrow beside the Mirror () button in the Sketcher Tools toolbar (Fig.Pro/E-3.5). A flyout appears (Fig.Pro/E-3.53).
3. *Click* the **Scale and Rotate** () button, as shown in Fig.Pro/E-3.53:

Fig.Pro/E-3.53

As soon as, you click the **Scale and Rotate** button, the selected entity is enclosed in a boundary, as shown in Fig.Pro/E-3.54:

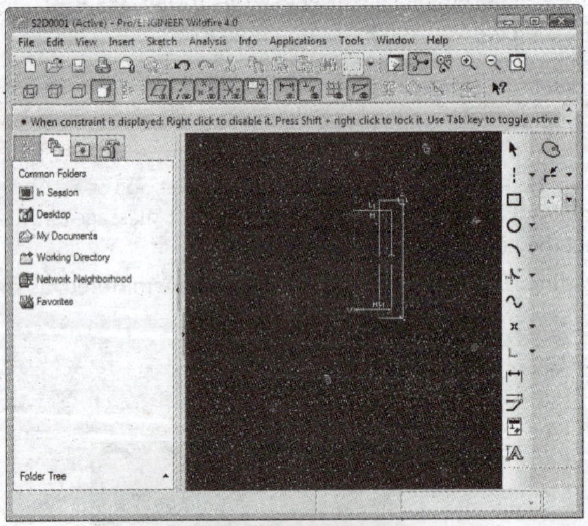

Fig.Pro/E-3.54

Apart from this, the **Scale Rotate** dialog box also appears, as shown in Fig.Pro/E-3.55:

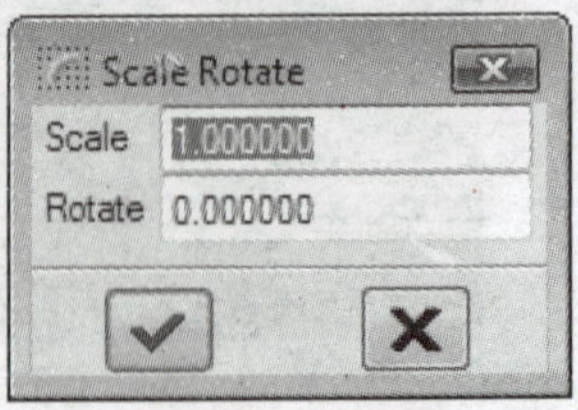

Fig.Pro/E-3.55

Now, you have following two approaches to scale and rotate the selected entity:

- First, modify the values in the text boxes beside the **Scale** and **Rotate** labels in the **Scale Rotate** dialog box. After modifying the values, click **Accept** (✔) button to accept the changes. As you click the accept button, changes are reflected in the drawing area.
- Second, use various handle available in Pro/ENGINEER Wildfire 4.0 to dynamically scale and rotate the selected entity. Following handles are available on the boundary of the selected entity to dynamically scale and rotate the selected entity:
  - **Rotate handle:** Used to dynamically rotate the selected entity.
  - **Base Point handle:** Used to move the selected entity from one place to another in the drawing area.
  - **Scale handle:** Used to scale the selected entity.

In our case, we have used the **Rotate** handle to rotate the selected entity and the resultant figure is shown in Fig.Pro/E-3.56:

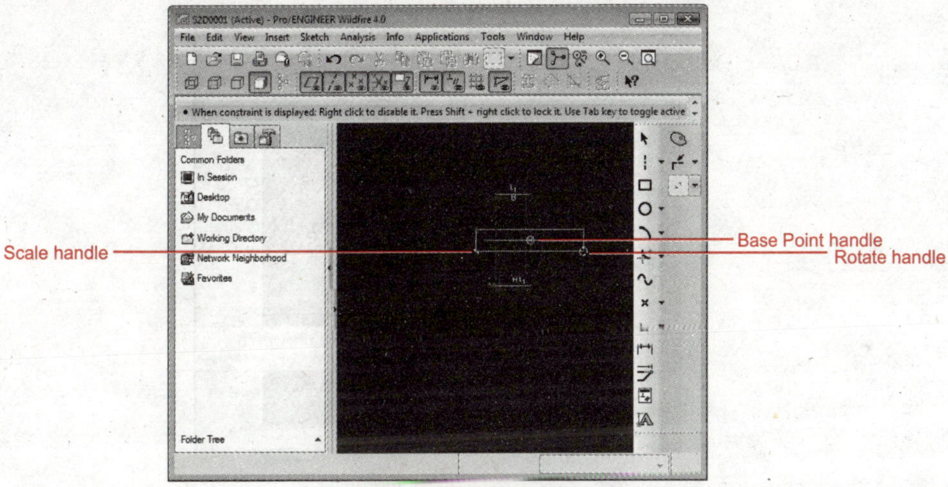

Fig.Pro/E-3.56

4. Now, *click* the **Accept** button in the **Scale Rotate** (Fig.Pro/E-3.55) dialog box to accept the changes. Changes are reflected in the drawing area, as shown in Fig.Pro/E-3.57:

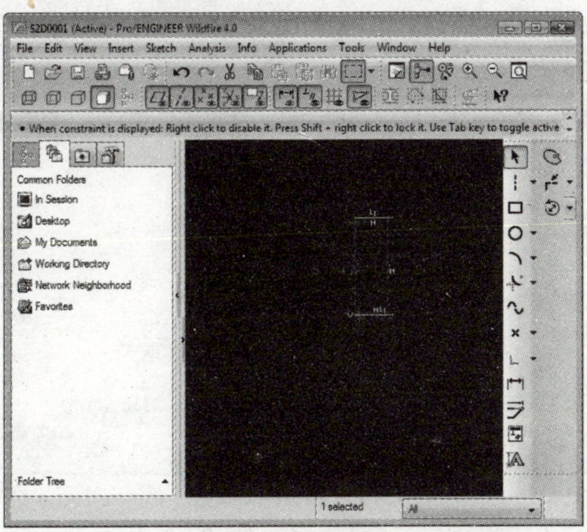

Fig.Pro/E-3.57

With this, we completed the discussion on scaling and rotating the entities. Now, we discuss the process to import an already created sketch.

Importing 2D Drawings

Sometimes, you may need to import an already created sketch to create another 2D sketch. For example, you have created a design of a pen few days ago but now you want to redesign a pen having different cap design. Now, you can import pre-existing sketch of the pen, change the cap design and the new pen is ready. This saves your time while creating new sketches from the pre-existing sketches. The Sketches/Drawings created in Sketcher mode are saved with the `.sec` extension so that they can be used to create other 2D sketches. The process to save a file has already depicted in chapter 2, *Exploring the User Interface*. However, to use already created 2D sketch, you have to import that sketch in the Sketcher mode. Following are the steps to import an already created sketch:

1. *Click* the **Sketch→Data from File→File System** as shown in Fig.Pro/E-3.58:

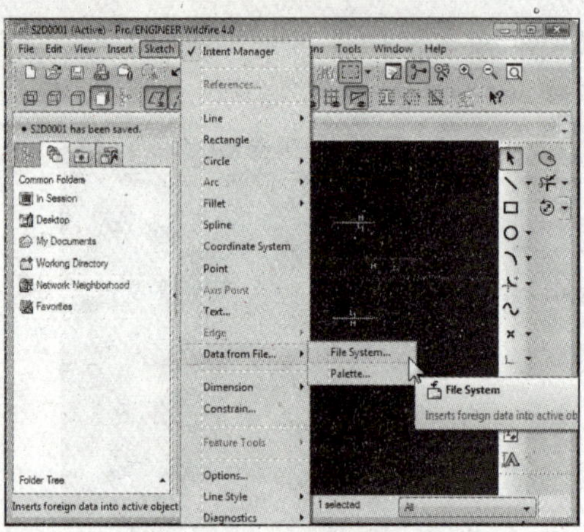

Fig.Pro/E-3.58

The **Open** dialog box appears (Fig.Pro/E-3.59).

2. Now, *select* the file, you want to open (Fig.Pro/E-3.59). In our case, we have selected the s2d0001.sec file.
3. Next, *click* the **Open** button, as shown in Fig.Pro/E-3.59:

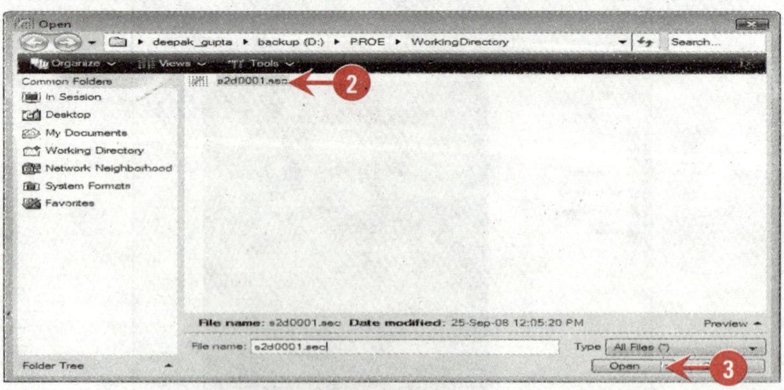

Fig.Pro/E-3.59

Selected file is opened in the drawing area, as shown in Fig.Pro/E-3.60:

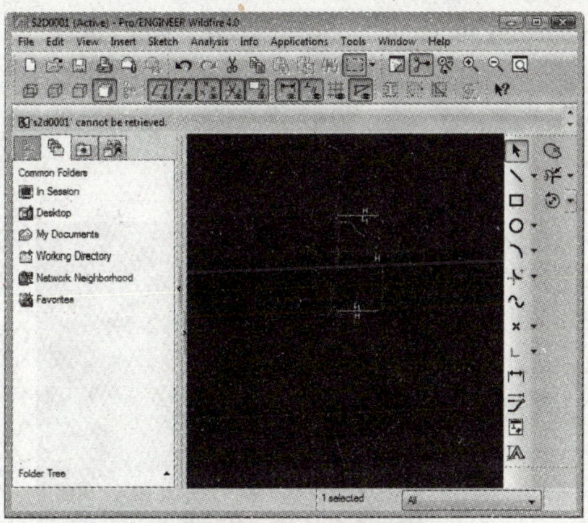

Fig.Pro/E-3.60

Now, you can use this file to create another 2D sketch. Let's now discuss the process to delete entities from a sketch.

Deleting Entities

Whenever creating a sketch in Pro/ENGINEER Wildfire 4.0, you may need to delete some extra entities to complete the sketch. In Pro/ENGINEER Wildfire 4.0, you have following two ways to delete entities:

- First, select the entity you want to delete and then press the **DELETE** key on your keyboard.

Note

To delete more than one entity, select multiple entities by holding the CTRL key.

- Second, by using the **Delete** option in the **Edit** menu.

In this chapter, we use the **Delete** option to delete the entities. For deleting entities, we use the sketch shown in Fig.Pro/E-3.59. Let's now follow the steps given here to delete the entities from a sketch:

1. *Select* an entity (Fig.Pro/E-3.60) that you want to delete.
2. Now, *click* the **Edit→Delete** option, as shown in Fig.Pro/E-3.61:

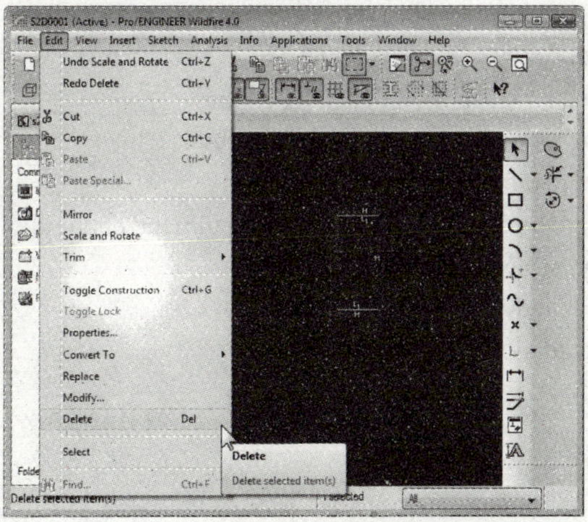

Fig.Pro/E-3.61

The selected entity is automatically deleted, as shown in Fig.Pro/E-3.62:

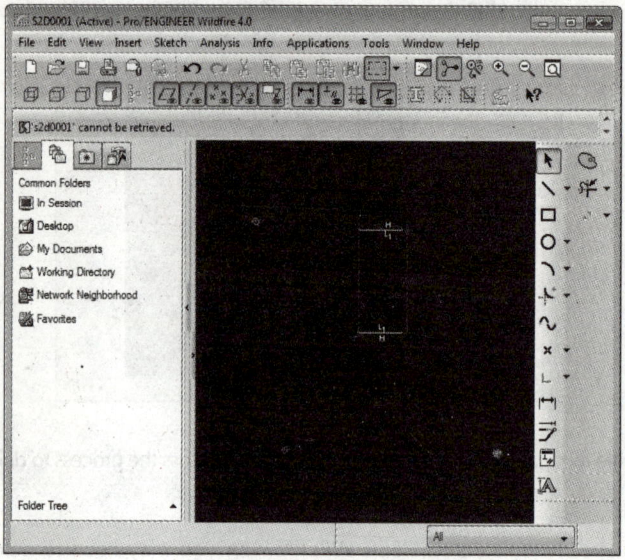

Fig.Pro/E-3.62

Note

> The Delete option is enabled when you select an entity. By default, the Delete option is disabled.

Here, we have completed the process to delete the entities. Apart from this, we have also reached to the end of this chapter. Before closing this chapter, let's have a look on the topic covered in this chapter.

Summary

In this chapter, you learned about:

- The process to start the Sketch mode
- The Sketcher Tools toolbar
- The processes to use button available on the Sketcher Tools toolbar to sketch various entities such as line, arc, and rectangle
- The process of applying dimensions on sketching entities
- The process of applying constraints on sketching entities
- Various options to modify dimensions
- The buttons available on the Sketcher Tools toolbar to trim the entities
- The processes to mirror, scale, and rotate the entities
- The process to import already created sketch files
- The process to delete entities from a sketch

Chapter 4

Exploring Pro/ENGINEER Wildfire 4.0 Part Mode

In this Section

- Starting the Part Mode
- Understanding Datums
- Creating an Extruded Feature
- Inducing the Hole Feature
- Creating a Cut Feature
- Inducing the Round Feature
- Inducing the Chamfer Feature
- Inducing a Shell Feature
- Inducing a Revolved Feature
- Working with the Pattern Feature
- Inducing the Sweep Feature
- Inducing the Blend Feature
- Inducing the Rib Feature
- Modifying Features
- Deleting a Feature

In Pro/ENGINEER Wildfire 4.0 the process of a model creation pass through- the **Sketch** mode, the **Part** mode, the **Assembly** mode, and the **Drawing** mode. The **Sketch** mode has already been discussed in chapter 3 Exploring the Pro/ENGINEER Wildfire 4.0 Sketch Mode. In this chapter, we discuss the **Part** mode. The **Part** mode in `Pro/ENGINEER Wildfire 4.0` is used to build three-dimensional (3D) solid features, such as holes, cuts, and rounds and has properties such as volume, mass, and surface area.

A feature in `Pro/ENGINEER Wildfire 4.0` is considered as a basic entity. When multiple features are combined together, (by keeping some design intent in mind,) a component is created. Further, when multiple components are combined, a 3D model is created.

Therefore, to create a component of the model, you first have to build a base feature. The base feature is the first solid feature of a model. However, to create features in `Pro/ENGINEER Wildfire 4.0`, you need to start the **Part** mode of `Pro/ENGINEER Wildfire 4.0`. Therefore, we start this chapter by understanding the process to start the **Part** mode. Next, we discuss the processes to create features, such as cuts, holes, rounds, and chamfers. Finally, after creation, we learn to modify and delete features with regard to other features. Now, let's start studying this chapter.

First, we learn the process to start the Part mode.

Starting the Part Mode

The **Part** mode is used to build solid features of a model. To start the **Part** mode, you first have to start the `Pro/ENGINEER Wildfire 4.0`. The process to start `Pro/ENGINEER Wildfire 4.0` has already been discussed in chapter 2, *Exploring the User Interface*. Now, let's follow the steps given here to start the **Part** mode:

1. *Select* the **File** menu on the menu bar in the `Pro/ENGINEER Wildfire 4.0` main window, as shown in Fig.Pro/E-4.1. The **File** menu options appear (Fig.Pro/E-4.1).
2. Now, *click* the **New** option, as shown in Fig.Pro/4.1:

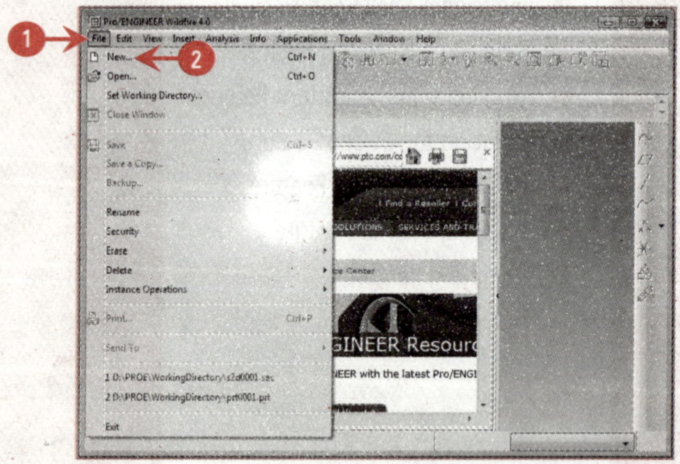

Fig.Pro/E-4.1

The **New** dialog box appears (Fig.Pro/E-4.2). By default, the the **Part** radio button is selected.

3. Assign a name for the **Part** you have to create. The part either accepts the default name or you can type any other name in the text box beside the **Name** label. In our case, we have accepted the default name, prt0001, as shown in Fig.Pro/E-4.2.
4. Next, *click* the **OK** button, as shown in Fig.Pro/E4.2:

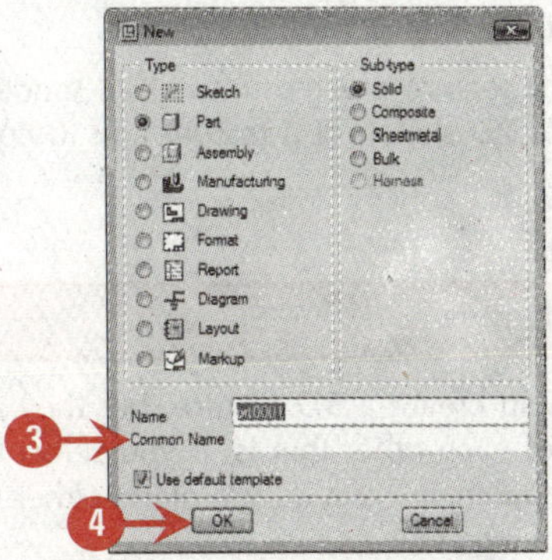

Fig.Pro/E-4.2

The `Pro/ENGINEER Wildfire 4.0` window with the **Part** mode appears, as shown in Fig.Pro/E-4.3:

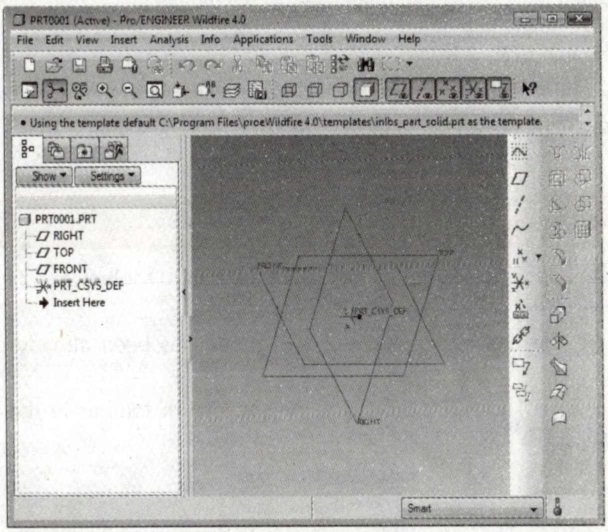

Fig.Pro/E-4.3

Now, the **Part** mode is ready to use. Fig.Pro/E-4.3 is the initial window, which appears in the **Part** mode. This initial window shows three datum planes that are created by default when you start the **Part** mode. **Datum** planes are important because they help in creating a base feature. Therefore, before proceeding ahead in this chapter with feature creation, we should understand datums.

Understanding Datums

Datums are imaginary features known as features having no mass, volume, and surface area. It is important to understand **Datums** because they work as reference for sketching a feature, helps in orienting components of a model, assembling components, and so on. Since, complex models contain many components encapsulating multiple features; **Datums** help in creating complex models. However, every model, simple or complex consists of a base feature on which other features are added. It is not easy to add all the features on a single plane. Therefore, to add other features, you may either select other available planes or create additional planes. By default, the following datum planes are created when you enter the **Part** mode:

- **Right:** Represents the right view of a feature or model.
- **Top:** Represents the top view of a feature or model.

- **Front:** Represents the front view of a feature or model.

These default datum planes are the first feature created in the **Part** mode and all are perpendicular to each other. These planes are used to draw a 2D sketch of a feature and to convert that sketch into a 3D feature through extrusion.

Note

> We will study about extrusion in detail, later in this chapter.

Though, these default datum planes help create a 3D feature, but they are not inter-related to each other. These datum planes are independent of each other that is, if you select the **Front** plane then only the front plane is selected, not all. However, you can select a datum plane either by selecting its name in the Model Tree or by clicking the plane in the drawing area.

Apart from using the default datum planes to group multiple features, you can also create your own datum planes as per your requirements. Next, let's learn to create datum planes.

Creating Datum Planes

Datum planes can be created by:

- Selecting the **Plane** option in the **Model Datum** option available in the **Insert** menu
- Using the **Plane** tool available on the **Datum** toolbar

Note

> The Datum toolbar has been already explained in Chapter 2, Exploring the User Interface.

In this chapter, we create additional datum planes using the datum toolbar. Let's follow the steps given here to create a datum plane:

1. *Start* the **Part** mode. The process to start the **Part** mode has been already discussed in the previous section.
2. Now, *click* the **Plane** (▱) tool (Fig.Pro/E-4.3) on the datum toolbar in the **Part** mode window. The DATUM **PLANE** dialog box appears, as shown in Fig.Pro/E-4.4:

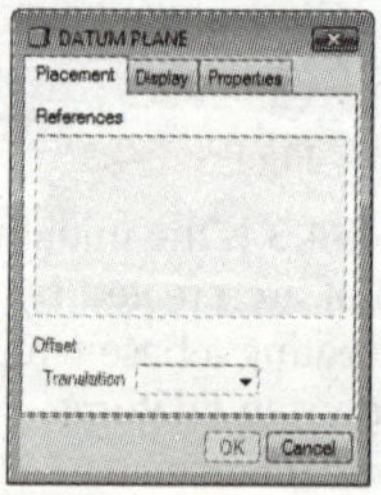

Fig.Pro/E-4.4

The DATUM PLANE dialog box contains following tabs to create a datum plane:

- **Placement Tab:** When the **DATUM PLANE** dialog box appears, then the **Placement** tab is already selected. Following are the options that the Placement tab contains:
    - **Reference area:** Reference area (Fig.Pro/E-4.4) displays the already existing datum plane, which is used to create a new datum plane. When you select a datum plane in the drawing area, its configuration appears in the **References** area. Various constraints are automatically applied on the new datum plane depending upon the constraint of the reference plane. These constraints are appeared on the right of the reference plane. Following are various constraints that are available with the reference planes:

- **Offset:** This is a default constraint of a datum plane. It is used to create a datum plane at an offset from the selected reference.
- **Through:** This constraint helps in creating the new datum plane through the selected reference.
- **Parallel:** This constraint helps in creating the new datum plane parallel to the selected reference.
- **Normal:** This constraint helps in creating the new datum plane normal to the selected reference.
- **Tangent:** This constraint helps in creating the new datum plane tangent to the cylindrical features.

- **Offset area:** This area is visible only when the constraint of the reference plane is the **Offset** constraint.

- **Display Tab:** The options available in the **Display** tab help flip the datum plane and set the size of the datum plane. The Display tab is shown in Fig.Pro/E-4.5:

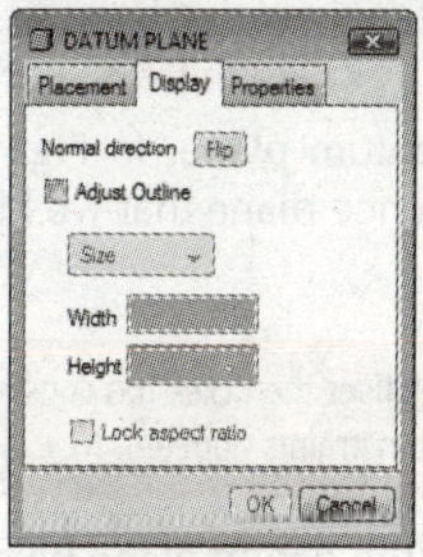

Fig.Pro/E-4.5

Following are the options available on the Display tab:

- **Flip:** This button helps in changing the normal direction of the datum plane which is to be created
- **Adjust Outline:** This option helps in adjusting the perimeter of the datum plane being created. This option is available when the checkbox (Fig.Pro/E-4.5) beside the **Adjust Outline** label is selected. When it is selected, then the text boxes beside the **Width** and **Height** labels are available to specify the width and height of the datum plane. Apart from this the combo-box named **Size**, also enables.
- **Lock aspect ratio:** This option is available only when the checkbox beside the **Adjust Outline** label is selected. This option helps in managing the proportion between the width and the height of the datum plane outline display.

- **Properties Tab:** This tab shows the name of the datum plane being created. By default, **DTM1** is assigned to the first datum plane created. When more than one datum plane is created, they are successively numbered. This tab also allows you to assign a new name to the datum plane being created. The Properties tab is shown in Fig.Pro/E-4.6:

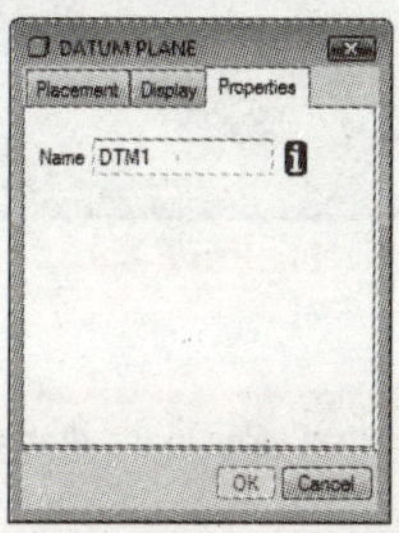

Fig.Pro/E-4.6

This is all about the **DATUM PLANE** dialog box and the constraints available to create a datum plane. As discussed, we have various constraints that help us to create a datum plane. However, in this chapter, we use the **Offset** constraint to create a new datum. Therefore, let's continue creating a datum plane.

3. *Select* an existing datum plane (Fig.Pro/E-4.3). The selected datum plane appears in the References area in the **DATUM PLANE** dialog box, as shown in Fig.Pro/E-4.7:

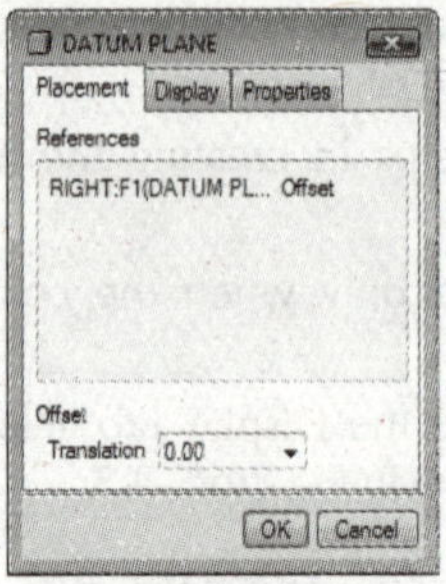

Fig.Pro/E-4.7

In our case, we have selected the **Right** datum plane. In Fig.Pro/E-4.7, you can see that the **Offset** area is enabled because the constraint of the reference plane that we have selected is Offset.

Note

> In case, the constraint of the referenced plane is not Offset then click the constraint appearing on the right of the selected datum plane in the DATUM PLANE dialog box. A drop-down list of constraints appears. Now, select the Offset constraint from the list.

Now, you have to set the offset distance of the datum plane being created.

4. To set the offset distance, *Click* and *drag* the offset handle to move the datum plane to the required distance, as shown in Fig.Pro/E-4.8:

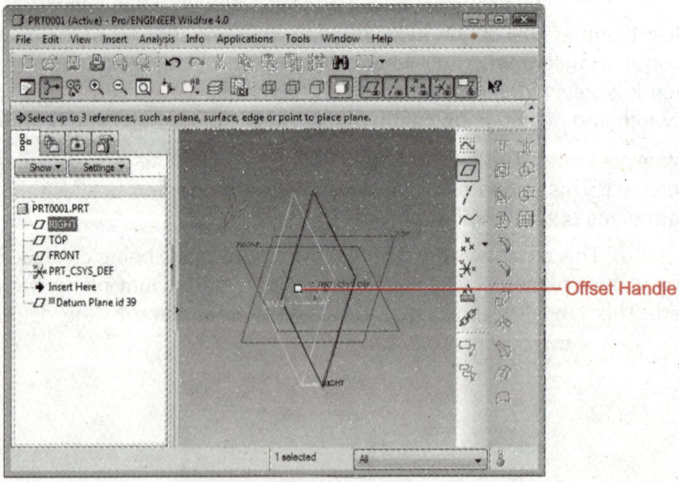

Fig.Pro/E-4.8

Note

> You can also set the offset distance by specifying the offset value in the drop-down list beside the Translation label (Fig.Pro/E-4.7) or select any value from the recently used values.

In Fig.Pro/E-4.8, the yellow colored datum plane is the new datum plane created. The value of the offset distance is automatically set in the drop-down list beside the Translation label in the DATUM PLANE dialog box, as shown in Fig.Pro/E-4.9.

5. Now, *click* the OK button (Fig.Pro/E-4.9) to complete the datum plane creation with the default name:

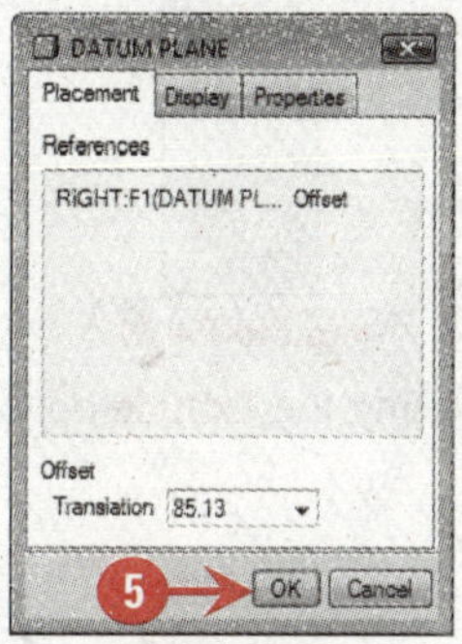

Fig.Pro/E-4.9

Note

> However, if you want to assign a new name to the datum plane, then click the Properties tab in the DATUM PLANE dialog box. Specify the new name to the datum plane and click the OK button. In our case, we have accepted the default name.

As soon as you click the OK button, the new datum plane appears in the drawing area and also in the Model Tree, as shown in Fig.Pro/E-4.10:

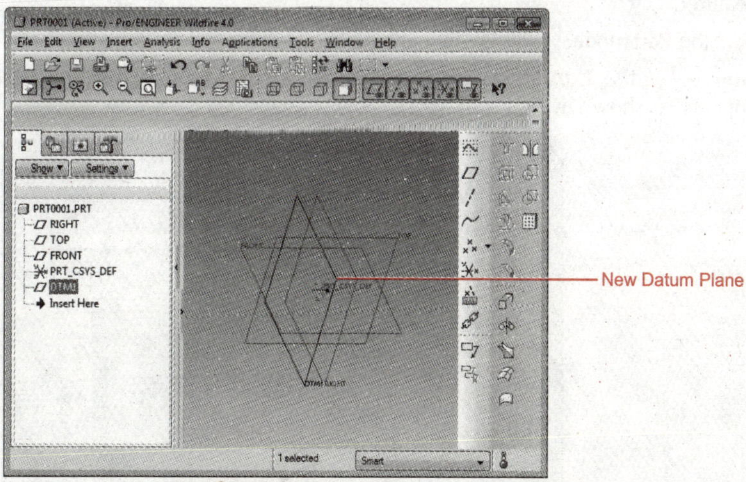

Fig.Pro/E-4.10

In Fig.Pro/E-4.10, you can see the new datum plane with the name DTM1 appearing in the Model Tree as well as the drawing area. This is all about creating a datum plane in Pro/ENGINEER Wildfire 4.0. Now, we start creating features.

Creating an Extruded Feature

Extrusion is a process of creating a 3D model of a 2D sketch. To create an extruded feature, the Extrude ( ) button available on the Base Features toolbar in the Part mode window can be used. The Base Features toolbar is shown in Fig.Pro/E-4.11:

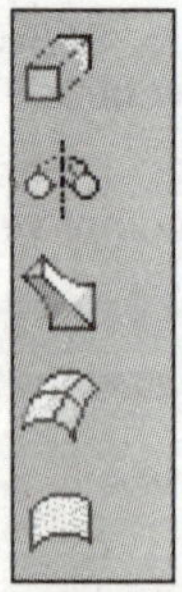

Fig.Pro/E-4.11

An extruded feature can also be created using the **Extrude** option in the **Insert** menu. The **Extrude** button is generally used for two purposes:

- Creating a Base Feature
- Creating a Protrusion

Let's now discuss these purposes one by one.

Creating a Base Feature

A base feature, as known, is the basic entity of a model because the shape of a model depends on its base. For example, if the base feature is rectangular, then a rectangular model is created and if the base feature is circular, it results in a circular model. All other features such as protrusions, cuts, and holes are added on the base feature to complete a model. A base feature varies according to the design intent of the model being designed. In our case, a rectangular slot is being created. Let's now follow the steps given here to create a base feature:

1. *Start* the **Part** mode.
2. Now, *click* the **Extrude** button in the **Base Features** toolbar (Fig.Pro/E-4.11). The **Extrude** dashboard appears, as shown in Fig.Pro/E-4.12:

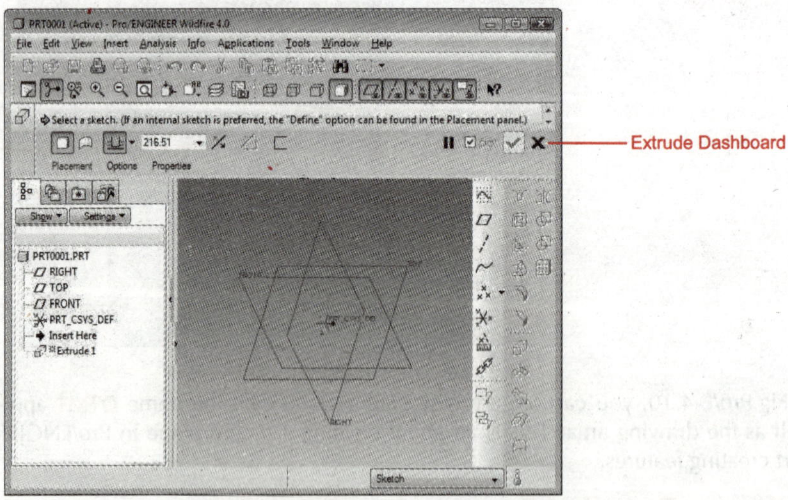

Fig.Pro/E-4.12

Let's have a brief idea about the buttons and tabs available in the Extrude dashboard, before moving ahead with the steps in the section. The Extrude dashboard contains the following buttons and tabs:

- **Extrude as solid Button** ( ): Used to create a solid feature.
- **Extrude as surface Button** ( ): Used to create the surface of a model.
- **Depth Options** ( ): Used to constrain the depth of a feature. When you click the down arrow beside in the Extrude dashboard then various depth options project out. You can use any of the depth option to constrain the depth of the feature being created.
- **Depth Box** (149.52): Specifies the depth of the feature being created. You can either select the available depth by clicking the down arrow beside (149.52) in the Extrude dashboard or specify any other required depth.
- **Flip Button** ( ): Flips the direction of the feature being created with respect to the sketching plane. This button is enabled only when the feature creation is completed.
- **Remove Material Button** ( ): Removes the material from the feature. This button is enabled only when the feature creation is completed.
- **Thicken Sketch Button** ( ): Assigns specific thickness to the sketch outline.
- **Pause Button** ( ): Used to pause the process of current feature creation.
- **Resume Button** ( ): Resumes the process of feature creation. This button is enabled only when the **Pause** button is clicked.
- **Preview Button** ( ): This button is used to preview the feature being created.
- **Build feature Button** ( ): This button is u**sed to complete** the feature creation.
- **Close Button** ( ): Aborts the process of feature creation.
- **Placement Tab** (Placement): Defines the sketching plane.
- **Options Tab** (Options): Specifies whether an extrusion is to be created on one side of the sketching plane or on both the sides.
- **Properties Tab** (Properties): Edits the feature name. This tab can also be used to know the feature details.

This is a brief about the **Extrude** dashboard. Let's now move ahead with the process of creating the base feature.

3. *Select* the **Placement** tab (Fig.Pro/E-4.13) to define the sketching plane. A slide-down panel, as shown in Fig.Pro/E-4.13 appears:

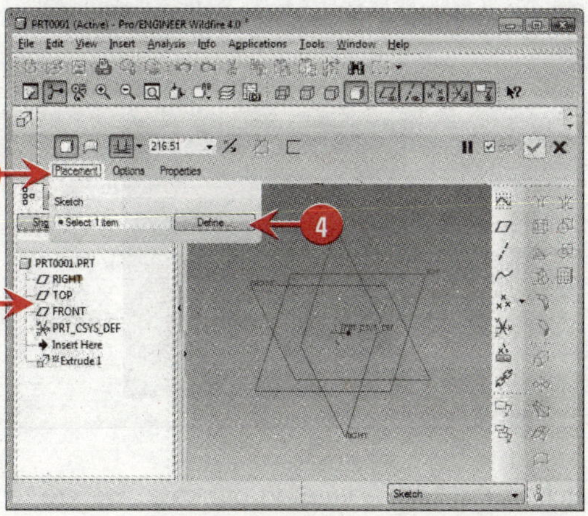

Fig.Pro/E-4.13

4. *Click* the **Define** button in the slide-down panel (Fig.Pro/E-4.13). The **Sketch** dialog box, as shown in Fig.Pro/E-4.14 appears:

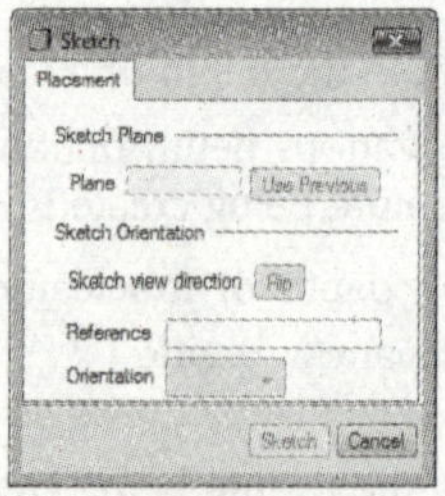

Fig.Pro/E-4.14

5. *Select* a datum plane either in the **Model Tree** or in the drawing area (Fig.Pro/E-4.13). Once you have selected the datum plane, the selected datum plane, its reference plane, and orientation is automatically specified in the **Sketch** dialog box, as shown in Fig.Pro/E-4.15.

In our case, we have selected the **Front** datum plane. In Fig.Pro/E-4.15, you see that the selected datum plane is specified in the text box beside the **Plane** label, its reference is specified in the text box beside the **Reference** label, and the orientation appears beside the Orientation label.

6. Next, *click* the **Sketch** button (Fig.Pro/E-4.15) to confirm the selection, as shown in Fig.Pro/E-4.15:

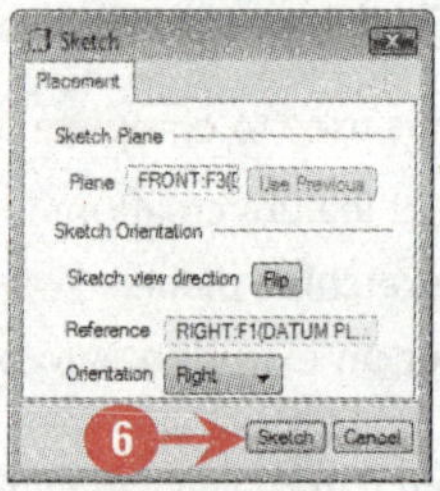

Fig.Pro/E-4.15

Now, the **Sketcher** tools toolbar appears, as shown in Fig.Pro/E-4.16:

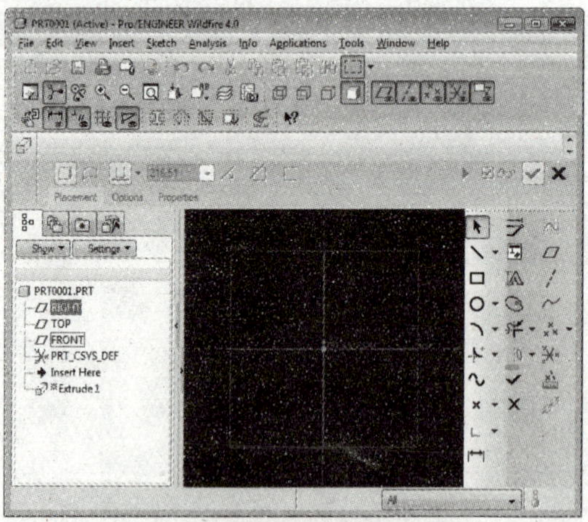

Fig.Pro/E-4.16

Now, you can create a sketch of the base feature. In our case, we created a rectangle. After creating a rectangle apply dimensions on it.

Note

> The process to sketch a rectangle and applying dimensions on it has been already depicted in chapter 3, Exploring the Pro/ENGINEER Wildfire 4.0 Sketch Mode.

Once the dimensions are applied on the rectangle, it appears as shown in Fig.Pro/E-4.17.

7. Now, *click* the **Done** (✓) button in the **Sketcher tools toolbar** (Fig.Pro/E-4.17):

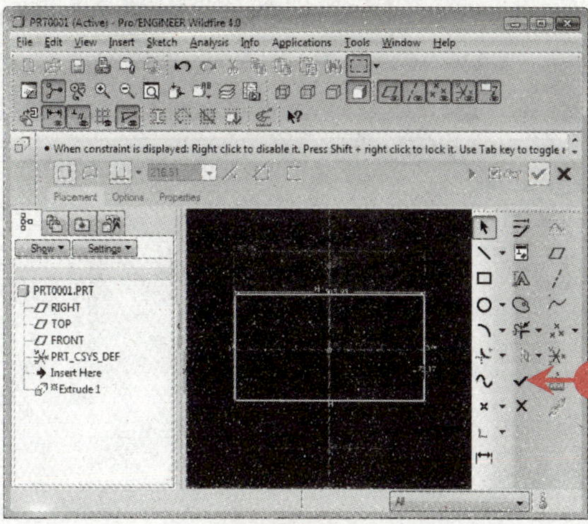

Fig.Pro/E-4.17

The sketch appears in yellow (Fig.Pro/E-4.18). Now, you can view the sketch in 3D mode using various orientations such as **Left**, **Right**.

8. To view the sketch in 3D format using different orientations, *click* the **Named View List** ( ) button in the View toolbar, as shown in Fig.Pro/E-4.18:

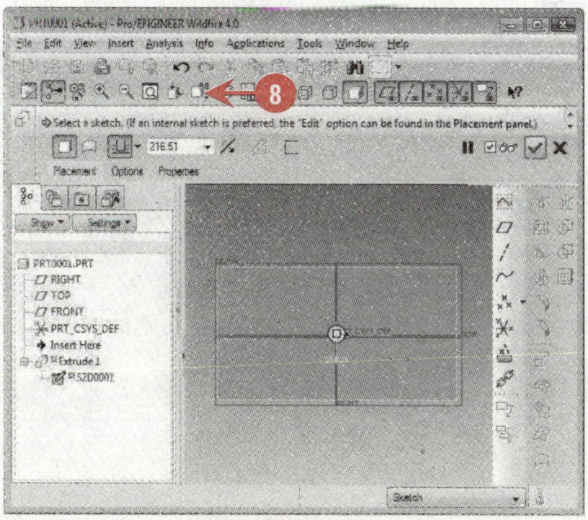

Fig.Pro/E-4.18

A list of various orientations appears (Fig.Pro/E-4.19).

9. *Select* any of the orientation to view the sketch in 3D format. In our case, we have selected the **Default Orientation**, as shown in Fig.Pro/E-4.19:

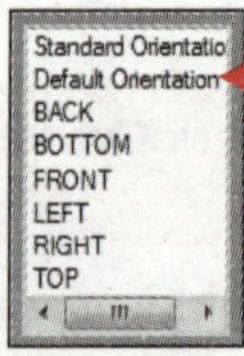

Fig.Pro/E-4.19

Once the default orientation has been selected, the rectangular slot created appears, as shown in Fig.Pro/E-4.20:

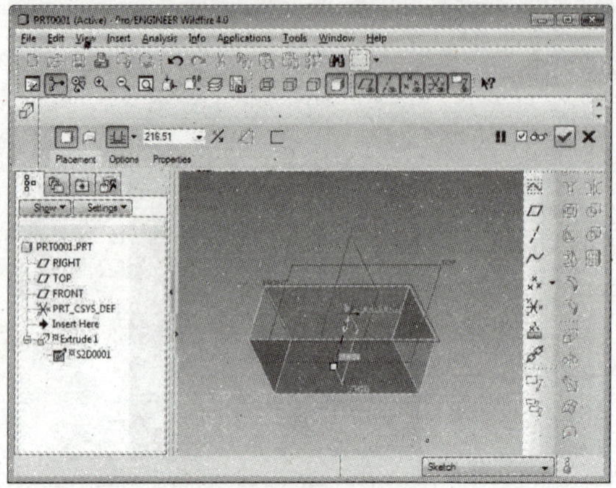

Fig.Pro/E-4.20

10. Now, *specify* its depth. To specify its depth, either select a previously used value from the **Depth** box or enter a new value in the **Depth** box. In our case, we entered 100.00 as the new value.
11. *Press* the **ENTER** key. The 3D view of the rectangular slot appears (Fig.Pro/E-4.21).
12. *Click* the **Preview button** ( ) in the **Extrude** dashboard to preview the rectangular slot being created, as shown in Fig.Pro/E-4.21:

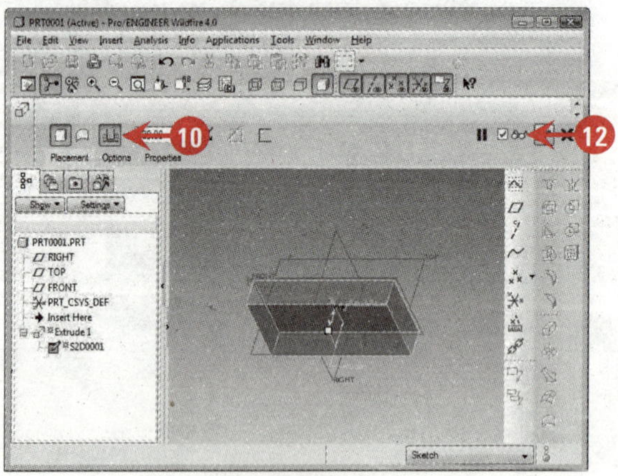

Fig.Pro/E-4.21

The preview of the rectangular slot appears (Fig.Pro/E-4.22).

13. *Click* the **Build feature** ( ) button in the **Extrude** dashboard to complete the base feature creation, as shown in Fig.Pro/E-4.22:

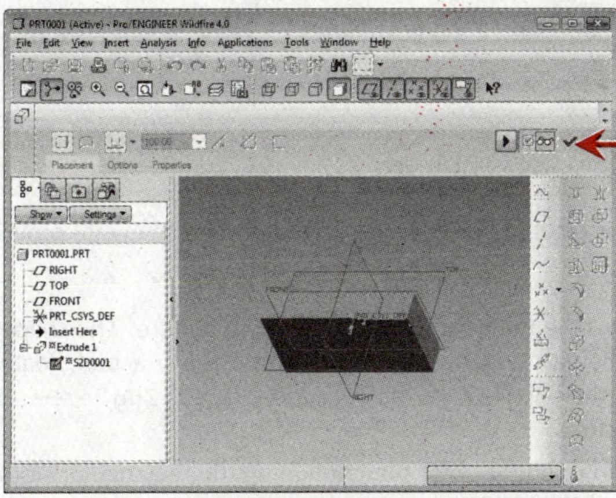

Fig.Pro/E-4.22

As soon as, you click the **Build feature** button, the rectangular slot appears, as shown in Fig.Pro/E-4.23:

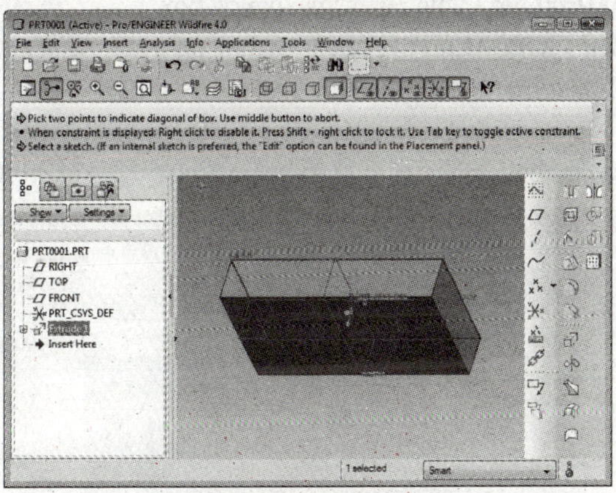

Fig.Pro/E-4.23

14. Now, *click* anywhere on the drawing area (Fig.Pro/E-4.23). The final view of the base feature appears, as shown in Fig.Pro/E-4.24:

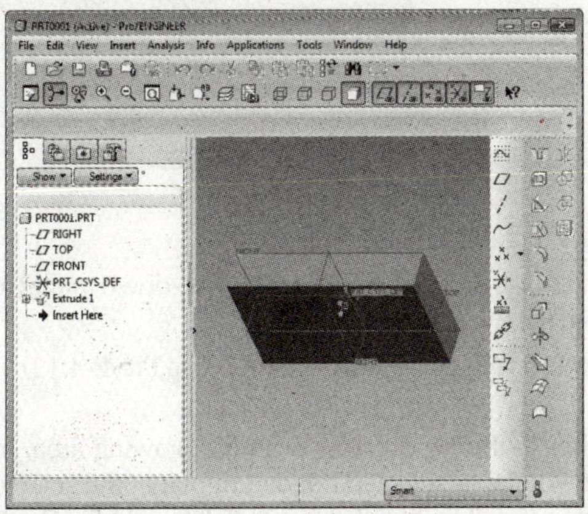

Fig.Pro/E-4.24

15. *Save* the feature. The process to save a feature is already depicted in chapter 2, *Exploring the User Interface*.

Note

> A feature created in the part mode is saved as a .PRT file.

In this way, a base feature is created. Next, we learn the process to create protrusions.

Creating a Protrusion

Protrusion can be defined as the material added to the base feature. The shape of the protrusion depends on the design of the model being created. To create a protrusion, a base feature must exist. In our case, a rectangular slot created, is the base feature in the section, `Creating a Base Feature`. Let's now follow the steps to create a protrusion:

1. *Start* the **Part** mode and open the .PRT file containing the rectangular slot to create a protrusion on it. In our case, the file name is PRT0001.prt.
2. To open the PRT0001.prt, *click* the **File→Open**. The **File Open** dialog box appears.
3. Now, *select* the PRT0001.prt file in the **File Open** dialog box.
4. *Click* the **OK** button. The PRT0001.prt file opens in the Part mode, as shown in Fig.Pro/E-4.25.

In Fig.Pro/E-4.25, you can view the rectangular slot created as a base feature in the section, `Creating a Base Feature`.

5. *Click* the **Extrude** button in the **Base Features** toolbar, as shown in Fig.Pro/E-4.25:

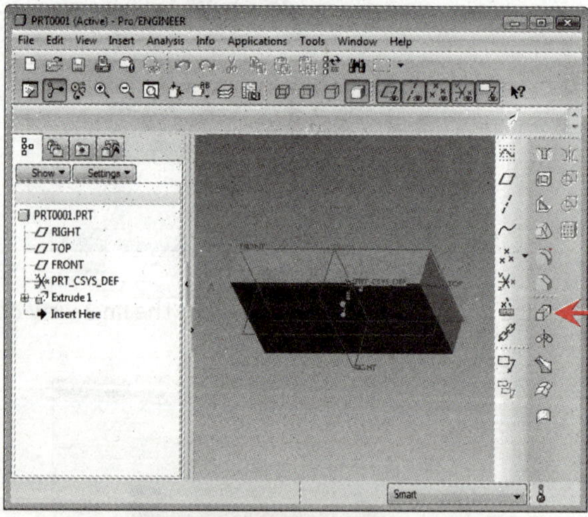

Fig.Pro/E-4.25

The **Extrude** dashboard appears (Fig.Pro/E-4.12).

6. Next, *select* the **Placement** tab (Fig.Pro/E-4.12). A slide-down panel that lets you define a sketching plane appears (Fig.Pro/E-4.13).
7. *Click* the **Define** button on the slide-down panel (Fig.Pro/E-4.13).The **Sketch** dialog box appears (Fig.Pro/E-4.14).
8. *Select* a datum plane either in the **Model Tree** or in the drawing area. In our case, we have selected the **Front** datum plane. Once you have selected the datum plane, the selected datum plane and its reference is automatically specified in the **Sketch** dialog box, as shown in Fig.Pro/E-4.26.

9. *Click* the **Sketch** button (Fig.Pro/E-4.26) to confirm selection:

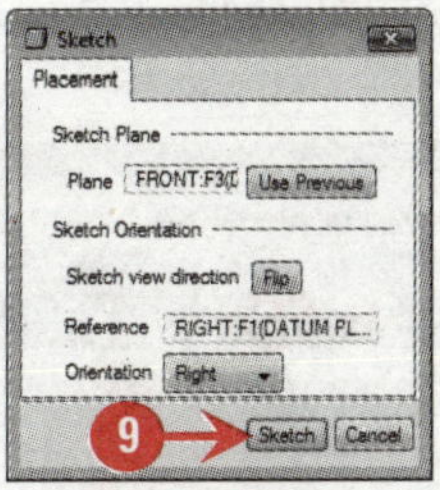

Fig.Pro/E-4.26

Now, the **Sketcher tools toolbar** is enabled, as shown in Fig.Pro/E-4.27:

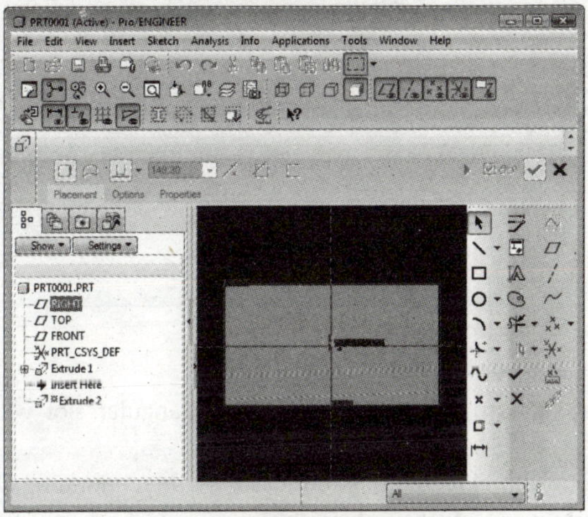

Fig.Pro/E-4.27

10. *Draw* a small rectangle at the upper right corner of the rectangular slot (Fig.Pro/E-4.28).
11. *Click* the **Done** (✔) button on the **Sketcher** tools toolbar, as shown in Fig.Pro/E-4.28:

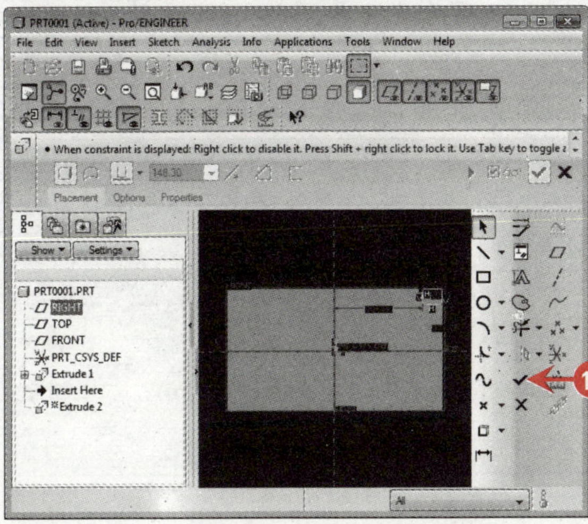

Fig.Pro/E-4.28

The preview of the newly created sketch appears in yellow (Fig.Pro/E-4.29).

12. *Click* the down arrow beside the **Named View List** (button icon) button in the view toolbar to view the feature in 3D mode, as shown in Fig.Pro/E-4.29:

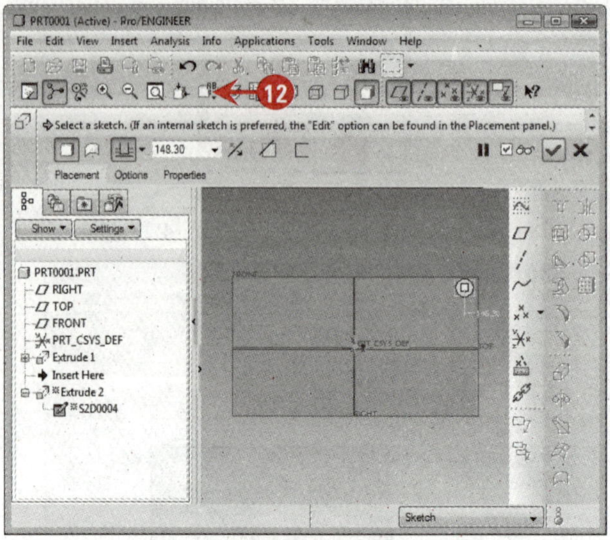

Fig.Pro/E-4.29

A list of various orientations appears (Fig.Pro/E-4.19).

13. *Select* the **Default Orientation** (Fig.Pro/E-4.19). The rectangular slot with the protrusion appears, as shown in Fig.Pro/E-4.30.

Modify the depth of the protrusion, if required. In our case, we are continuing with the default values.

14. *Click* the **Build feature** button in the **Extrude** dashboard, as shown in Fig.Pro/E-4.30:

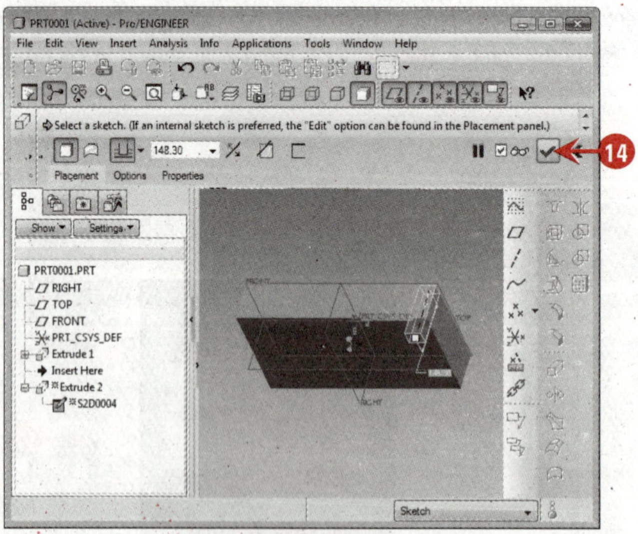

Fig.Pro/E-4.30

The rectangular slot with a protrusion appears, as shown in Fig.Pro/E-4.31:

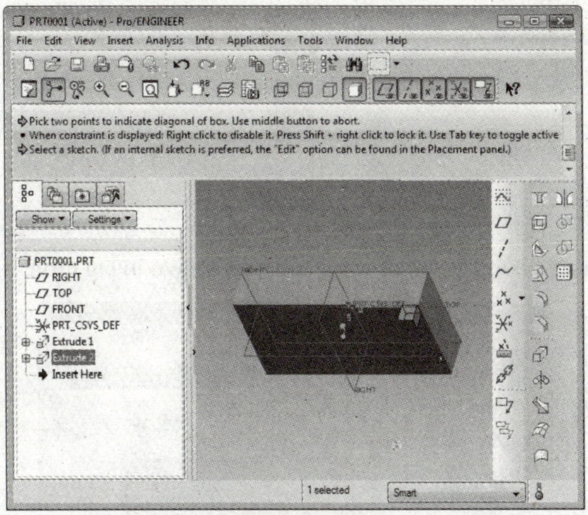

Fig.Pro/E-4.31

15. Now, *click* anywhere in the drawing area (Fig.Pro/E-4.31). The final view of the base feature along with the protrusion appears, as shown in Fig.Pro/E-4.32:

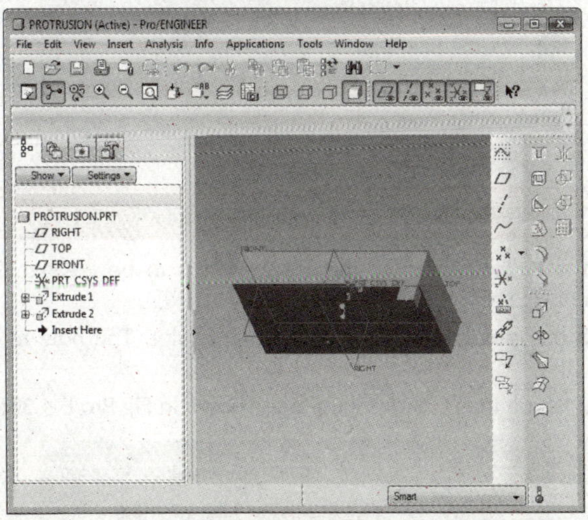

Fig.Pro/E-4.32

In this way, a protrusion is created. Let's move ahead by adding another feature in the base feature.

Inducing the Hole Feature

In `Pro/ENGINEER Wildfire 4.0` simple and industry-standard holes can be created. An industry-standard hole is the hole that has some industry standards to define its diameter. In this section, we see the process to create a simple hole. To create a simple hole, we need a base feature on which the hole is to be created. Therefore, we use the rectangular slot created in the section, `Creating a Base feature` as the base feature on which a hole is created. To create a hole, follow the steps given:

1. *Start* the **Part** mode and *open* the PRT0001.prt file.
2. *Click* the **Hole** ( ) button in the **Engineering Features** toolbar in right tool chest. The **Engineering Features** toolbar is shown in Fig.Pro/E-4.33:

Fig.Pro/E-4.33

The **Hole** dashboard appears on top of the drawing area, as shown in Fig.Pro/E-4.34:

Fig.Pro/E-4.34

Note

As soon as the Hole dashboard appears, the Engineering Features toolbar is disabled.

3. *Select* an area on the rectangular slot where the hole is to be created. The preview of the hole is displayed, as shown in Fig.Pro/E-4.35.

In our case, we selected the middle area of the rectangular slot. The hole to be created is previewed in yellow color (Fig.Pro/E-4.35).

4. *Select* the **Placement** tab on the **Hole** dashboard, as shown in Fig.Pro/E-4.35:

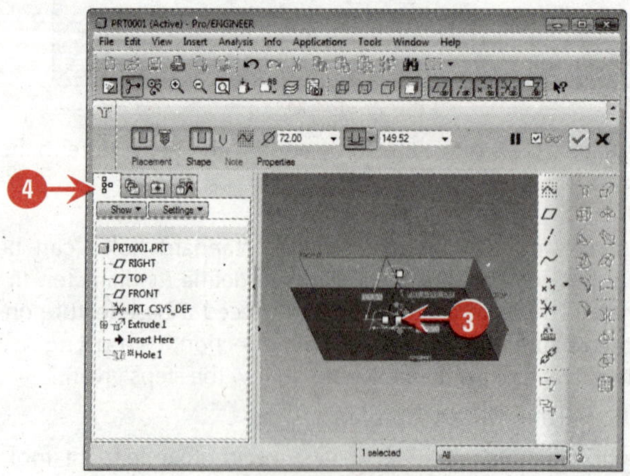

Fig.Pro/E-4.35

The **Placement** tab is selected to choose the sketching plane and its offset references. The **Placement** slide-down panel appears, as shown in Fig.Pro/E-4.36.

In Fig.Pro/E-4.36, the plane selected on the rectangular slot is appearing in the text box in the **Placement** option.

5. *Click* inside the text area in the **Offset References** option to specify the offset references, as shown in Fig.Pro/E-4.36:

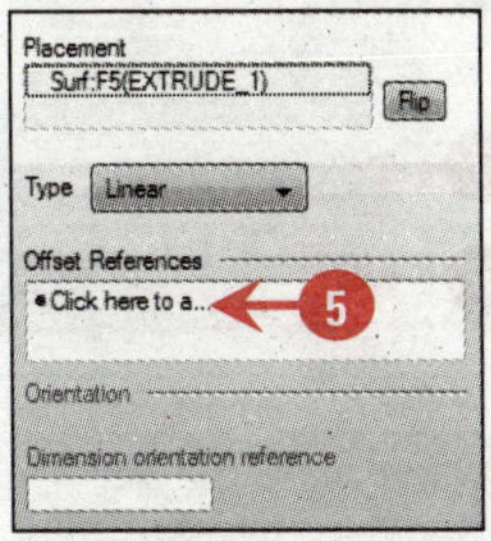

Fig.Pro/E-4.36

In the text area, a message Select 2 items is flagged. This message prompts you to select two items to specify the offset references, as shown in Fig.Pro/E-4.37:

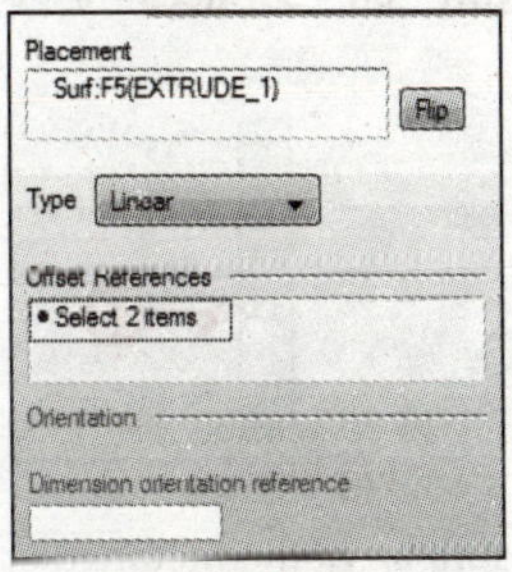

Fig.Pro/E-4.37

6. *Select* the left edge of the rectangular toolbar. The configuration of the selected edge appears in the **Offset References** edit box. Apart from that, you are also prompted to select the remaining item as an offset reference, as shown in Fig.Pro/E-4.38:

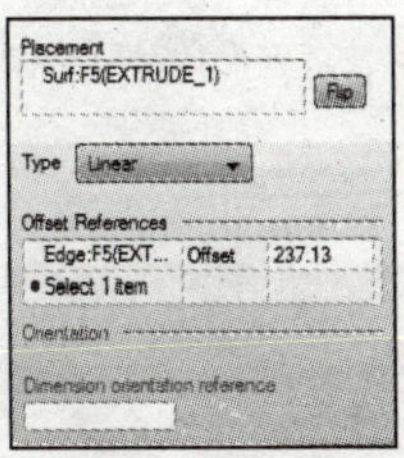

Fig.Pro/E-4.38

7. *Press* **CTRL** key and *select* the top edge of the rectangular slot. Both the selected edges appear in the edit box in the **Offset Reference** option. Apart from that, both edges are highlighted in red color (Fig.Pro/E-4.39).
8. *Select* the **Placement** tab once again (Fig.Pro/E-4.39) to close the **Placement** slide-down panel.
9. *Click* the down arrow in the Depth box to *select* the predefined depth option, as shown in Fig.Pro/E-4.39:

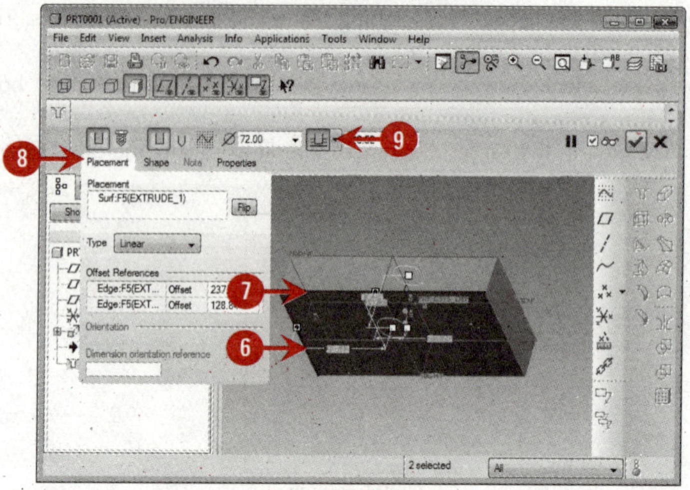

Fig.Pro/E-4.39

A list of available depth options appears (Fig.Pro/E-4.40).

10. *Select* the **Drill to intersect with all surfaces** ( ) option from the list, as shown in Fig.Pro/E-4.40:

Fig.Pro/E-4.40

This option creates a hole by drilling the surface of the rectangular slot.

11. *Click* the **Build feature** ( ) button (Fig.Pro/E-4.39) to complete the hole creation. A hole, as shown in Fig.Pro/E-4.41 is created:

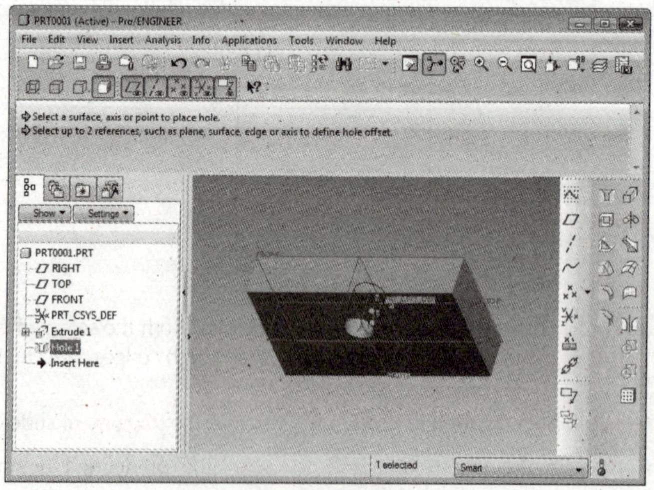

Fig.Pro/E-4.41

12. *Click* anywhere in the drawing area (Fig.Pro/E-4.41). The final view of the base feature along with the hole feature appears, as shown in Fig.Pro/E-4.42:

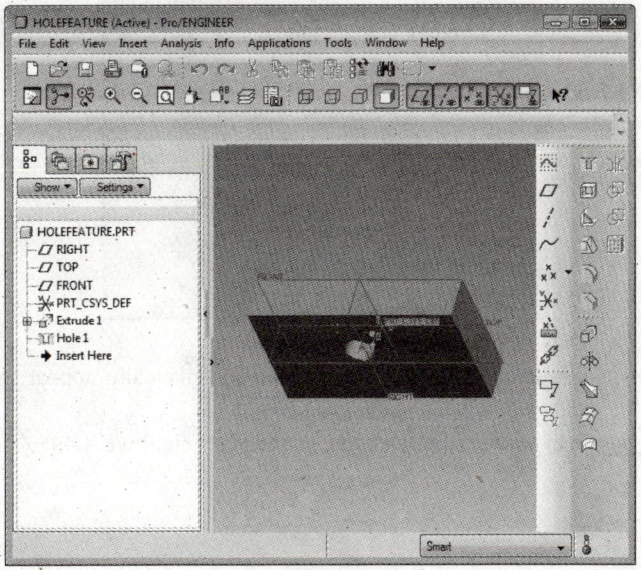

Fig.Pro/E-4.42

13. *Save* the file. In our case, we have saved the file with the name **HOLEFEATURE.prt**.

This is how a hole is created on a feature in `Pro/ENGINEER Wildfire 4.0`. Let's now move further and depict the process to create a feature called **cut**.

Creating a Cut Feature

A cut in `Pro/ENGINEER Wildfire 4.0` can be rectangular, tri-angular, or semi-circular. The shape of a cut depends on the sketch created on a feature. A cut in Pro/ENGINEER Wildfire 4.0 is created using the **Extrude** dashboard. However, to create a cut a 3D feature is required. In our case, we use the **HOLEFEATURE.prt** file created in the preceding section as the base feature. Now, let's follow the steps given here to create a cut feature:

1. *Start* the **Part** mode and *open* the HOLEFEATURE.prt file (Fig.Pro/E-4.42).
2. *Click* the **Extrude** ( ) button in the **Base Features** toolbar (Fig.Pro/E-4.11) in the right tool chest. The **Extrude** dashboard appears, as shown in Fig.Pro/E-4.43.

Note

> Fig.Pro/E-4.43 is also displaying the HOLEFEATURE.prt file.

3. *Click* the **Remove Material** ( ) button on the **Extrude** toolbar (Fig.Pro/E-4.43).
4. *Select* the **Placement** tab to define the sketching plane, as shown in Fig.Pro/E-4.43.

As soon as you select the **Placement** tab, a slide-down panel appears (Fig.Pro/E-4.13).

5. Next, *click* the **Define** button in the slide-down panel (Fig.Pro/E-4.13). The **Sketch** dialog box appears (Fig.Pro/E-4.14).
6. *Select* the datum plane or click on the feature's surface to create a cut. In our case, we have *clicked* the right side of the hole to specify the surface, as shown in Fig.Pro/E-4.43:

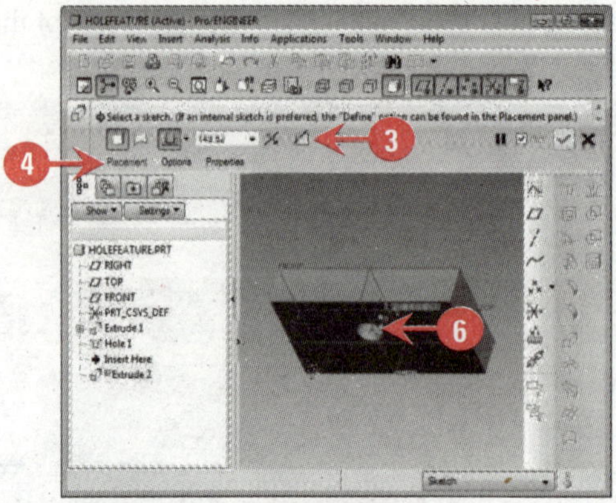

Fig.Pro/E-4.43

The selected surface, its reference plane, and orientation automatically appear in the **Sketch** dialog box (Fig.Pro/E-4.44).

7. *Click* the **Sketch** button to confirm the selection, as shown in Fig.Pro/E-4.44:

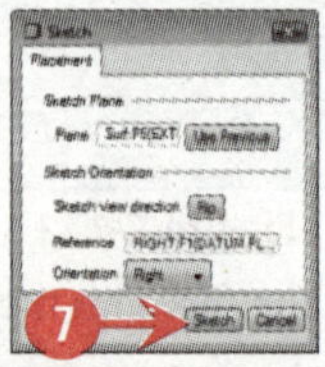

Fig.Pro/E-4.44

The Sketcher tools toolbar appears (Fig.Pro/E-4.45).

8. *Select* the **Rectangle** button on the **Sketcher tools toolbar** (Fig.Pro/E-4.45).
9. *Draw* another rectangle on the right side of the hole (Fig.Pro/E-4.45).
10. *Click* the **Done** (✔) button in the **Sketcher** tools toolbar, as shown in Fig.Pro/E-4.45:

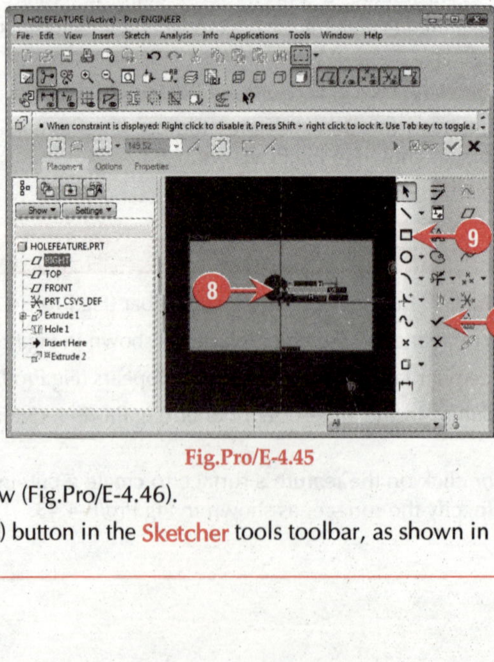

Fig.Pro/E-4.45

The sketch appears in yellow (Fig.Pro/E-4.46).

11. *Click* the **Preview** (☑ oo) button in the **Sketcher** tools toolbar, as shown in Fig.Pro/E-4.46:

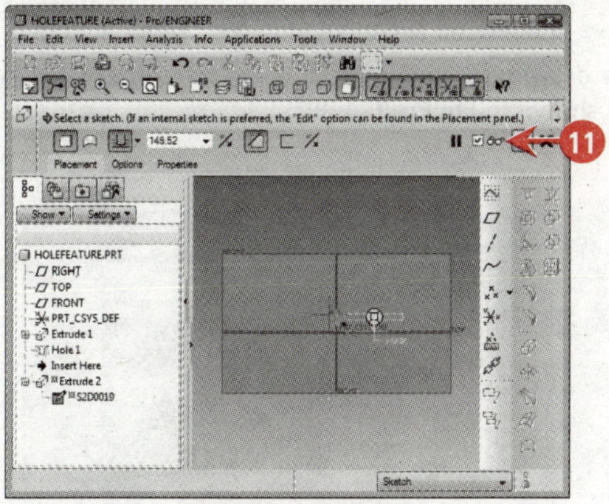

Fig.Pro/E-4.46

The cut feature is previewed (Fig.Pro/E-4.47).

12. *Click* the **Build feature** (☑) button in the **Extrude** dashboard (Fig.Pro/E-4.47) to complete the cut creation:

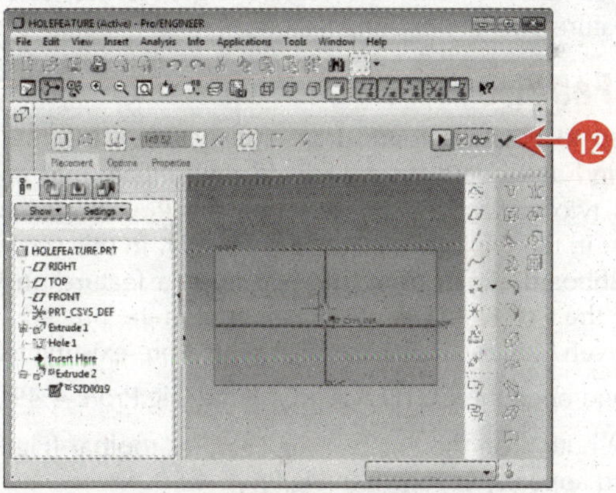

Fig.Pro/E-4.47

The final 3D cut feature is shown in Fig.Pro/E-4.48:

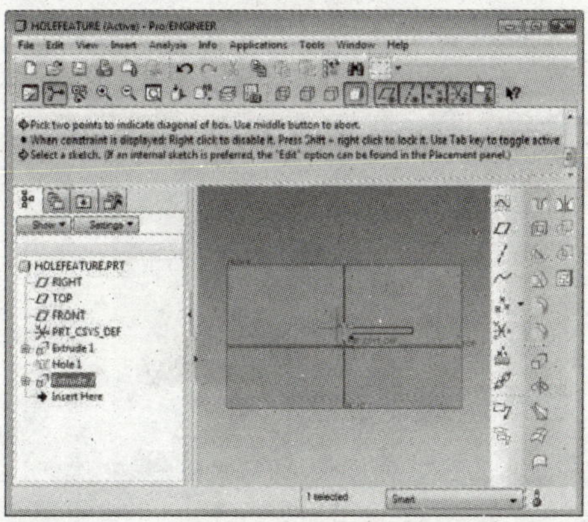

Fig.Pro/E-4.48

The edges of the cut we have created are highlighted in red color.

13. *Click* anywhere in the drawing area (Fig.Pro/E-4.48). The final view of the base feature along with the cut feature appears, as shown in Fig.Pro/E-4.49:

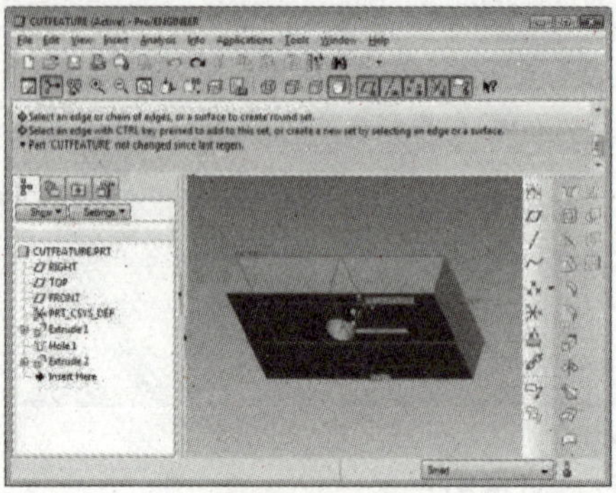

Fig.Pro/E-4.49

14. *Save* the file. In our case, we have saved the file with the name **CUTFEATURE.prt**

In this way, a cut feature is created on any other feature to get the desired 3D component. Now, we study the process to add a round feature on a pre-existing feature.

Inducing the Round Feature

The round feature is used to create a round transition between two adjacent edges. In `Pro/ENGINEER Wildfire 4.0` the **Round** dashboard is used to add round feature to an existing feature. You can invoke the **Round** dashboard in two ways. First, by selecting the **Round** option in the **Insert** menu; second, by clicking the **Round** button in the **Engineering Features** toolbar. In this section, we have used the second way to invoke the **Round** dashboard. Apart from that, we need a feature on which the round feature is to be added. Therefore, we use the **CUTFEATURE.prt** file created in the preceding section to add round feature on it. Let's follow the steps given here to add round feature on a pre-existing feature:

1. *Start* the **Part** mode and open the CUTFEATURE.prt file (Fig.Pro/E-4.49).
2. *Click* the **Round** ( ) button on the **Engineering Features** toolbar (Fig.Pro/E-4.33) in the right tool chest. The **Round** dashboard appears, as shown in Fig.Pro/E-4.50.

Apart from the **Round** dashboard, the part file CUTFEATURE.prt is also displayed in Fig.Pro/E-4.50.

3. *Select* the **Sets** tab in the **Round** dashboard, as shown in Fig.Pro/E-4.50:

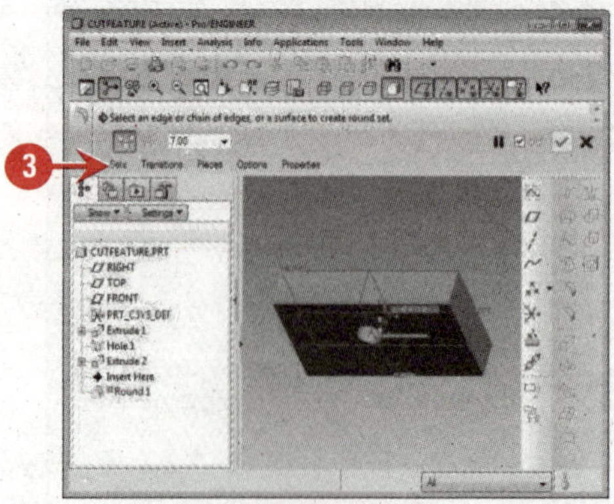

Fig.Pro/E-4.50

A slide-down panel appears, as shown in Fig.Pro/E-4.51:

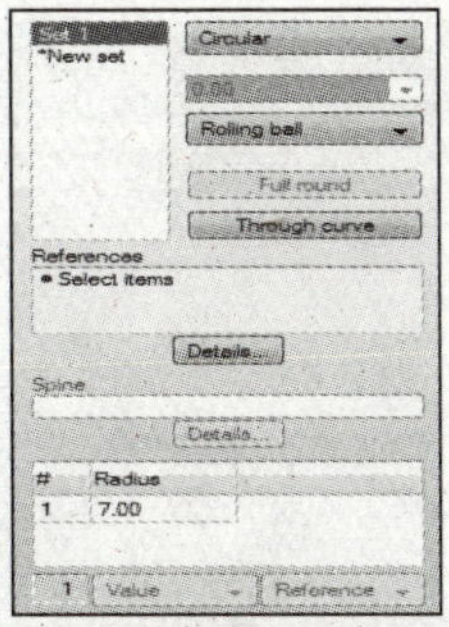

Fig.Pro/E-4.51

This slide-down panel allows you to select the edges of a feature on which the round feature has to be applied. The edges you select are grouped in sets and those sets are numbered as **Set1**, **Set2**, and so on. You can create multiple sets by clicking the ***New set** in the slide-down panel to group different edges. By default, **Set 1** is created and all the edges you select are grouped in this set. In our case, we group all the edges in the default set. The message, `Select items` in the **References** text box prompts you to select the edges that need to be rounded.

4. *Select* the first edge, you want to round. In our case, we have selected the edge of the circle in the base feature, as shown in Fig.Pro/E-4.52:

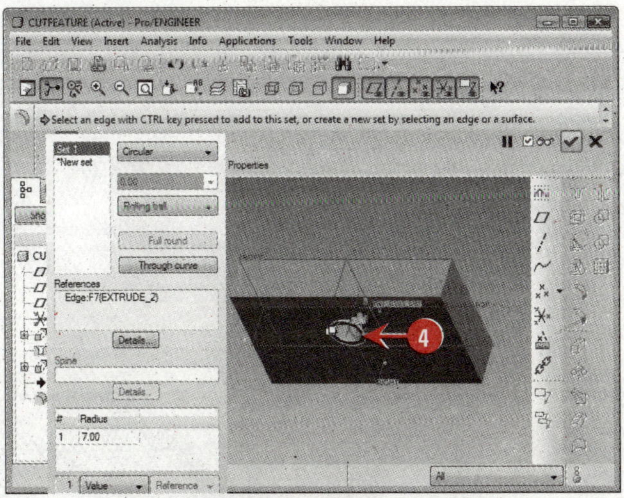

Fig.Pro/E-4.52

The selected edge appears in the **References** text box (Fig.Pro/E-4.52).

5. To select more edges, *hold* the **CTRL** key and select edges. In our case, we have selected all the edges appearing in the default orientation of the base feature (Fig.Pro/E-4.53).

6. *Press* the middle mouse button and spin the base feature to select the remaining edges, if required. In our case, we have selected all the edges of the rectangular slot.

After selecting all the edges, *set* the radius of the round. By default, some value as the radius of the round appears. In our case 7.00 is the radius of the round, which can be viewed in the edit box of the **Radius** option.

7. *Modify* the radius of the round (Fig.Pro/E-4.53). In our case, we continue with the default value.

8. *Click* the **Build feature** (✓) button to complete adding the round feature, as shown in Fig.Pro/E-4.53:

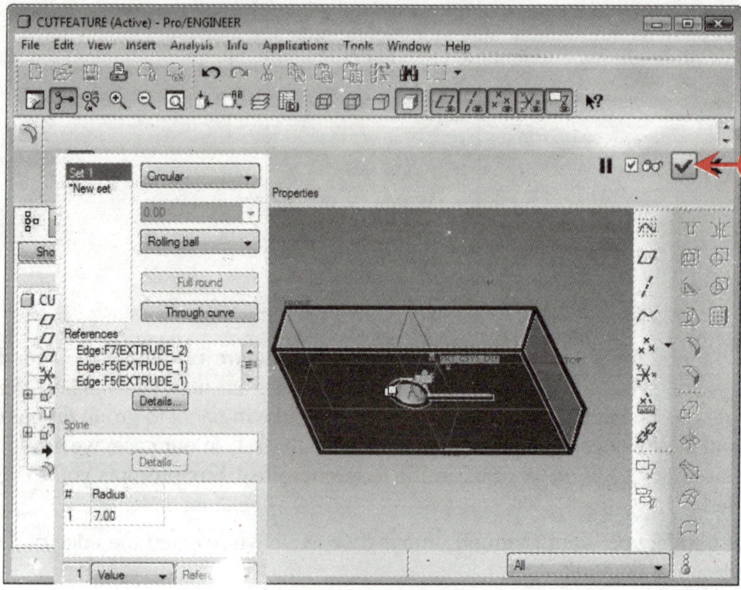

Fig.Pro/E-4.53

The base feature after adding the round feature appears, as shown in Fig.Pro/E-4.54:

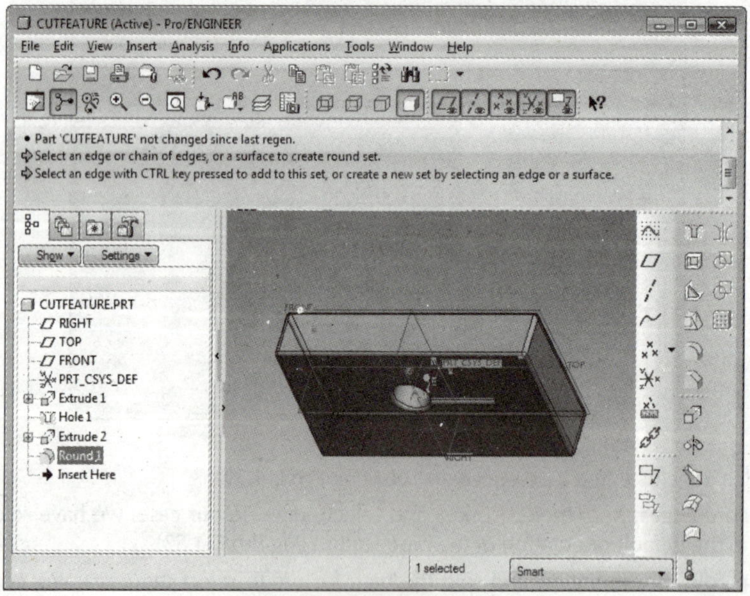

Fig.Pro/E-4.54

9. *Click* anywhere in the drawing area (Fig.Pro/E-4.54). The final view of the base feature along with the round feature appears, as shown in Fig.Pro/E-4.55:

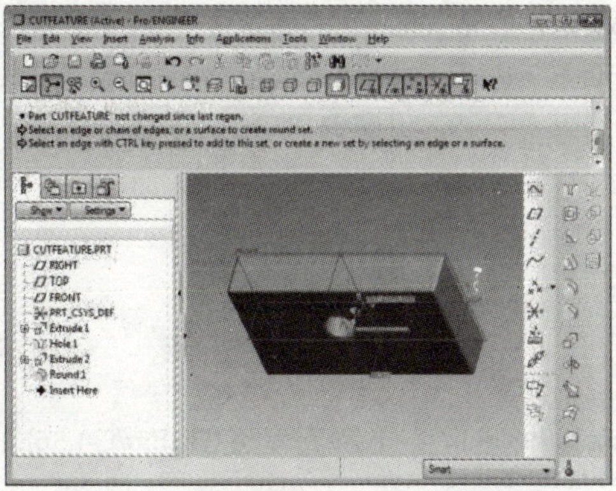

Fig.Pro/E-4.55

10. *Finally, save* the file. In our case, we have saved the file with the name ROUNDFEATURE.prt.

This is the process of adding a round feature on a base feature. Now, we learn another feature called chamfer.

Inducing the Chamfer Feature

The chamfer feature in `Pro/ENGINEER Wildfire 4.0` is used to create beveled surface around the selected edges or at the corners of selected edges. In Pro/ENGINEER Wildfire 4.0 two types of chamfer features can be added to a base feature:

- **Edge chamfer:** Bevels the selected edges
- **Corner chamfer:** Bevels the corners of the selected edges

In this section, we see the process to add edge chamfer on the edges of the base feature. Let's follow the steps given here to add edge chamfer to a feature:

1. *Start* the Part mode and *open* the CUTFEATURE.prt file (Fig.Pro/E-4.49).
2. *Click* the Edge Chamfer ( ) button in the Engineering Features toolbar (Fig.Pro/E-4.33) in the right tool chest. The Edge Chamfer dashboard appears, as shown in Fig.Pro/E-4.56.

Note

> You can also invoke the Edge Chamfer dashboard by selecting the Edge Chamfer option from the Chamfer option in the Insert menu.

3. *Select* the Sets tab in the Edge Chamfer dashboard, as shown in Fig.Pro/E-4.56:

Fig.Pro/E-4.56

A slide-down panel that prompts you to select the edges that need to be chamfered appears, as shown in Fig.Pro/E-4.57:

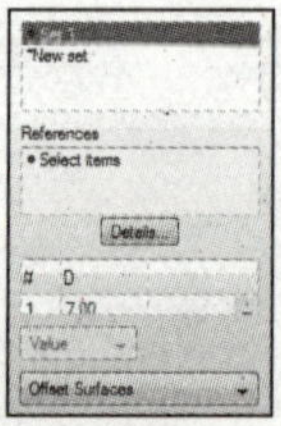

Fig.Pro/E-4.57

4. *Select* the edges you want to bevel in the base feature (Fig.Pro/E-4.56). In our case, we have selected all four top edges in the base feature (Fig.Pro/E-4.58).

In Fig.Pro/E-4.58, the number 7.00 is appearing in yellow color. This number is representing the distance of the edge chamfer from the selected edges. Now, you have two options, either continue with this distance value or modify it. In our case, we are modifying the value.

5. To modify the distance, *double-click* the number, as shown in Fig.Pro/E-4.58:

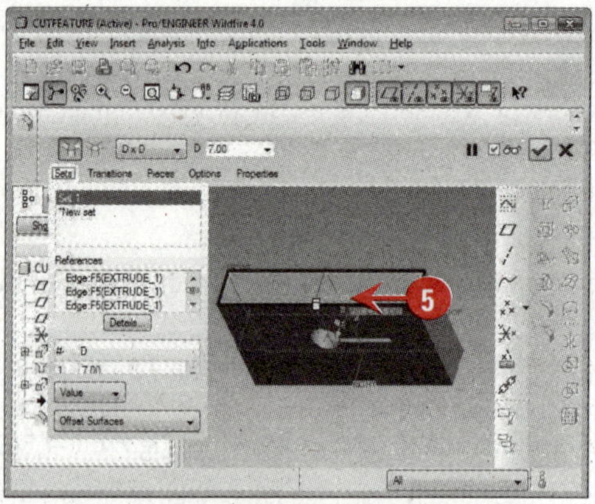

Fig.Pro/E-4.58

An edit box appears, as shown in Fig.Pro/E-4.59:

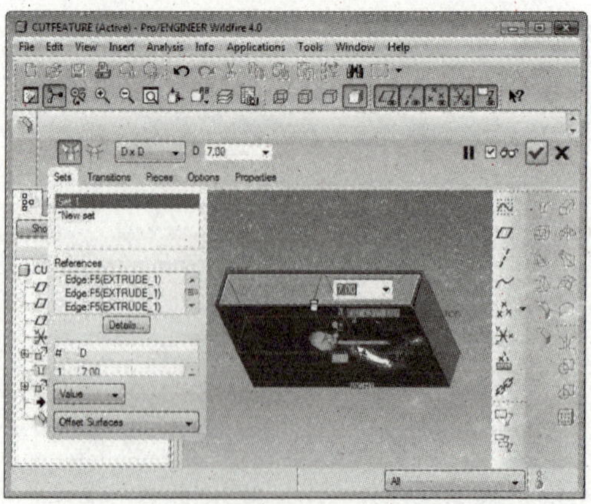

Fig.Pro/E-4.59

6. Either *enter* some new value in the edit box or select any previously used value. In our case, we have entered 15.00 as new value. However, to select any previously used value, *click* the down arrow beside the number and select the value.
7. *Press* the **ENTER** key. The edit box disappears (Fig.Pro/E-4.60). The new distance value is highlighted in yellow (Fig.Pro/E-4.60).
8. *Click* the **Build feature** (✓) button to complete adding edge chamfer on the base feature, as shown in Fig.Pro/E-4.60:

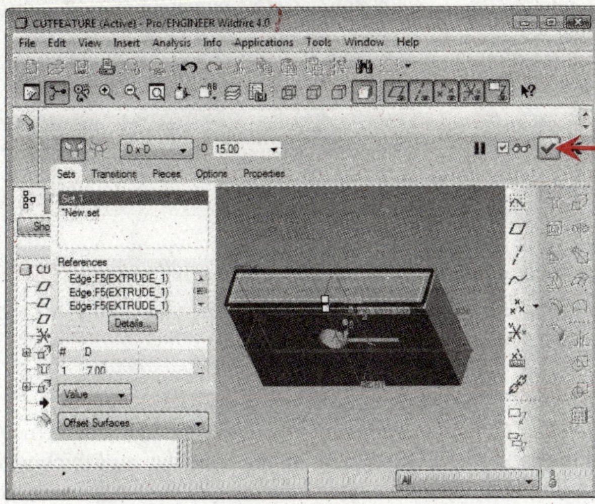

Fig.Pro/E-4.60

The slide-down panel disappears and the base feature along with the edge chamfer feature appears, as shown in Fig.Pro/E-4.61:

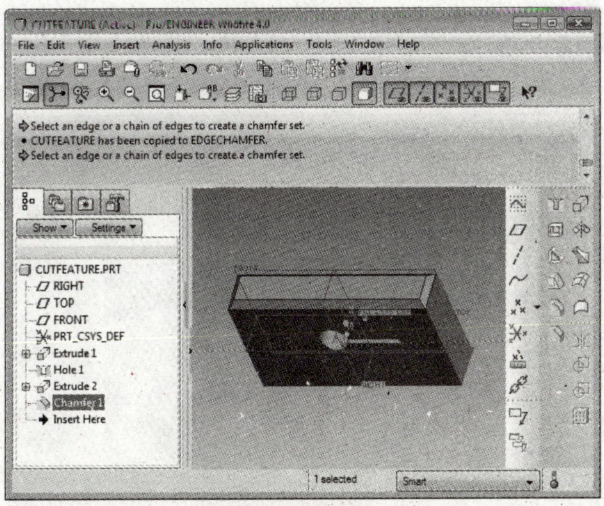

Fig.Pro/E-4.61

9. *Click* anywhere in the drawing area. The final view of the base feature along with the edge chamfer feature appears, as shown in Fig.Pro/E-4.62:

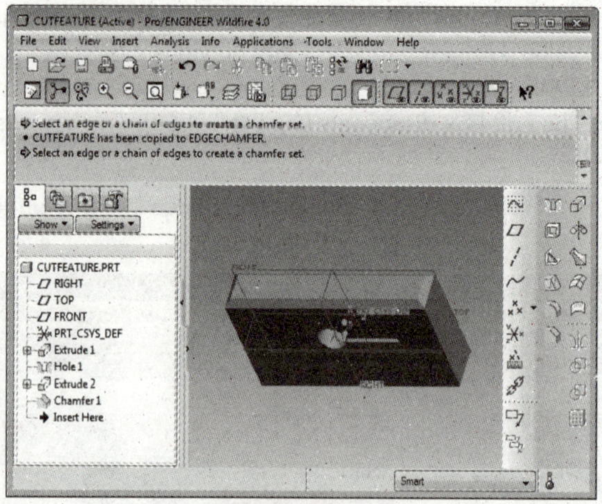

Fig.Pro/E-4.62

10. Finally, *save* the file. In our case, we have saved the file with the name **EDGEFEATURE.prt**.

With this, we e completed learning the process of adding chamfer feature on a base feature. Now, we will discuss another feature called **shell**.

Inducing a Shell Feature

The shell feature in `Pro/ENGINEER Wildfire 4.0` creates a shell by removing a surface from a solid object. The shell created in Pro/ENGINEER Wildfire 4.0 has specific wall thickness. A shell in Pro/ENGINEER Wildfire 4.0 is created using the **Shell** dashboard. However, to create a shell we need a base feature; therefore, in this section we use PRT0001.prt file as the base feature. This file contains a rectangular slot. Now, let's follow the steps given here to create a shell:

1. *Start* the **Part** mode and *open* the **PRT0001.prt** file (Fig.Pro/E-4.24).
2. *Click* the **Shell** ( ) button on the **Engineering Features** toolbar (Fig.Pro/E-4.33) in the right tool chest. The **Shell** dashboard appears, as shown in Fig.Pro/E-4.63:

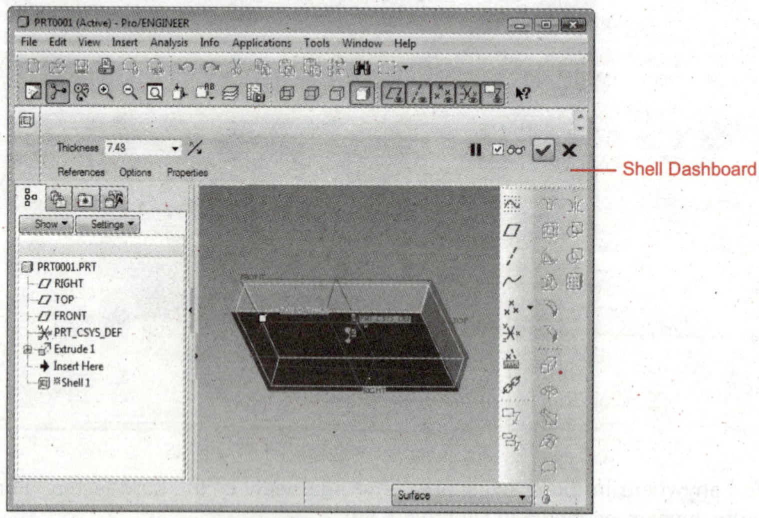

Fig.Pro/E-4.63

As soon as the **Shell** dashboard is invoked, all the surfaces of the base feature get highlighted in yellow. Apart from that, the thickness of the walls of the surfaces also appears beside the **O_THICK** label on the base feature in the drawing area. In our case, the thickness of the walls of all the surfaces is 7.48. The thickness of the walls can also be viewed in the drop-down edit box beside the **Thickness** label in the **Shell** dashboard (Fig.Pro/E-4.63).

Note

> If the thickness of the walls varies then the thickness of the walls appears individually.

3. *Select* the surface that has to be removed from the rectangular slot to create a shell. In our case, we have selected the front surface. As soon as you select the surface, the color of the selected surface changes (Fig.Pro/E-4.64).

Now, we have two options; either we can modify the thickness of the walls of the rectangular slot or, continue the shell creation with the default thickness.

In our case, we are modifying the thickness value. To modify the thickness, again we have two ways:

- *Double-click* the thickness appearing on the base feature in the drawing area (Fig.Pro/E-4.64).
- Enter new thickness value in the drop-down edit box beside the **Thickness** label in the **Shell** dashboard (Fig.Pro/E-4.64).

4. *Double-click* the thickness appearing on the base feature, as shown in Fig.Pro/E-4.64:

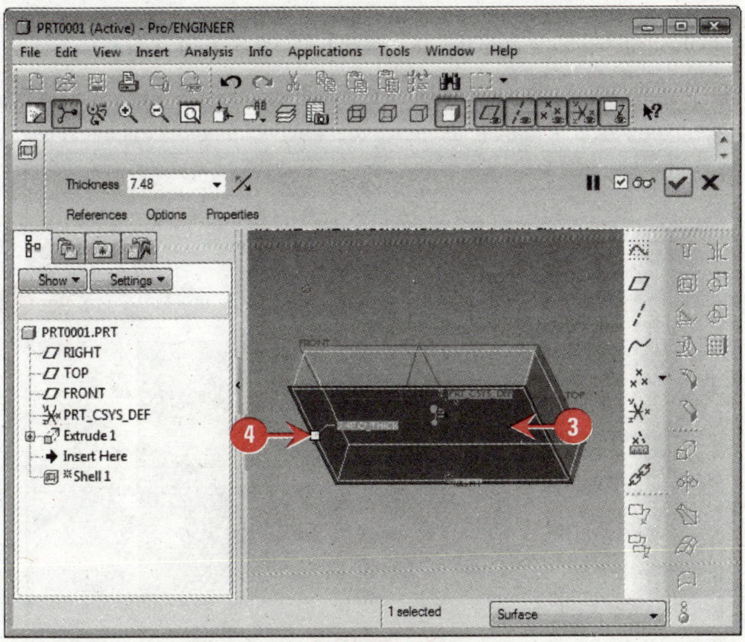

Fig.Pro/E-4.64

The thickness of the edit box appears (Fig.Pro/E-4.65).

5. Now, *specify* a new thickness in the edit box (Fig.Pro/E-4.65). In our case, we entered 25.25 as new value.

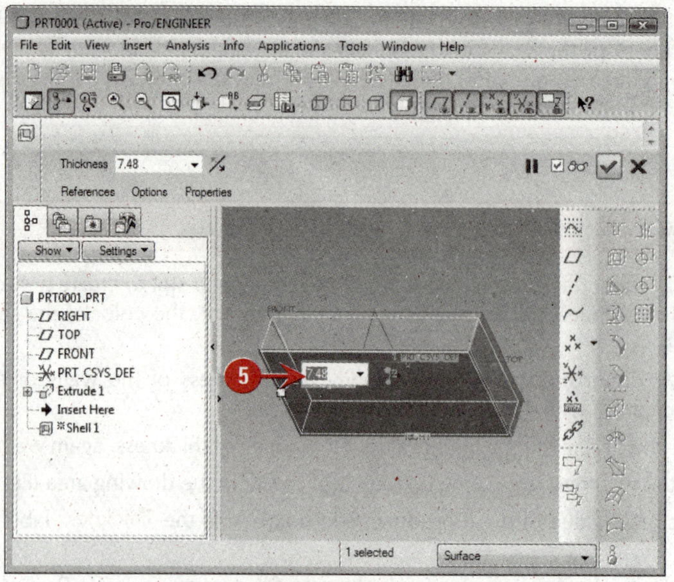

Fig.Pro/E-4.65

6. *Press* the **ENTER** key. The new thickness is previewed in the drawing area and also in the drop-down edit box beside the **Thickness** label in the **Shell** dashboard (Fig.Pro/E-4.66).
7. *Click* the **Build feature** (✓) button to complete the shell creation, as shown in Fig.Pro/E-4.66:

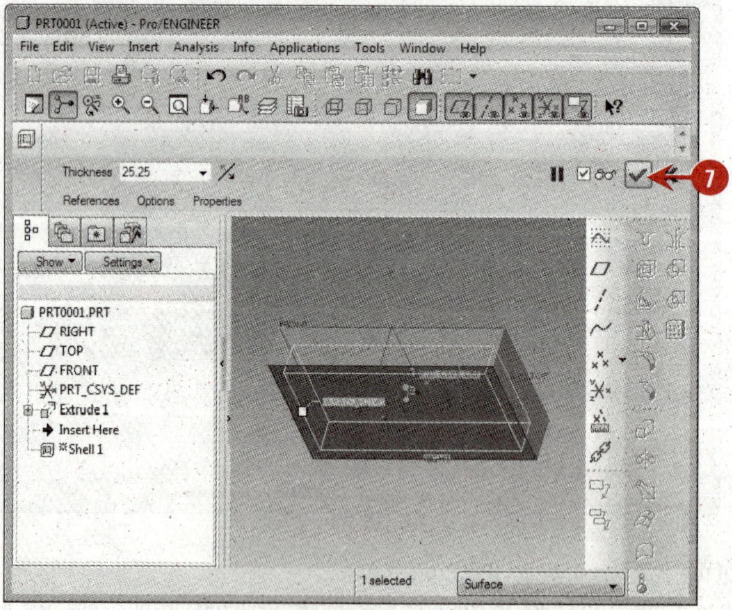

Fig.Pro/E-4.66

The shell appears, as shown in Fig.Pro/E-4.67:

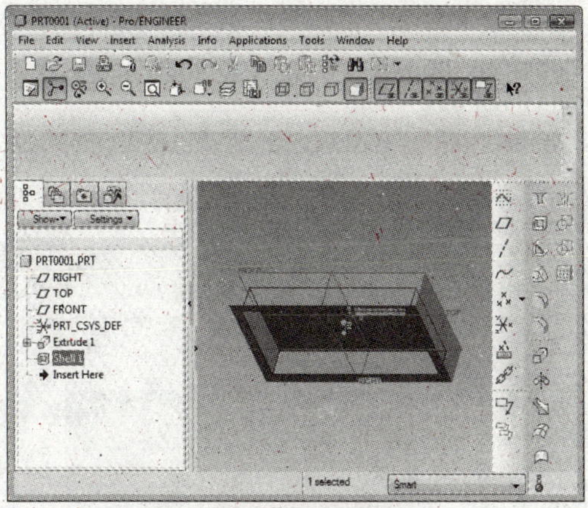

Fig.Pro/E-4.67

The boundary of the shell is highlighted in red color (Fig.Pro/E-4.67).

Note

As soon as the Build feature button is clicked the Shell dashboard disappears (Fig.Pro/E-4.67).

8. *Click* anywhere in the drawing area. The final view of the shell appears, as shown in Fig.Pro/E-4.68:

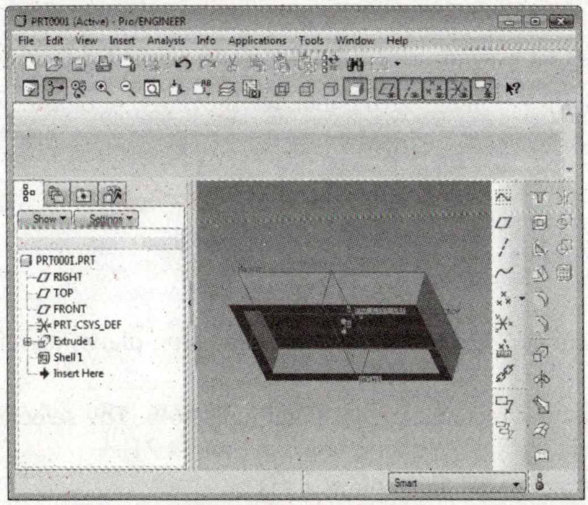

Fig.Pro/E-4.68

9. Finally, *save* the file. In our case, the file is saved with the name **SHELLFEATURE.prt**.

Here, we have completed the process of creating a shell on a base feature. Now, we move ahead and understand the process of creating a revolved feature.

Inducing a Revolved Feature

The revolved feature in `Pro/ENGINEER Wildfire 4.0` is used to create features by revolving a sketch around the centerline. To create a revolved feature, the **Revolve** dashboard is used, which helps you create features, such as revolved protrusions, revolve cuts, and revolved surfaces. However, to create a revolved feature, a base feature is required. Therefore, in this section, we use the **PRT0001.prt** file as the base feature, which contains a rectangular slot. On that rectangular slot, we create a revolved protrusion. Let's follow the steps given here to create a revolved protrusion:

1. *Start* the **Part** mode and open the **PRT0001.prt** file (Fig.Pro/E-4.24).
2. *Click* the **Revolve** ( ) button in the **Base Features** toolbar (Fig.Pro/E-4.11) in the right tool chest. The **Revolve** dashboard appears (Fig.Pro/E-4.69).
3. *Select* the **Placement** tab in the **Revolve** dashboard, as shown in Fig.Pro/E-4.69:

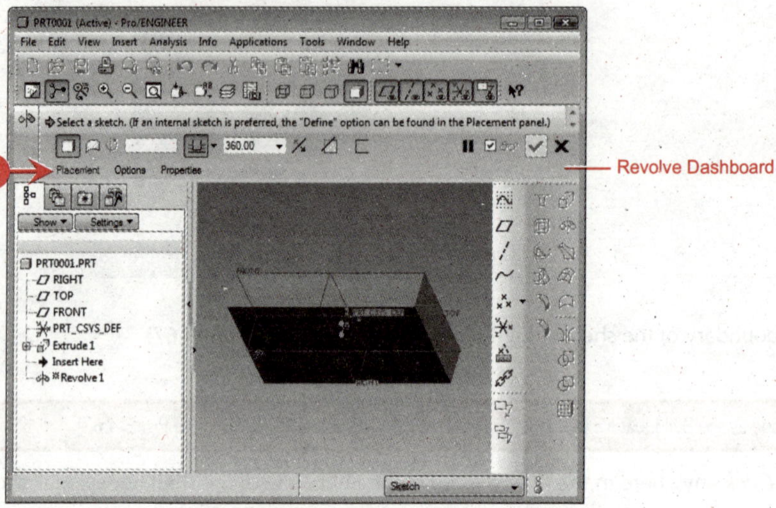

Fig.Pro/E-4.69

A slide-down panel appears that lets you define the sketching plane (Fig.Pro/E-4.70).

4. *Click* the **Define** button on the slide-down panel, as shown in Fig.Pro/E-4.70:

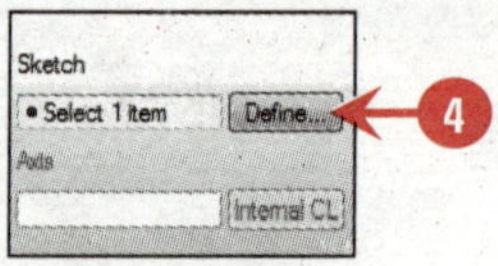

Fig.Pro/E-4.70

The **Sketch** dialog box where you have to specify the sketching plane, its reference plane, and orientation appears (Fig.Pro/E-4.14).

5. *Select* the top surface of the rectangular slot (Fig.Pro/E-4.69). The selected surface, its reference plane, and orientation appear in the **Sketch** dialog box (Fig.Pro/E-4.71).
6. *Click* the **Sketch** button, as shown in Fig.Pro/E-4.71:

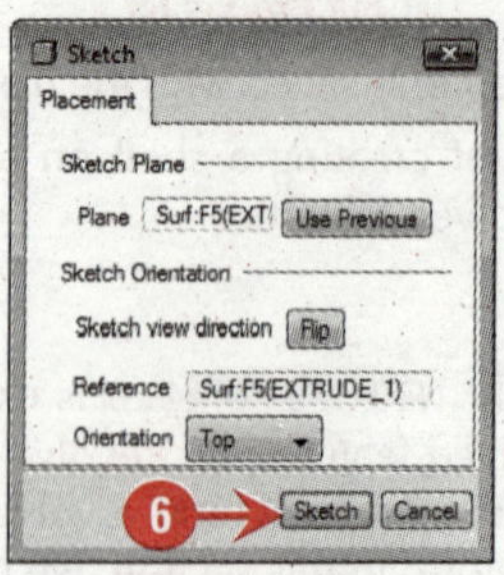

Fig.Pro/E-4.71

The **Sketcher tools toolbar** appears to let you draw the sketch of the protrusion, as shown in Fig.Pro/E-4.72:

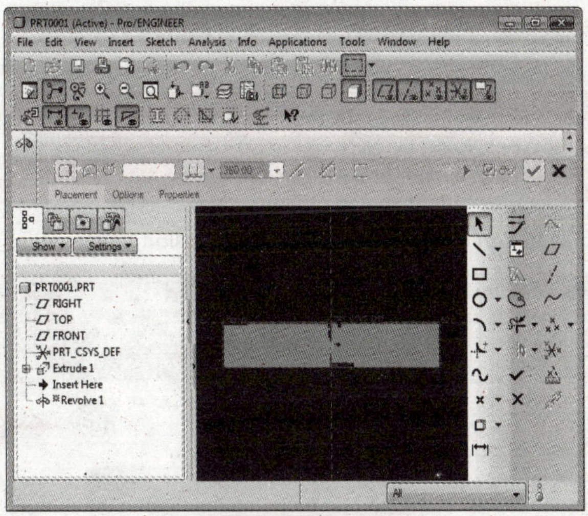

Fig.Pro/E-4.72

7. Now, *draw* a sketch. In our case, we have drawn a circle, as shown in Fig.Pro/E-4.73.

Note

The process to sketch a circle has been already depicted in chapter 3, Exploring the Pro/ENGINEER Wildfire 4.0 Sketch Mode.

8. *Draw* a centerline around the circle to use it as an axis of revolution (Fig.Pro/E-4.73).

Note

The process to sketch a centerline has already depicted in chapter 3, Exploring the Pro/ENGINEER Wildfire 4.0 Sketch Mode.

9. *Click* the **Done** (✓) button, as shown in Fig.pro/E-4.73:

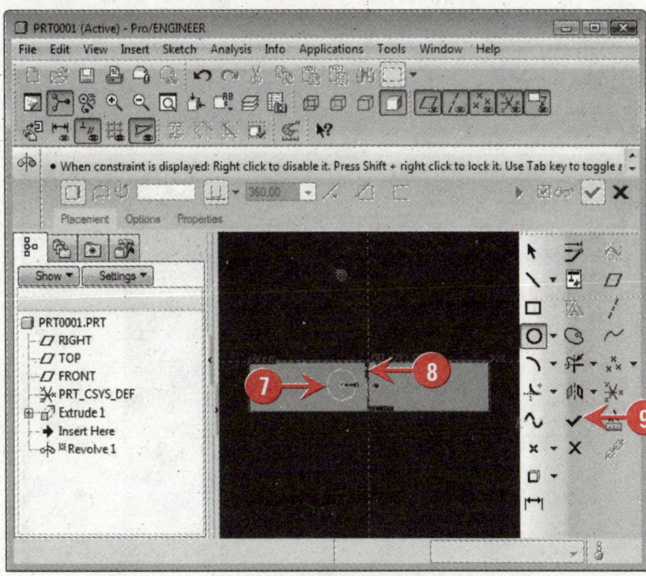

Fig.Pro/E-4.73

The sketch of the revolved protrusion is created, as shown in Fig.Pro/E-4.74.

Note

> The revolved protrusion is created according to the default setting of Revolved dashboard. For example, the default angle for the protrusion is 360 °; however, if you want to modify the angle then click the down arrow beside [360.00 ▾] and select another angle. You can also enter an angle rather than selecting an angle from the drop-down edit list.

10. *Click* the **Preview** (☑ 👓) button to verify the revolved protrusion, as shown in Fig.Pro/E-4.74:

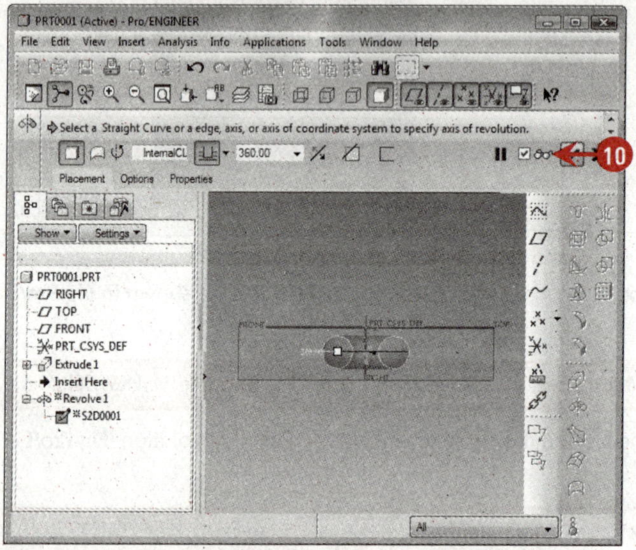

Fig.Pro/E-4.74

The preview of the revolved protrusion appears, as shown in Fig.Pro/E-4.75:

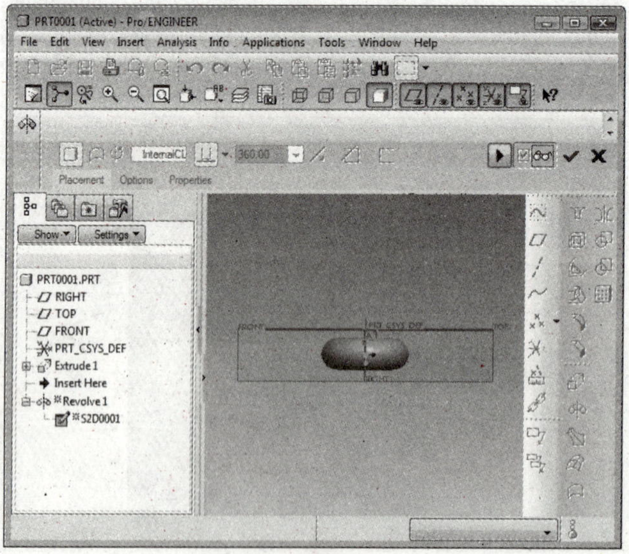

Fig.Pro/E-4.75

11. *Hold* the middle mouse button and spin the rectangular slot to have a 3D view of the protrusion, as shown in Fig.Pro/E-4.76.

12. *Click* the **Build feature** (✓) button to complete the revolved protrusion, as shown in Fig.Pro/E-4.76:

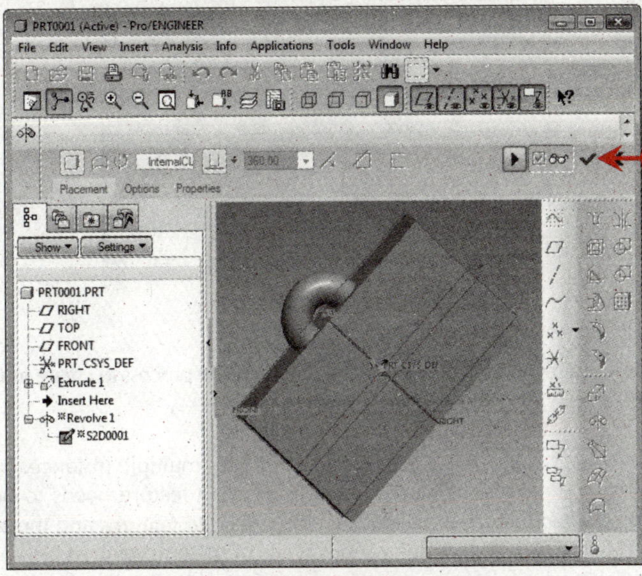

Fig.Pro/E-4.76

The revolved protrusion is highlighted, as shown in Fig.Pro/E-4.77:

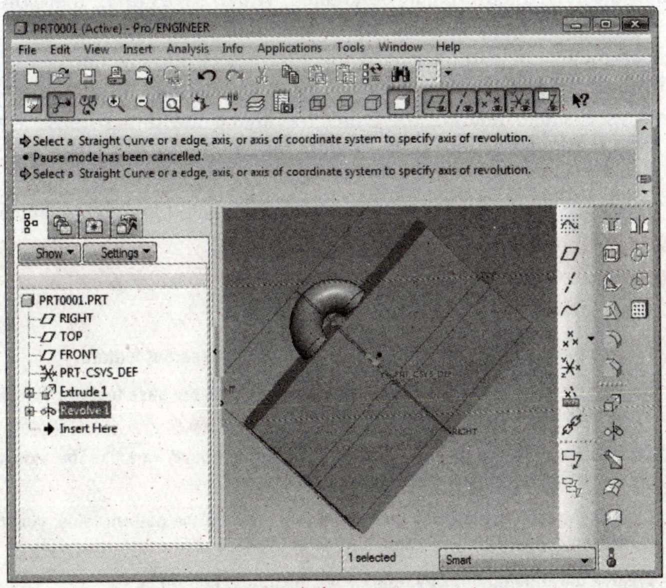

Fig.Pro/E-4.77

13. *Click* anywhere in the drawing area. The final view of the revolved protrusion is shown in Fig.Pro/E-4.78:

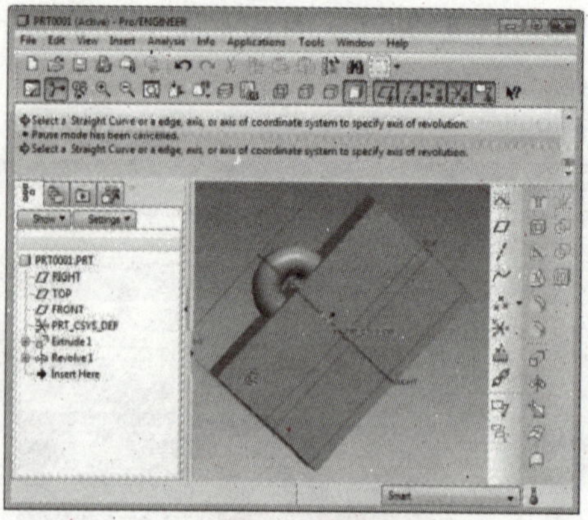

Fig.Pro/E-4.78

14. Finally, *save* the file.

This is the process of creating a revolved protrusion. Now, we study the process to create multiple copies of a feature.

Working with the Pattern Feature

The pattern feature in `Pro/ENGINEER Wildfire 4.0` creates multiple instances of a selected feature. The pattern feature speeds up the process of model creation where a feature needs to be created several times. Multiple instances of a selected feature are called **patterns** or **child features** and the selected feature is called **parent** feature. In Pro/ENGINEER Wildfire 4.0 various types of patterns such as **Dimension** pattern, **Direction** pattern can be created. However, before understanding pattern types, we discuss the benefits of creating patterns. Following are the benefits of creating patterns:

- Multiple copies of a selected feature can be created, which saves time; therefore, the process of model creation speeds up.
- All patterns are parameterized and modifying the parameters, such as number of child features, spacing between child features are easy.
- All patterns work as a single feature; therefore; modifying the dimensions of all the copies is easier because if you modify the dimension of the parent feature then the change is reflected in all the child features.

Let's see Table 4.1, which lists the type of patterns that can be created in Pro/ENGINEER Wildfire 4.0:

Table 4.1: Pattern Types Available in Pro/ENGINEER Wildfire 4.0

Pattern Type	Explanation
Dimension	Creates the child features using the dimensions of the parent feature.
Direction	Creates the child features by specifying a direction on the base feature. The direction can be specified by selecting a sketching plane or an edge.
Axis	Creates the child features by specifying an angle around an axis. The axis should be of the feature around which the child features have to be created.
Table	Creates the child features by using a pattern table. In the pattern table, you must need to specify the dimension of every child feature.
Reference	Creates the child features by referencing the parent feature.
Fill	Creates the child features by filling the sketched area with a selected feature.
Curve	Creates the child features by specifying the numbers of the child features along a curve.

There are the various types of patterns that can be created in Pro/ENGINEER Wildfire 4.0. However, in this section, we see the process to create the child features using the **Axis** pattern in the following steps:

1. *Start* the **Part** mode and *open* the **HOLEFEATURE.prt** file (Fig.Pro/E-4.42).
2. *Create* a hole of diameter 30.00 adjacent to the hole in the base feature. Once the hole has been created, the base feature appears (Fig.Pro/E-4.79).
3. Now, *select* the hole whose patterns need to be created, as shown in Fig.Pro/E-4.79:

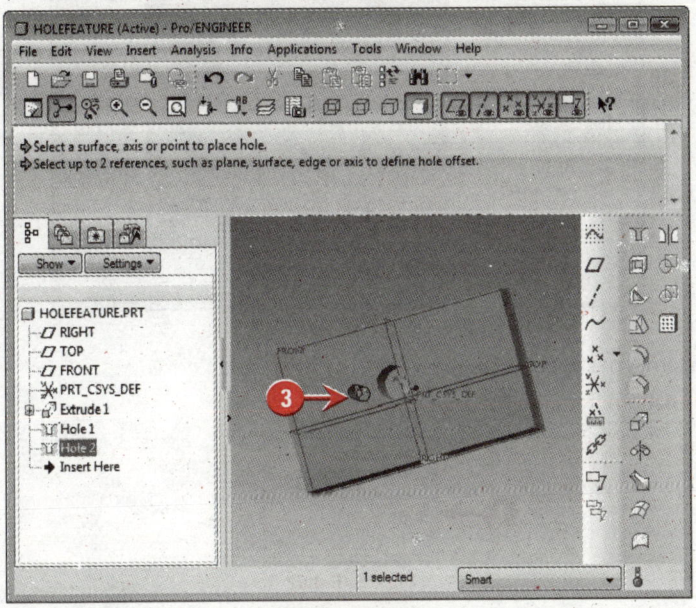

Fig.Pro/E-4.79

In our case, the hole on the left side (Fig.Pro/E-4.79) is selected.

4. *Click* the **Pattern** ( ) button in the **Edit Features** toolbar, as shown in Fig.Pro/E-4.80:

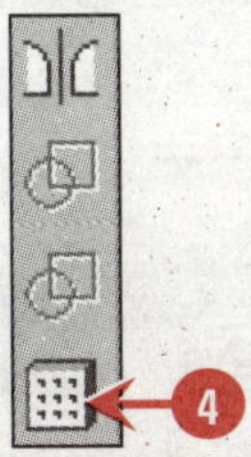

Fig.Pro/E-4.80

Note

> By default, all buttons in the Edit Feature toolbar are disabled. These buttons get enabled depending on their requirements. For example, the Pattern button is enabled when a feature, which is to be patterned, is selected.

The **Pattern** dashboard appears (Fig.Pro/E-4.81).

5. *Click* the down-arrow beside Dimension, as shown in Fig.Pro/E-4.81:

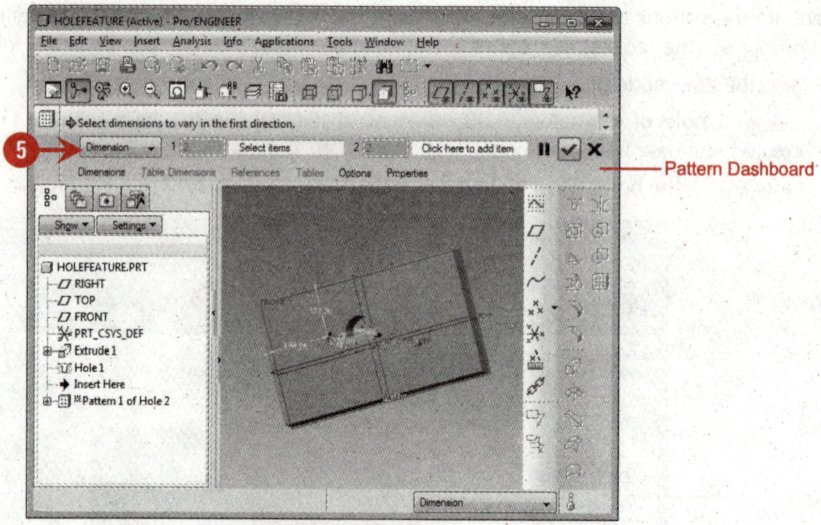

Fig.Pro/E-4.81

A list of various pattern types appears, as shown in Fig.Pro/E-4.82.

6. *Select* the **Axis** pattern from the list (Fig.Pro/E-4.82):

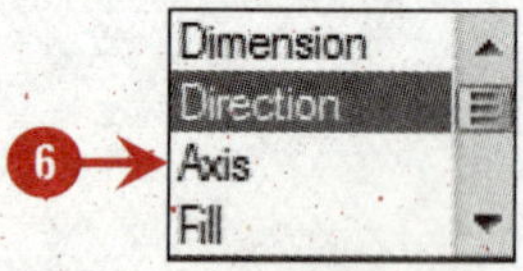

Fig.Pro/E-4.82

The layout of the **Pattern** dashboard changes (Fig.Pro/E-4.83).

7. *Select* the axis, A_1, of the hole, as shown in Fig.Pro/E-4.83:

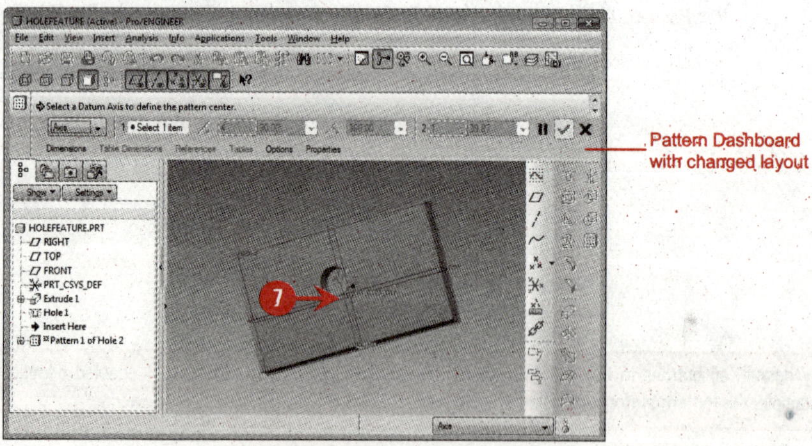

Fig.Pro/E-4.83

Note

The layout of the Pattern dashboard depends on the type of pattern selected.

The selected axis turns red, as shown in Fig.Pro/E-4.84:

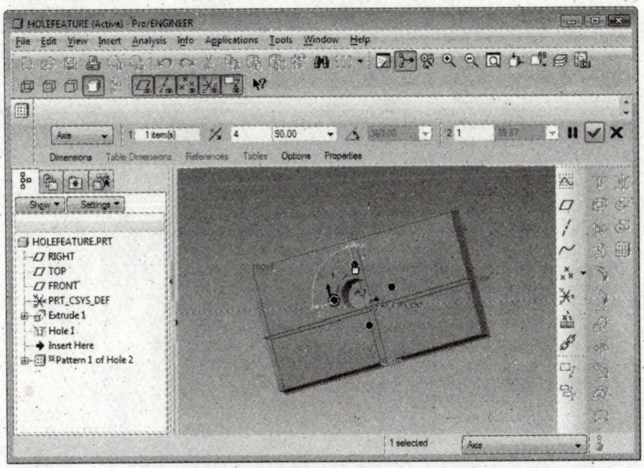

Fig.Pro/E-4.84

Apart from that, a preview of the patterns also appears (Fig.Pro/E-4.84). You see four holes around the selected axis (Fig.Pro/E-4.84); this is because by default, the numbers of patterns are 4 and the angle between these patterns is 90 degrees.

8. To modify the total number of the patterns, *specify* a new number in the text box beside the **Flip** ( ) button. In our case, we have entered 6.

Note

The Flip button is used to flip the angular direction of the pattern.

9. To modify the angle between the patterns, either select an angle from the list of pre-defined angles by clicking the down-arrow beside [90.00 ▾] or enter any other angle. In our case, we have selected 60 ° from the list of given angles.
10. *Press* the **ENTER** key. Changes are reflected in the drawing area, as shown in Fig.Pro/E-4.85.
11. *Click* the **Build feature** ( ) button, as shown in Fig.Pro/E-4.85:

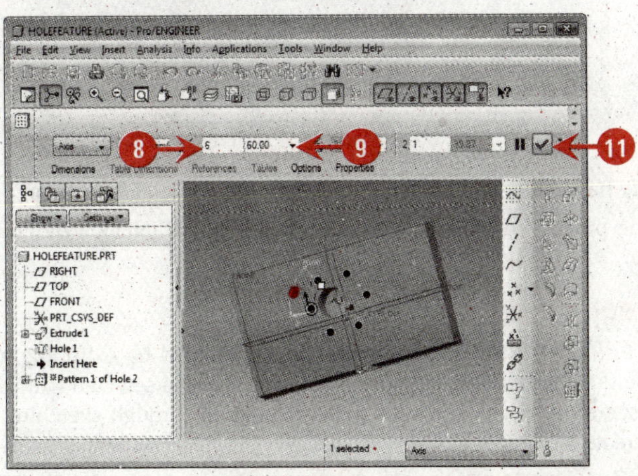

Fig.Pro/E-4.85

All the patterns are highlighted in red color, as shown in Fig.Pro/E-4.86:

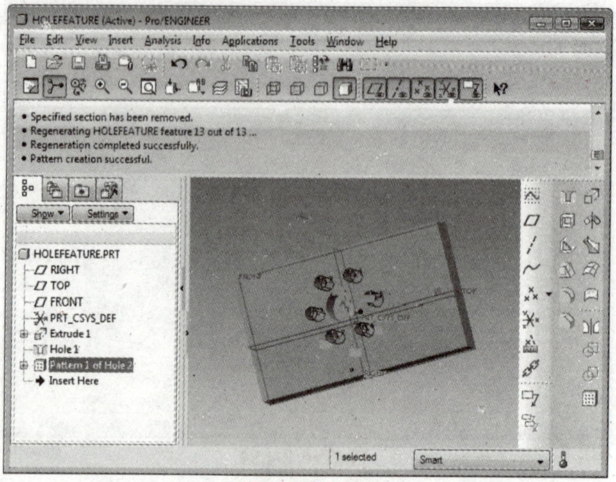

Fig.Pro/E-4.86

12. *Click* anywhere in the drawing area to have a look at the final view of the patterns. The base feature along with the patterns appears, as shown in Fig.Pro/E-4.87:

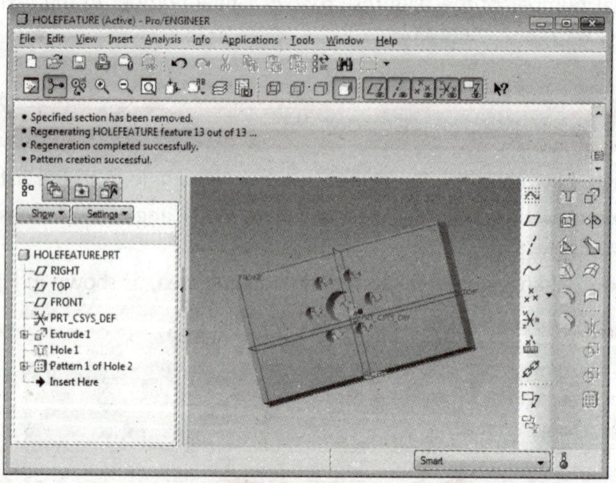

Fig.Pro/E-4.87

13. *Save* the file. We have saved the file with the name, PATTERNFEATURE.prt

Now, we learned the process to create Axis patterns in Pro/ENGINEER Wildfire 4.0. Next, we will study a feature called sweep.

Inducing the Sweep Feature

A sweep feature in `Pro/ENGINEER Wildfire 4.0` is created by extruding a section along a trajectory. The trajectory can be sketched during a feature creation or can be selected within an already created feature. In this section, we see the process to create a sweep feature through sketching a section. Let's follow the steps given here to create a sweep feature:

1. *Start* the **Part** mode. *Select* **Insert→Sweep→Protrusion**, as shown in Fig.Pro/E-4.88:

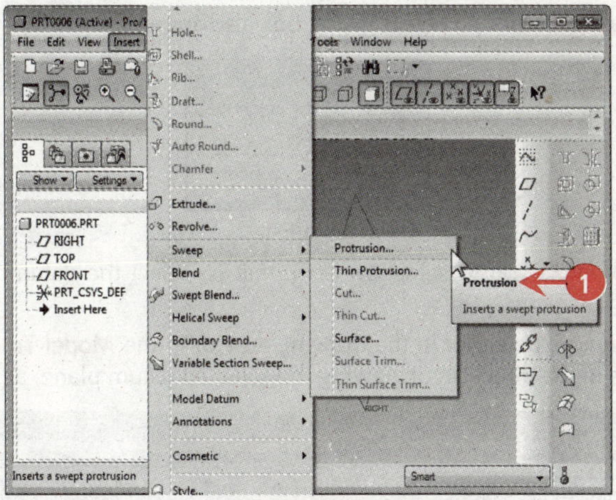

Fig.Pro/E-4.88

The **PROTRUSION: Sweep** dialog box appears, as shown in Fig.Pro/E-4.89:

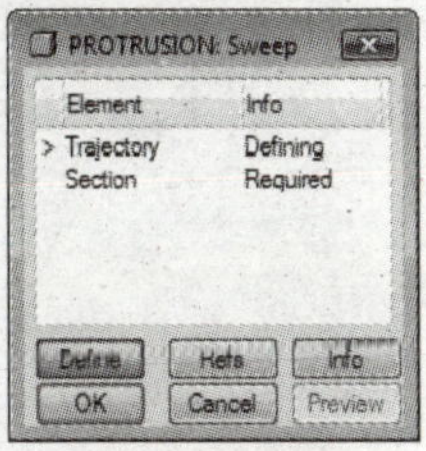

Fig.Pro/E-4.89

Apart from **PROTRUSION: Sweep** dialog box, a **Menu Manager** menu also appears (Fig.Pro/E-4.90).

The **Menu Manager** menu provides you two options:

- To sketch a trajectory
- To select a trajectory from the existing feature

2. *Select* the **Sketch Traj** option, as shown in Fig.Pro/E-4.90:

Fig.Pro/E-4.90

The appearance of the **Menu Manager** menu changes, as shown in Fig.Pro/E-4.91:

Fig.Pro/E-4.91

Apart from the **Menu Manager** menu, a **Select** dialog box also appears, as shown in Fig.Pro/E-4.92:

Fig.Pro/E-4.92

The **Select** dialog box prompts you to select an item, that is, select the datum plane to define a trajectory along the selected datum plane.

3. *Select* the **Front** datum plane either in the drawing area or in the **Model Tree** in the **Part** mode window. A **red** arrow appears in the direction of viewing the selected datum plane, as shown in Fig.Pro/E-4.93:

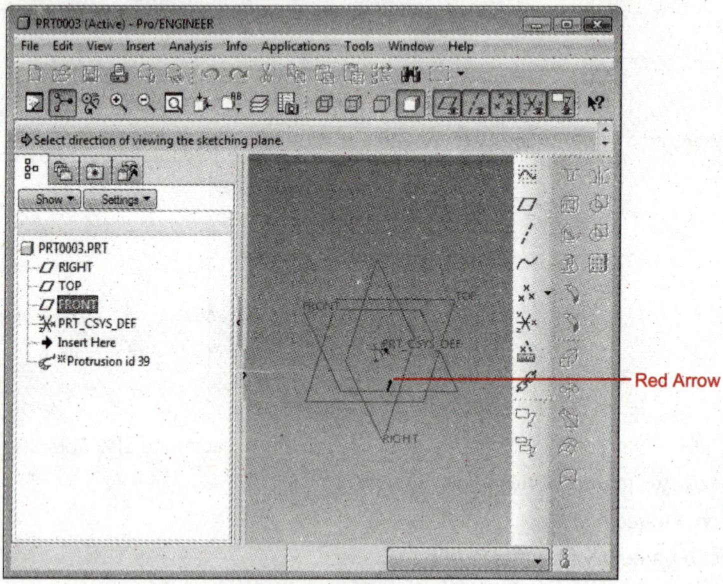

Fig.Pro/E-4.93

Apart from this, the appearance of the **Menu Manager** menu changes once again (Fig.Pro/E-4.94).

4. *Select* the **Okay** option to continue creating a sweep feature, as shown in Fig.Pro/E-4.94:

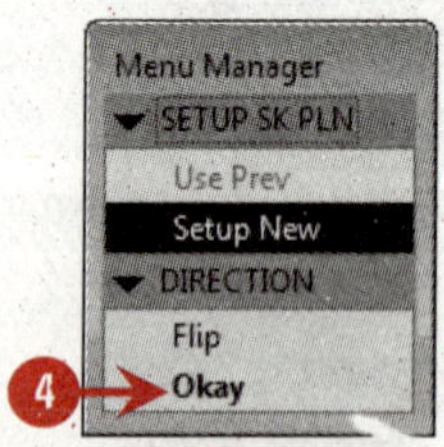

Fig.Pro/E-4.94

The appearance of the **Menu Manager** menu changes again (Fig.Pro/E-4.95). Now, you have to select the orientation view of the sweep feature from the options given in the **SKET VIEW** sub-menu of the **Menu Manager** menu (Fig.Pro/E-4.95).

5. *Select* the **Top** option in the **SKET VIEW** sub-menu, as shown in Fig.Pro/E-4.95:

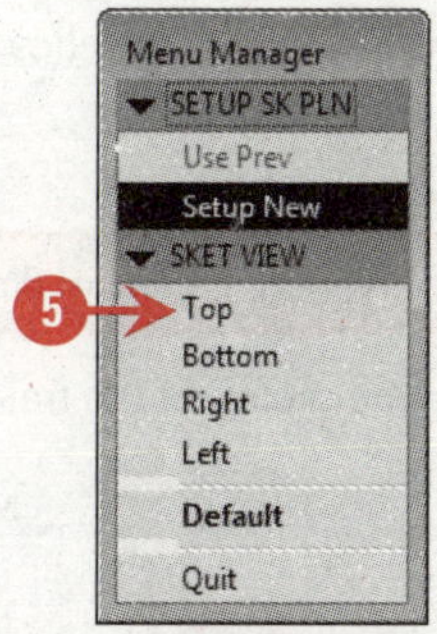

Fig.Pro/E-4.95

The appearance of **Menu Manager** menu changes again, as shown in Fig.Pro/E-4.96:

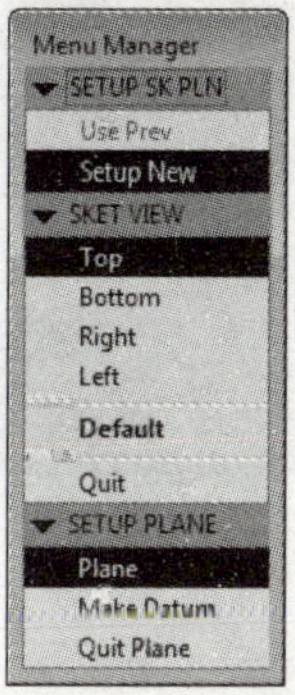

Fig.Pro/E-4.96

Apart from this, the **Select** dialog box (Fig.Pro/E-4.92) appears. Both, the **Menu Manager** menu (Fig.Pro/E-4.96) and the **Select** dialog box prompt you to select the reference plane of the **Front** datum plane in the part mode window.

6. *Select* the **Top** datum plane in the **Model Tree**. The **Sketcher tools toolbar** appears to let you sketch the trajectory (Fig.Pro/E-4.97).
7. *Select* the **3 Point/Tangent End** ( ) button in the **Sketcher tools toolbar** (Fig.Pro/E-4.97) to draw an arc, as shown in Fig.Pro/E-4.97:

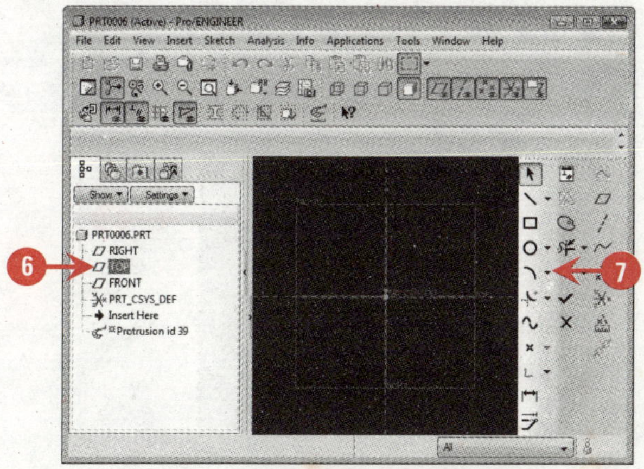

Fig.Pro/E-4.97

A **yellow** arrow is attached to the arc (Fig.Pro/E-4.98). This yellow arrow represents the starting point of the trajectory and it points in the direction of the sweep.

Note

The process of drawing an arc has already been depicted in Chapter 3, Exploring Pro/ENGINEER Wildfire 4.0 Sketch Mode.

8. *Click* the **Done** (✓) button to complete the sketch of the trajectory, as shown in Fig.Pro/E-4.98:

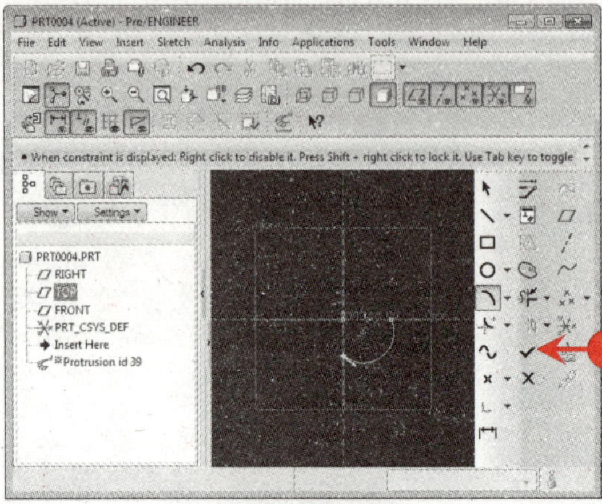

Fig.Pro/E-4.98

The view of the drawing mode is changed. This lets you create a section for sweep, as shown in Fig.Pro/E-4.99.

In Fig.Pro/E-4.99, the intersection point of two infinite yellow lines represents the starting point of the trajectory. The starting point of the section, which is to be created for a sweep, should lie at the start point of the trajectory.

9. *Click* the **Rectangle** (□) button (Fig.Pro/E-4.99) in the **Sketcher tools toolbar** to draw a rectangle as a section for the sweep.

Note

The process to draw a rectangle has already been depicted in Chapter 3, Exploring Pro/ENGINEER Wildfire 4.0 Sketch Mode.

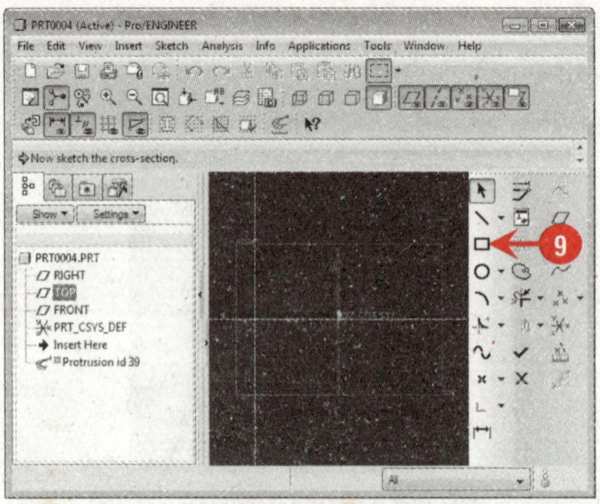

Fig.Pro/E-4.99

The rectangle appears in yellow (Fig.Pro/E-4.100).

10. *Click* the **Done** (✔) button in the **Sketcher** tools toolbar, as shown in Fig.Pro/E-4.100:

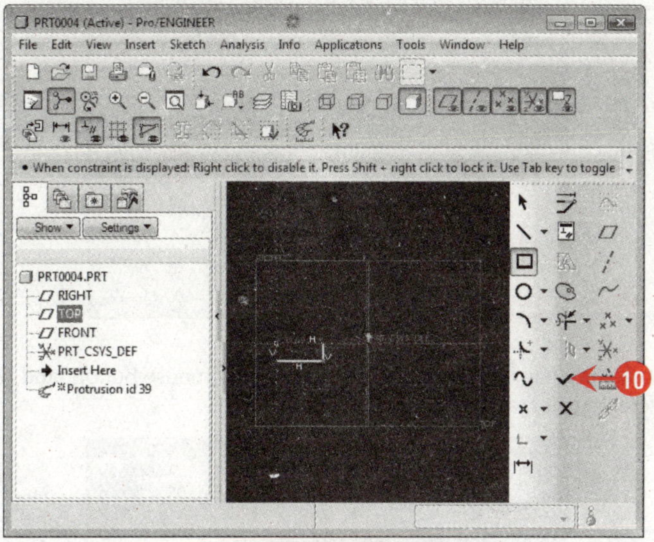

Fig.Pro/E-4.100

The sweep feature is created, as shown in Fig.Pro/E-4.101:

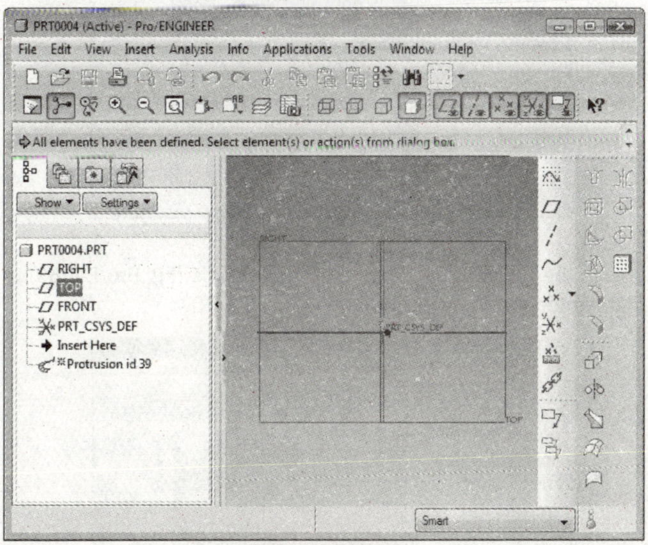

Fig.Pro/E-4.101

In Fig.Pro/E-4.101, the datum planes, along with the sweep feature are displayed.

11. *Click* the **Preview** button in the **PROTRUSION: Sweep** dialog box (Fig.Pro/E-4.89) to view the solid sweep feature. The sweep feature appears, as shown in Fig.Pro/E-4.102:

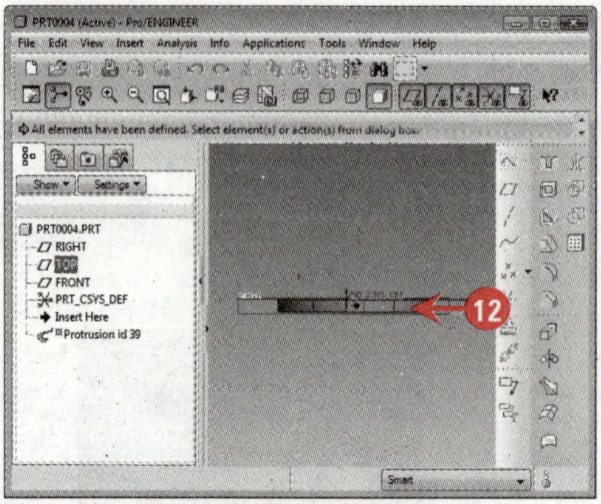

Fig.Pro/E-4.102

12. *Select* the sweep feature (Fig.Pro/E-4.102), *hold* the middle mouse button, and *spin* the sweep feature to view it from different angles, as shown in Fig.Pro/E-4.103:

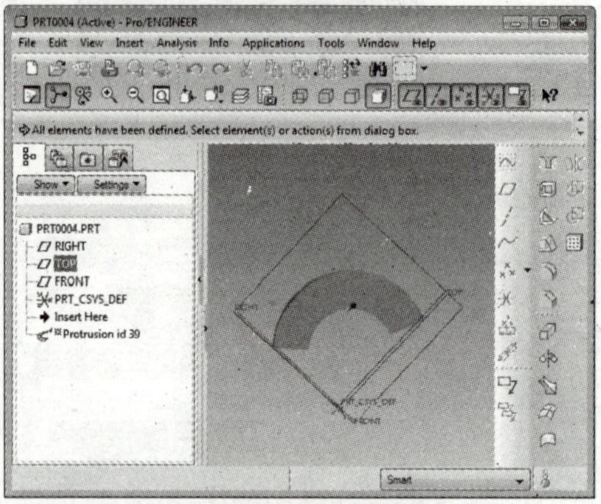

Fig.Pro/E-4.103

13. *Click* the **OK** button in the **PROTRUSION: Sweep** dialog box (Fig.Pro/E-4.89) to finish the process. The sweep feature is highlighted in red, as shown in Fig.Pro/E-4.104:

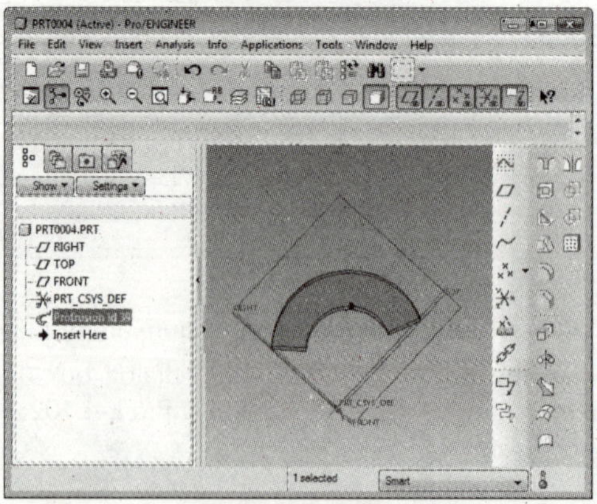

Fig.Pro/E-4.104

14. *Save* the file.

Presently, we studied the process to create a sweep feature. Now, we will study another feature, which blends two sections of different features to create a single feature.

Inducing the Blend Feature

A blended feature in Pro/ENGINEER Wildfire 4.0 is created by joining two or more sections at their edges through transitional faces. The blending of two or more sections results in a continuous feature. However, an important point to remember while creating a blended feature is that the total number of entities in all the sections must be the same. For example, a rectangle has 4 entities (4 lines); therefore, it can either be blended with another rectangle or with any section having 4 entities. In Pro/ENGINEER Wildfire 4.0, different blend types are used to blend different sections. The different blend types are:

- **Parallel:** This type is used to blend the sections that are drawn parallel to each other.
- **Rotational:** This type is used to blend the sections that are rotated about the Y-axis up to an angle of maximum 120 °. The angle between the sections is known as the rotational blend angle. If the angle between the sections is 0 ° then this option works as the **Parallel** blend option. All the sections that are blended using this option are sketched individually and aligned using the co-ordinate system of the sections.
- **General:** This type is used to blend the sections that can be rotated and translated on the X, Y, and Z axis. All the sections blended using this option, are also sketched individually and aligned using the co-ordinate system of the sections.

In this section, we use the **Parallel** blend option to blend two rectangular sections. Following are the steps that help you blending two rectangular sections:

1. *Start* the **Part** mode.
2. *Select* **Insert→Blend→Protrusion**, as shown in Fig.Pro/E-4.105:

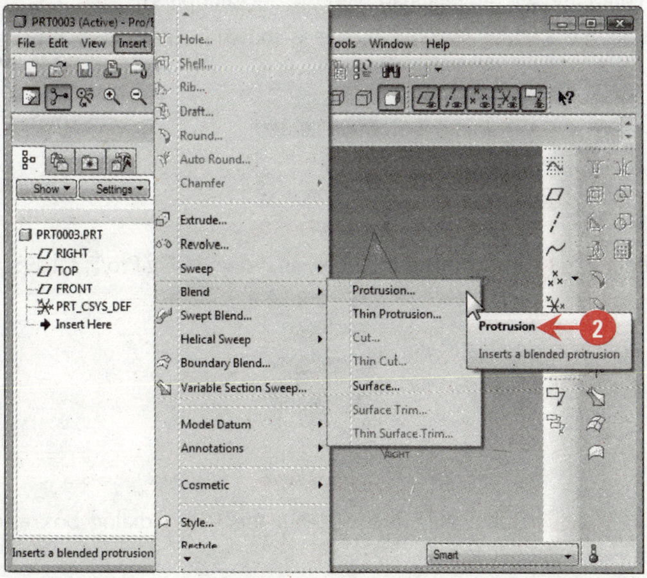

Fig.Pro/E-4.105

The **Menu Manager** menu appears, as shown in Fig.Pro/E-4.106.

3. *Select* **Done** in the **Menu Manager** menu, as shown in Fig.Pro/E-4.106:

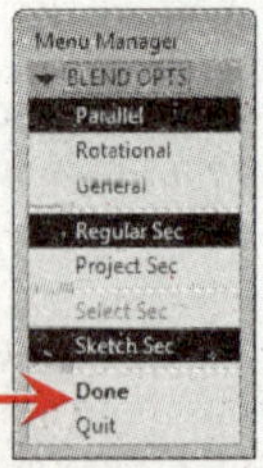

Fig.Pro/E-4.106

The **PROTRUSION: Blend**, **Parallel**, **Regular Sections** dialog box appears, as shown in Fig.Pro/E-4.107:

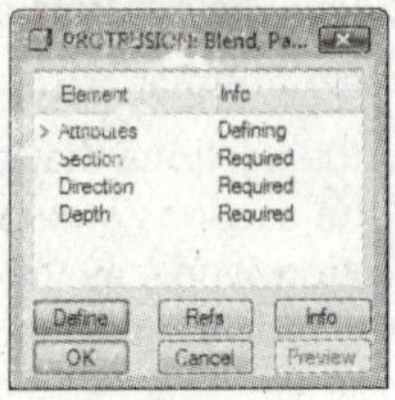

Fig.Pro/E-4.107

Apart from the **PROTRUSION: Blend**, **Parallel**, **Regular Sections** dialog box, the appearance of the **Menu Manager** menu has changed (Fig.Pro/E-4.108).

This time the **Menu Manager** menu appears with the following attributes:

- **Straight:** This attribute joins the sections with straight lines.
- **Smooth:** This attribute joins the sections with curves.

By default, the **Straight** attribute is selected. Now, either select anyone of the attribute or move ahead with the default option. In our case, we move ahead with the default option.

4. *Select* **Done** option in the **Menu Manager** menu, as shown in Fig.Pro/E-4.108:

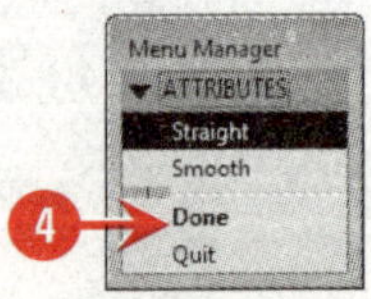

Fig.Pro/E-4.108

The appearance of the **Menu Manager** menu changes, as shown in Fig.Pro/E-4.109:

Fig.Pro/E-4.109

Apart from the **Menu Manager** menu (Fig.Pro/E-4.109), the **Select** dialog box also appears, as shown in Fig.Pro/E-4.110:

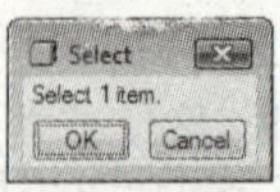

Fig.Pro/E-4.110

Both, the **Menu Manager** menu (Fig.Pro/E-4.109) and the **Select** dialog box (Fig.Pro/E-4.110) prompt you to select the datum plane along with the sections that are to be drawn.

5. *Select* the **Front** datum plane either in the **Model Tree** or in the drawing area. A red arrow appears on the **Front** datum plane, which points in the direction of feature creation, as shown in Fig.Pro/E-4.111:

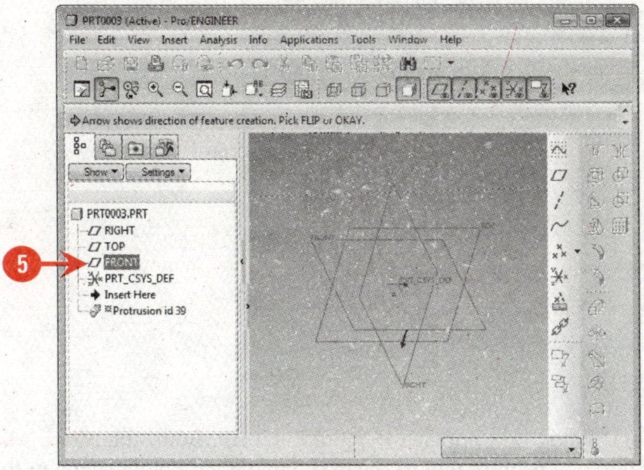

Fig.Pro/E-4.111

Apart from this, the appearance of the **Menu Manager** menu changes once again, as shown in Fig.Pro/E-4.112.

Now, either flip the direction of the feature creation by selecting the **Flip** option in the **Menu Manager** menu (Fig.Pro/E-4.112) or accept the default direction by selecting the **Okay** option. In our case, we are selecting the **Okay** option.

6. *Select* the **Okay** option in the **Menu Manager** menu, as shown in Fig.Pro/E-4.112:

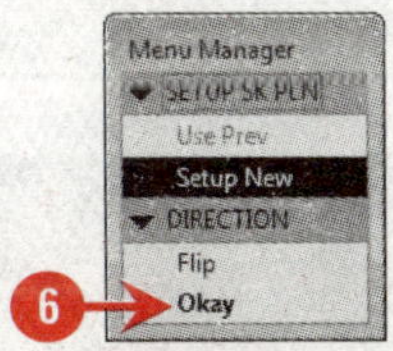

Fig.Pro/E-4.112

The appearance of the **Menu Manager** menu changes once again (Fig.Pro/E-4.113). Now, you need to select anyone of the option given in the **SKET VIEW** sub-menu of the **Menu Manager** menu to set the orientation of the section.

7. In our case, we are *selecting* the **Top** view, as shown in Fig.Pro/E-4.113:

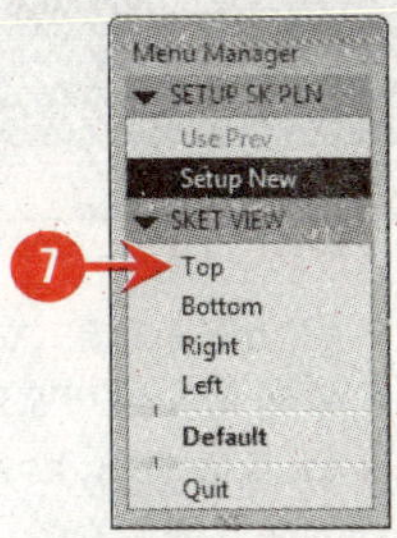

Fig.Pro/E-4.113

The appearance of the **Menu Manager** menu changes again, as shown in Fig.Pro/E-4.114:

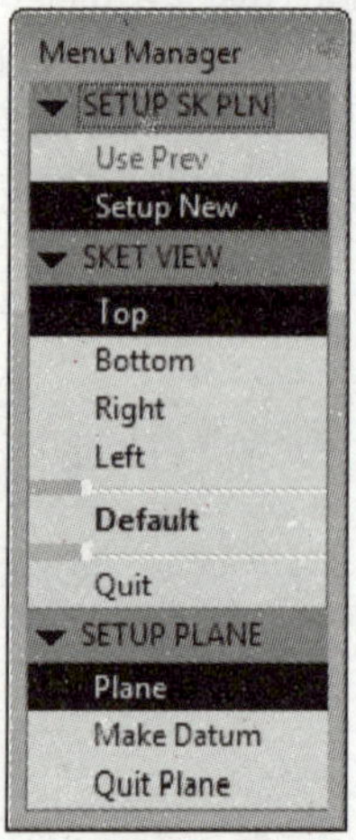

Fig.Pro/E-4.114

Apart from this, the **Select** dialog box appears (Fig.Pro/E-4.110). Both, the **Menu Manager** menu (Fig.Pro/E-4.114) and the **Select** dialog box prompt you to select the reference datum plane of the **Front** datum plane.

8. *Select* the **Top** datum plane (Fig.Pro/E-4.115) in the **Model Tree** as the reference datum plane. The **Sketcher tools toolbar** appears, as shown in Fig.Pro/E-4.115:

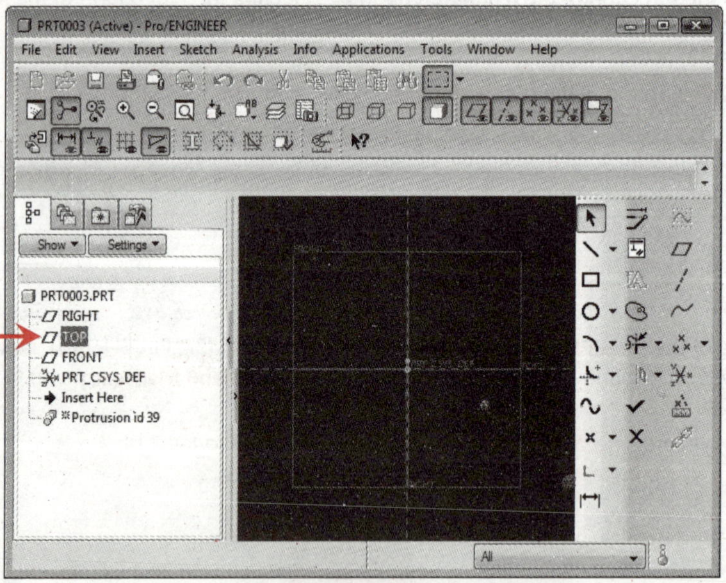

Fig.Pro/E-4.115

9. Now, draw a rectangle, as shown in Fig.Pro/E-4.116. A yellow arrow is attached to the rectangle (Fig.Pro/E-4.116). This yellow arrow is showing the starting point of the rectangle.
10. *Click* the **Done** (✔) button in the **Sketcher** tools toolbar, as shown in Fig.Pro/E-4.116:

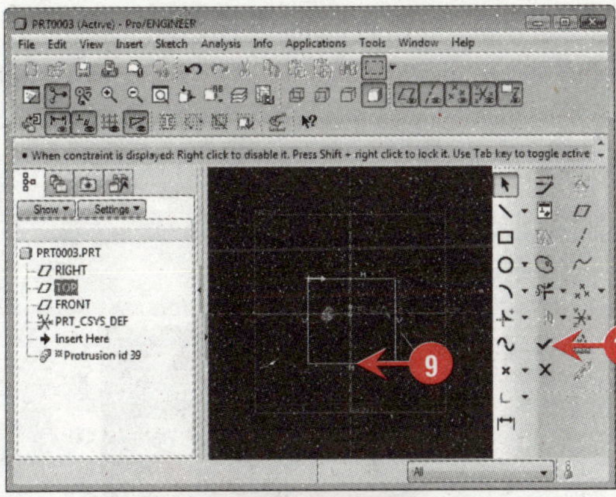

Fig.Pro/E-4.116

Now, create another section. However, to create another section, you must invoke the **Toggle Section** option from the **Sketch** menu. If the **Toggle Section** option is not invoked then another section cannot be created.

11. *Select* **Sketch→Feature Tools→Toggle Section**, as shown in Fig.Pro/E-4.117:

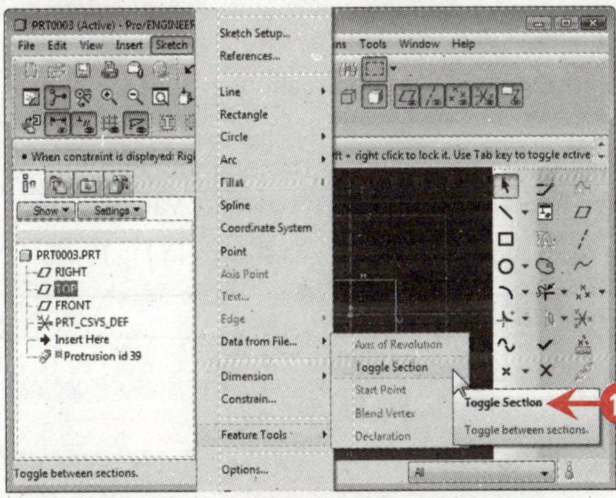

Fig.Pro/E-4.117

All inactive portions of the section turn gray, as shown in Fig.Pro/E-4.118:

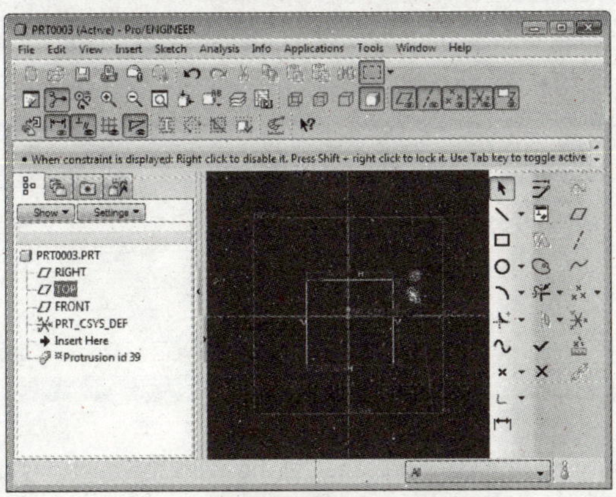

Fig.Pro/E-4.118

12. Now, *create* another rectangle, as shown in Fig.Pro/E-4.119.
13. *Click* the **Done** (✔) button in the **Sketcher tools toolbar** to complete the section, as shown in Fig.Pro/E-4.119:

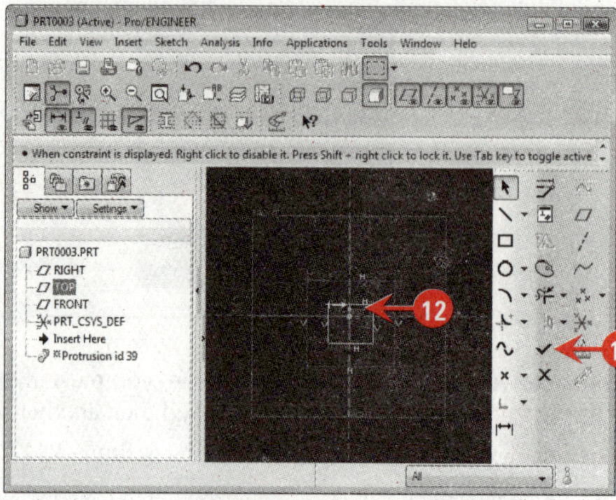

Fig.Pro/E-4.119

A text box that allows you to Enter DEPTH for section 2 appears above the drawing area, as shown in Fig.Pro/E-4.120.

14. Now, either specify depth value for section 2 or continue with the default value appearing in the text box (Fig.Pro/E-4.120). We are continuing with the default value.
15. *Click* the **Accept value** (☑) button to continue, as shown in Fig.Pro/E-4.120:

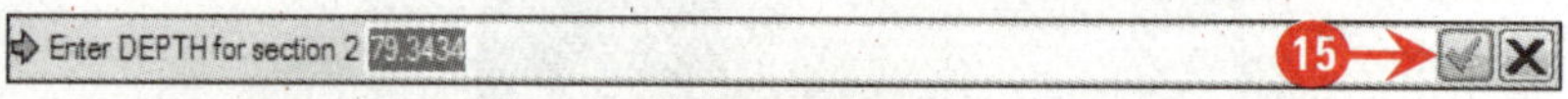

Fig.Pro/E-4.120

Both, the sections are blended, as shown in Fig.Pro/E-4.121:

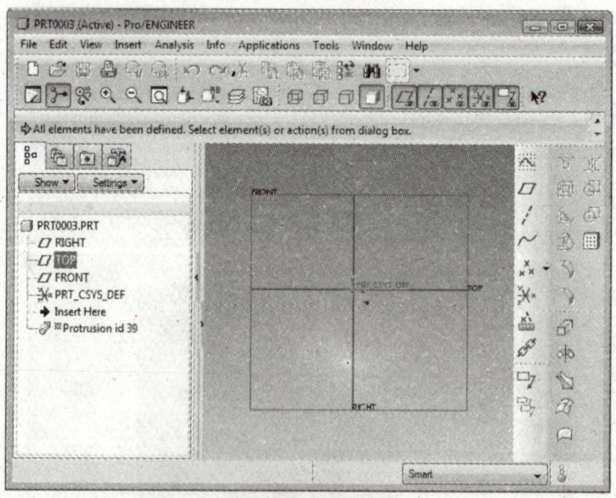

Fig.Pro/E-4.121

In Fig.Pro/E-4.121, you can view the datum planes along with the sections are blended. To have a look on the solid display of the blended section, you need to preview the blended sections.

16. *Click* the **Preview** button in the **PROTRUSION: Blend, Parallel, Regular Sections** dialog box (Fig.Pro/E-4.107). The preview of the blended sections appear, as shown in Fig.Pro/E-4.122:

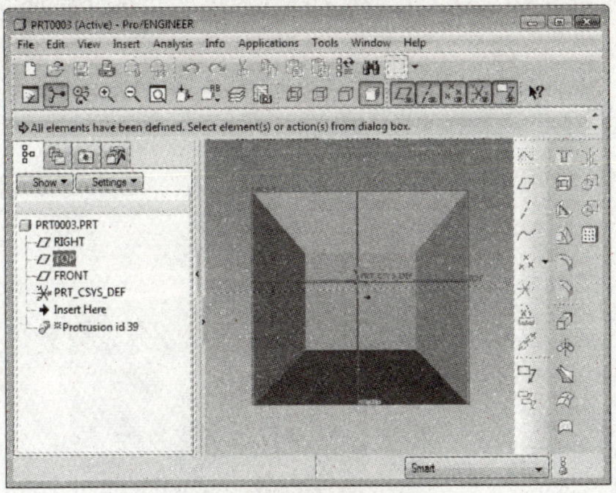

Fig.Pro/E-4.122

17. *Click* the **OK** button in the **PROTRUSION: Blend, Parallel, Regular Sections** dialog box (Fig.Pro/E-4.107) to finish the process. The blended sections appear in red, as shown in Fig.Pro/E-4.123:

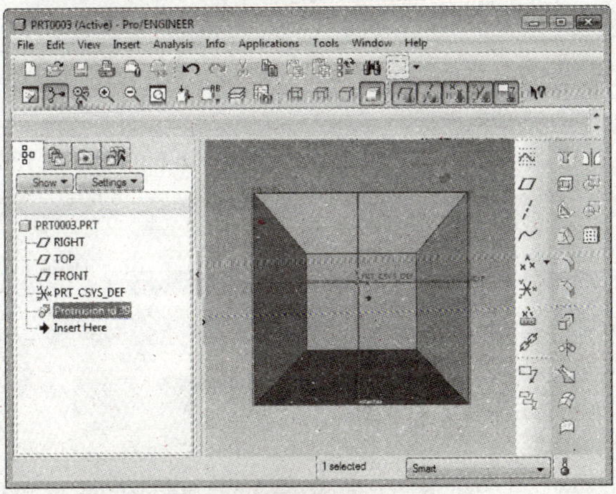

Fig.Pro/E-4.123

18. Finally, *save* the file.

With this, we complete the study of the blending feature. Now, we study another feature called the Rib feature, which is used to provide support to the features in a model.

Inducing the Rib Feature

A **Rib** is used to support and strengthen the weak features of a model. A rib in `Pro/ENGINEER Wildfire 4.0` is created using the **Rib** dashboard. However, a rib is created along with a feature; therefore, we first need a feature to create a rib. In this section, we follow the rib creation after the feature creation. Let's follow the steps given here to create a rib:

1. *Start* the **Part** mode and *create* a feature, as shown in Fig.Pro/E-4.124.

The feature shown in Fig.Pro/E-4.124 is created by drawing a horizontal rectangular slot and a vertical rectangular slot.

Note

The process to create a rectangular slot is similar to creating a base feature, which is already discussed in the section, Creating a Base Feature.

2. Next, *click* the **Rib** button in the **Engineering Features** toolbar to invoke the **Rib** dashboard, as shown in Fig.Pro/E-4.124:

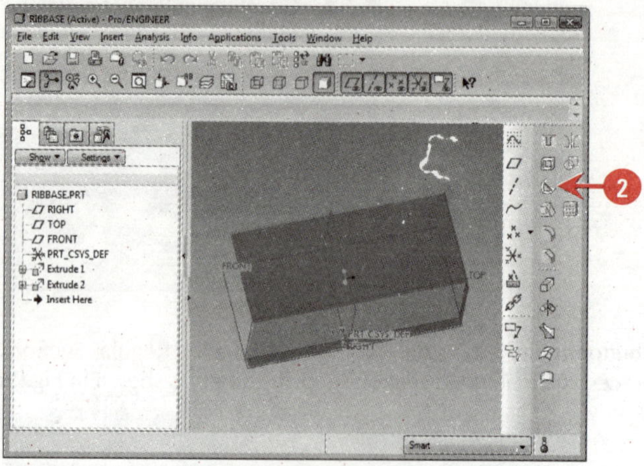

Fig.Pro/E-4.124

The **Rib** dashboard appears (Fig.Pro/E-4.125).

3. *Select* the **References** tab in the **Rib** dashboard, as shown in Fig.Pro/E-4.125:

Fig.Pro/E-4.125

A slide-down panel that helps select the sketch plane to draw the sketch of the rib appears (Fig.Pro/E-4.126).

4. *Click* the **Define** button in the slide-down panel, as shown in Fig.Pro/E-4.126:

Fig.Pro/E-4.126

The **Sketch** dialog box appears (Fig.Pro/E-4.127):

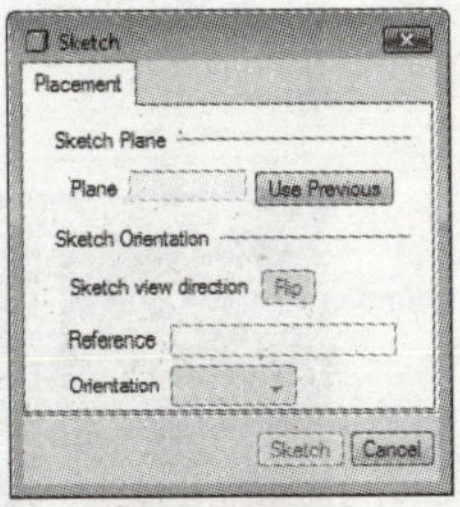

Fig.Pro/E-4.127

5. *Select* the **Right** datum plane, either in the **Model Tree** or in the drawing area. The **Right** datum plane and its reference datum plane along with the orientation automatically appear in the **Sketch** dialog box (Fig.Pro/E-4.128).

Note

> Select the orientation as Top, if the orientation is not already Top. *Select* the orientation by clicking the down arrow beside the Orientation drop-down list.

6. *Click* the **Sketch** button, as shown in Fig.Pro/E-4.128:

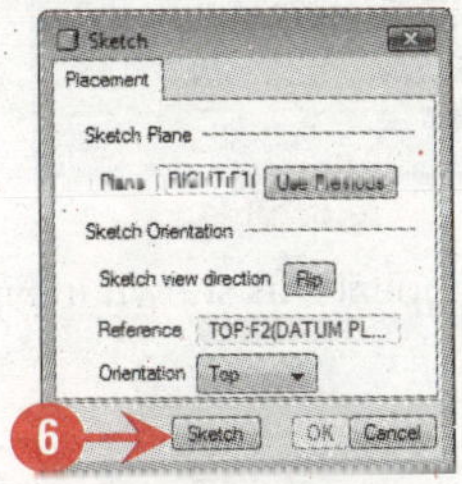

Fig.Pro/E-4.128

The **Sketcher tools toolbar** appears to let you create the sketch of the rib, as shown in Fig.Pro/E-4.129:

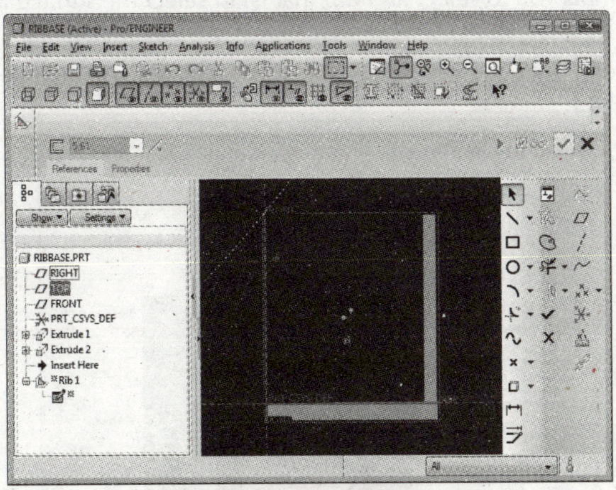

Fig.Pro/E-4.129

Now, you have to set the reference line along which the rib will be created.

7. *Select* the **References** option in the **Sketch** menu, as shown in Fig.Pro/E-4.130:

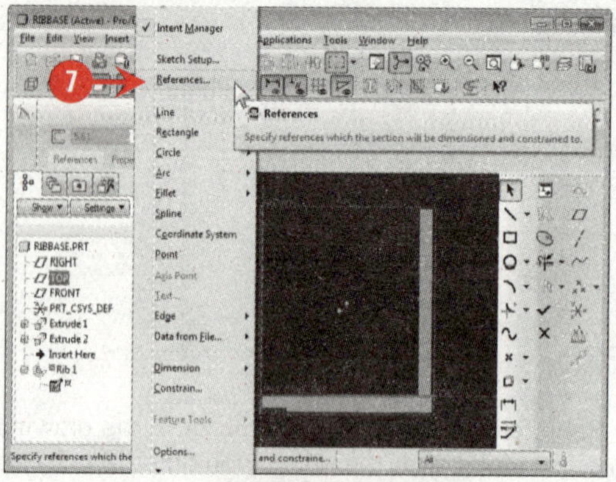

Fig.Pro/E-4.130

As soon as the **References** option is selected, the **References** window appears, as shown in Fig.Pro/E-4.131:

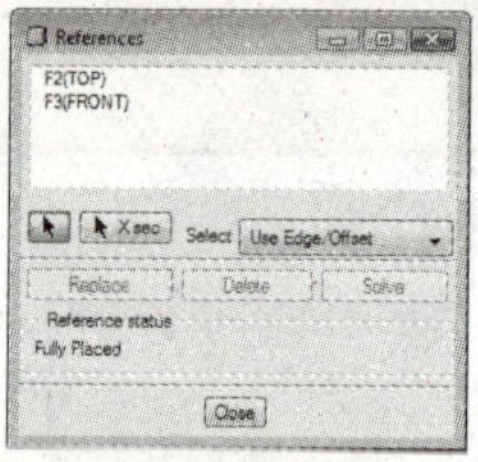

Fig.Pro/E-4.131

Apart from that, the **Select** dialog box also appears, as shown in Fig.Pro/E-4.132:

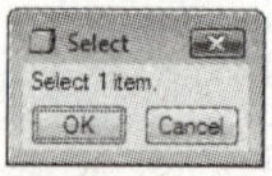

Fig.Pro/E-4.132

The **Select** dialog box prompts you to select an item that is, to select the surface along with the reference line that will be created.

8. *Select* the surface along with the reference line to be created. An infinite reference line appears, as shown in Fig.Pro/E-4.133:

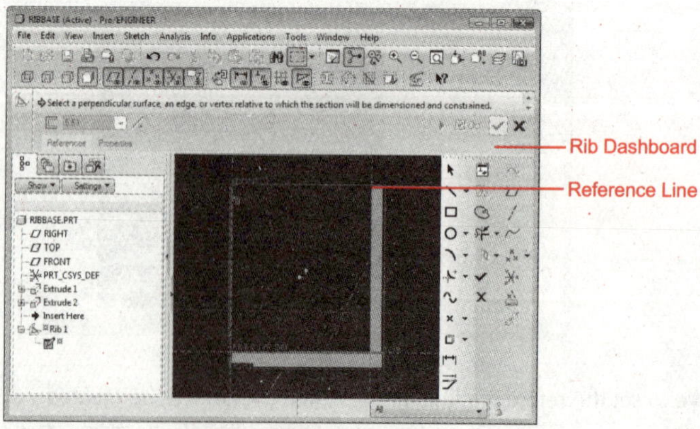

Fig.Pro/E-4.133

The selected surface is automatically added in the **References** window.

9. *Select* the cross-section of the two infinite lines along with the rib to be created.
10. Next, *draw* the sketch of the rib (Fig.Pro/E-4.134).
11. *Click* the **Done** (✓) button to complete the sketch and to exit from the **Sketcher tools toolbar** (Fig.Pro/E-4.134):

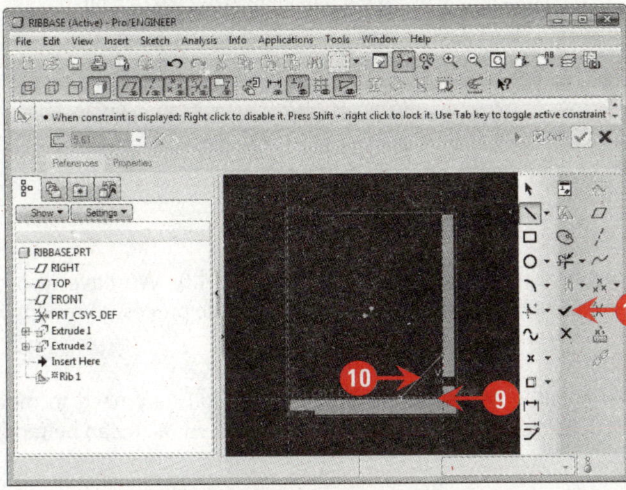

Fig.Pro/E-4.134

The rib is previewed, as shown in Fig.Pro/E-4.135.

12. *Enter* 8.50 in the **Dimension** box (5.61) in the **Rib** dashboard (Fig.Pro/E-4.135).
13. *Press*, the **ENTER** key to project the new value in the **Dimension** box (Fig.Pro/E-4.135).
14. *Click* the **Build feature** button (✓) in the **Rib** dashboard to complete the rib creation, as shown in Fig.Pro/E-4.135:

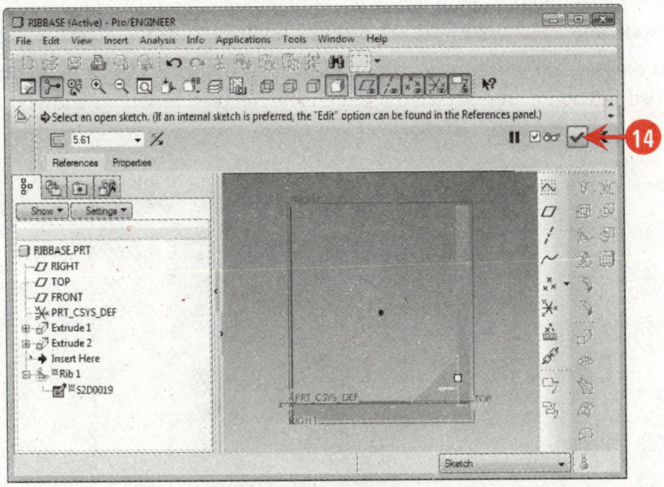

Fig.Pro/E-4.135

The rib is highlighted in red, as shown in Fig.Pro/E-4.136:

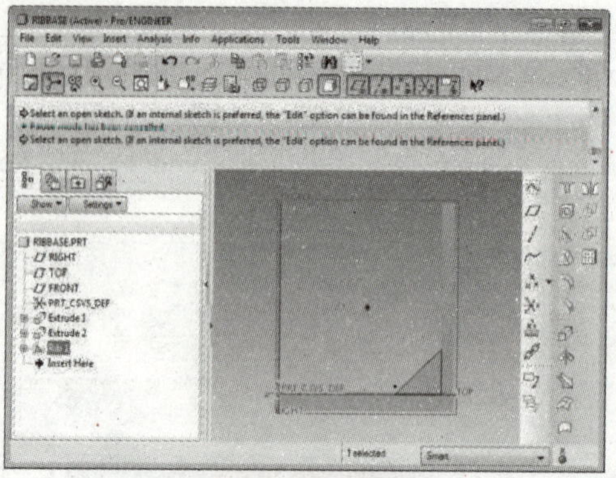

Fig.Pro/E-4.136

15. Finally, *save* the file.

This is the process to create a rib in Pro/ENGINEER Wildfire 4.0. We have also completed creating and inducing features in Pro/ENGINEER Wildfire 4.0. Now, we see the process to modify features.

Modifying Features

Sometimes during model creation or after creating a feature, you may need to modify a feature to get the desired feature or model. A feature in `Pro/ENGINEER Wildfire 4.0` can be modified by:

- Modifying Dimensions
- Modifying Definition

Let's first study modifying dimensions.

Modifying Dimensions

Features are generally modified by editing their dimensions. However, to learn the process of modifying a feature by editing its dimensions, we need a feature. Therefore, we use the **PRT0001.prt** file, which we created previously and contains a rectangular slot (Fig.Pro/E-4.24). Let's now follow the steps given here to understand the process of modifying dimensions:

1. *Start* the **Part** mode and *open* the PRT0001.prt file (Fig.Pro/E-4.24).
2. *Select* the feature whose dimensions need to be modified in the **Model Tree**. In our case, we have selected the feature named **Extrude 1**. The selected feature is highlighted in red, as shown in Fig.Pro/E-4.137:

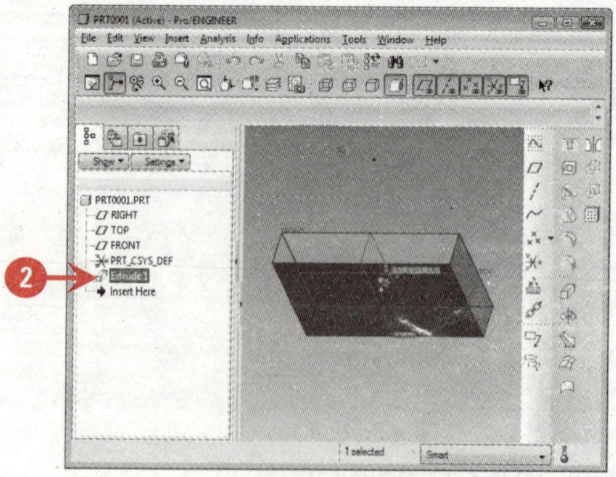

Fig.Pro/E-4.137

3. *Right-click* the selected feature in the **Model Tree** (Fig.Pro/E-4.137). A context menu appears (Fig.Pro/E-4.138).
4. *Select* the **Edit** option in the context menu, as shown in Fig.Pro/E-4.138:

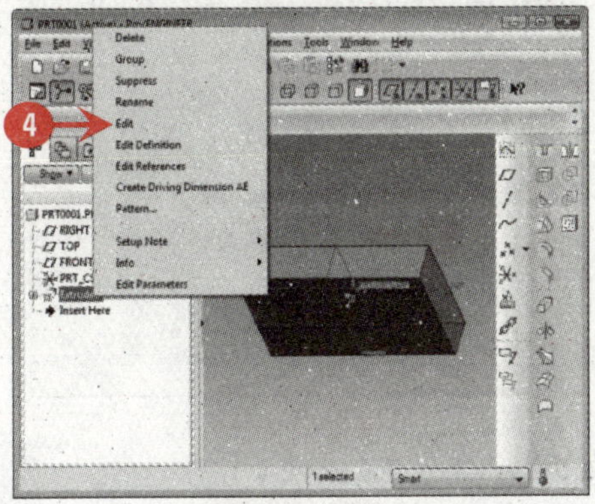

Fig.Pro/E-4.138

All the dimensions of the selected feature are highlighted in yellow, as shown in Fig.Pro/E-4.139:

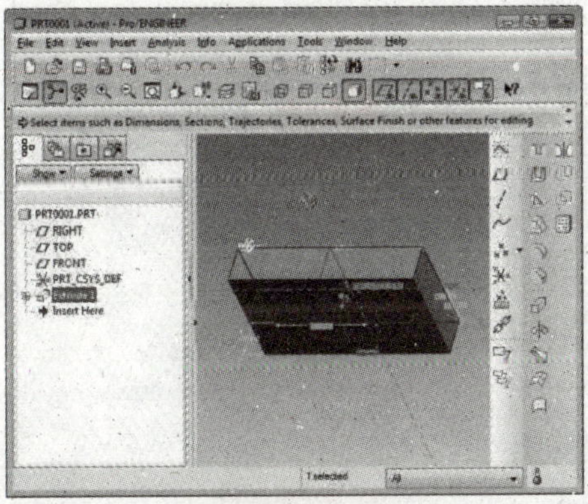

Fig.Pro/E-4.139

5. *Select* the dimension you want to modify. Selected dimension turns red, as shown in Fig.Pro/E-4.140:

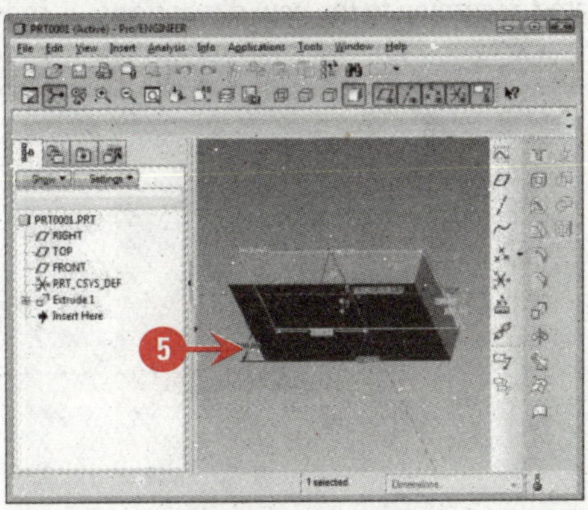

Fig.Pro/E-4.140

6. *Double-click* the dimension (Fig.Pro/E-4.140). An edit box appears, as shown in Fig.Pro/E-4.141:

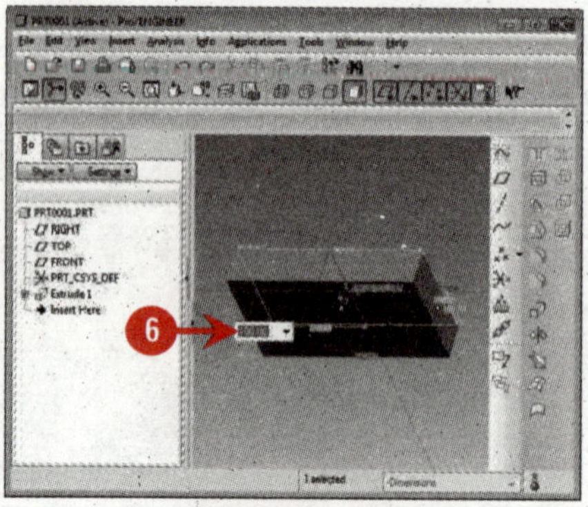

Fig.Pro/E-4.141

7. *Specify* a new dimension in the edit box (Fig.Pro/E-4.141).
8. After specifying the new dimension, *press* the **ENTER** key. The new dimension is highlighted in light green (Fig.Pro/E-4.142).
9. *Click* the **Regenerate** ( ) button which is placed on the top tool chest, as shown in Fig.Pro/E-4.142:

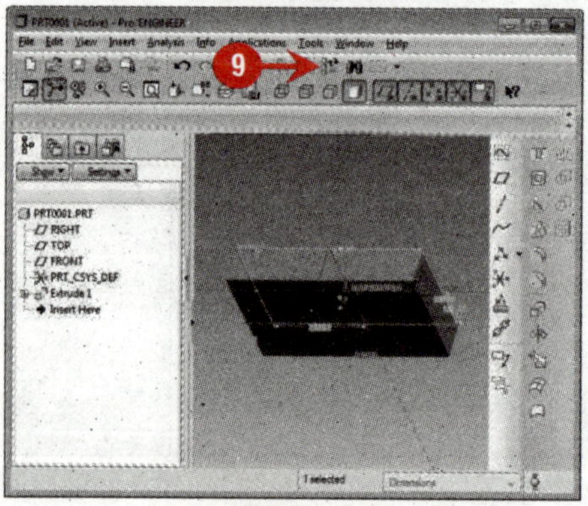

Fig.Pro/E-4.142

The selected feature is regenerated according to the new dimensions, as shown in Fig.Pro/E-4.143:

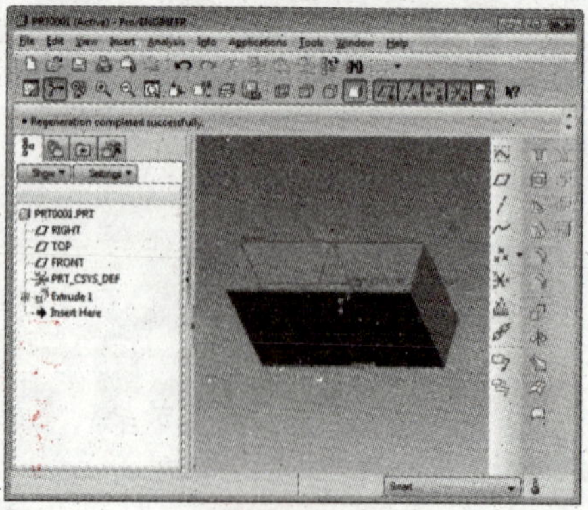

Fig.Pro/E-4.143

10. Finally, *save* the modified file.

Modifying Definition

Features can also be modified by editing their definition. Editing definition of a feature refers to modifying the parameters used to create that feature. In this section, we use the PATTERNFEATURE.prt file to understand the process of modifying a feature by editing its definition. This file contains a pattern of 6 holes around another hole on a rectangular slot. We will modify this pattern by editing its definition. Let's follow the steps given here to understand the process of modifying the definition of a feature:

1. *Start* the Part mode and *open* the PATTERNFEATURE.prt file (Fig.Pro/E-4.87).
2. *Select* the feature whose definition needs to be modified in the Model Tree. In our case, we have selected the pattern feature named Pattern 1 of Hole 2 (Fig.Pro/E-4.87). The selected feature is highlighted in red, as shown in Fig.Pro/E-4.144:

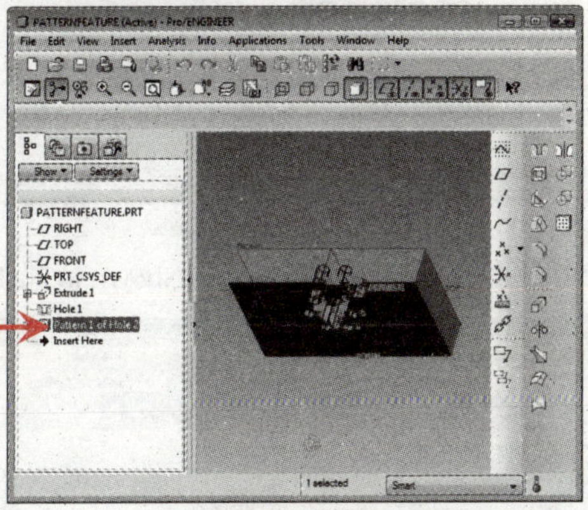

Fig.Pro/E-4.144

3. *Right-click* the selected feature (Fig.Pro/E-4.144). A context menu appears (Fig.Pro/E-4.145).
4. *Select* the Edit Definition option in the menu, as shown in Fig.Pro/E-4.145:

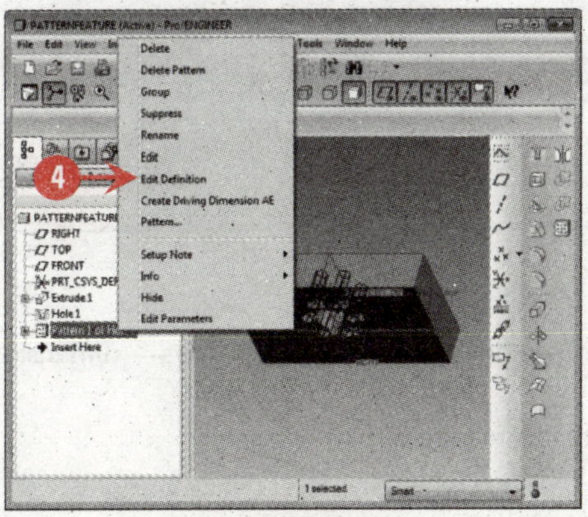

Fig.Pro/E-4.145

The dashboard corresponding to the selected feature appears. In our case, the Pattern dashboard appears, as shown in Fig.Pro/E-4.146.

5. *Replace* the number 6 in the Pattern dashboard (Fig.Pro/E-4.146) with the number 8 to increase the number of the holes in the pattern.

6. Next, *change* the angle between the holes from 60 ° to 45°.
7. *Click* the **Build feature** (✓) button, as shown in Fig.Pro/E-4.146:

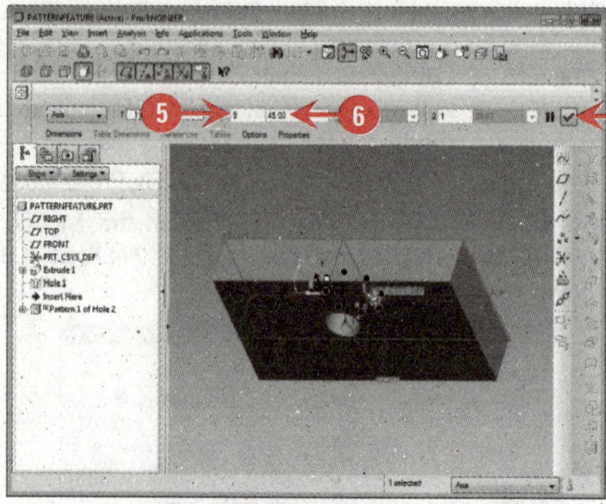

Fig.Pro/E-4.146

The modifications are reflected (in red) in the drawing area, as shown in Fig.Pro/E-4.147:

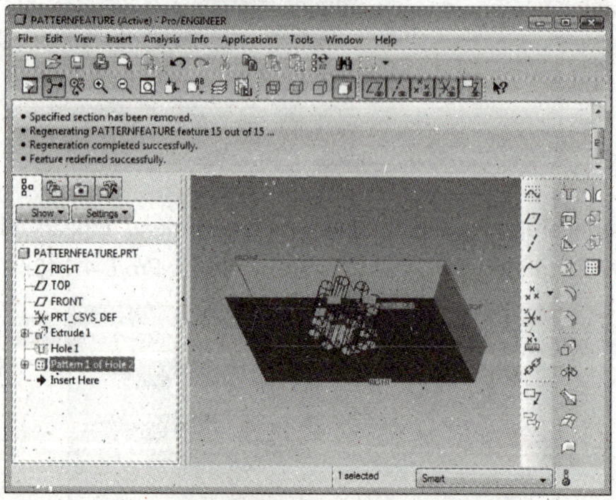

Fig.Pro/E-4.147

8. *Save* the file.

Here, we have discussed how to modify features in two ways— dimensions and definitions. With this, we complete modifying features.

Now, we study the process to delete features.

Deleting a Feature

During model creation, it may happen that you need to delete one or more features to get the desired model. To delete a feature from a component or a model, let's follow the steps given here:

1. *Start* the **Part** mode and *open* the **HOLEFEATURE.prt** file (Fig.Pro/E-4.42).
2. *Select* the feature to be deleted in the **Model Tree**. In our case, we have selected the feature named **Hole 1**. The selected feature is highlighted in red, as shown in Fig.Pro/E-4.148:

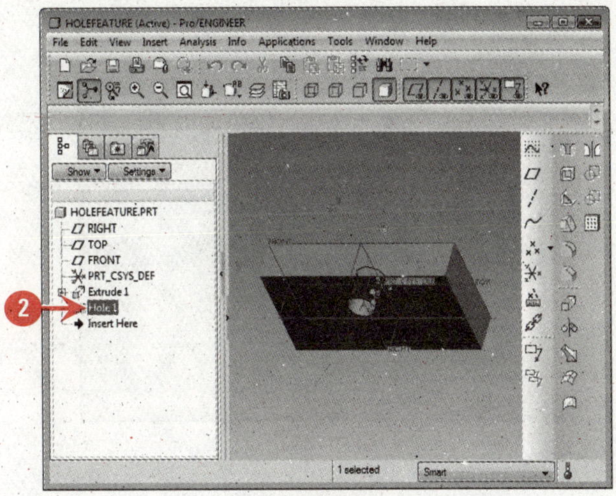

Fig.Pro/E-4.148

3. *Right-click* the selected feature (Fig.Pro/E-4.148). A context menu appears (Fig.Pro/E-4.149).
4. *Select* the **Delete** option, as shown in Fig.Pro/E-4.149:

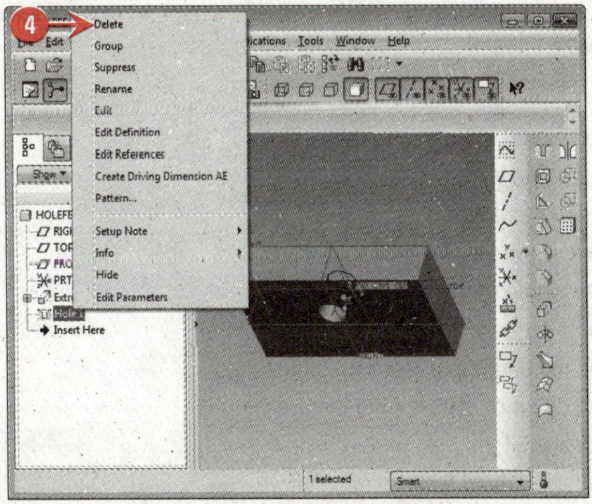

Fig.Pro/E-4.149

The **Delete** dialog box appears to confirm the deletion (Fig.Pro/E-4.150).

5. *Click* the **OK** button to delete the selected feature, as shown in Fig.Pro/E-4.150:

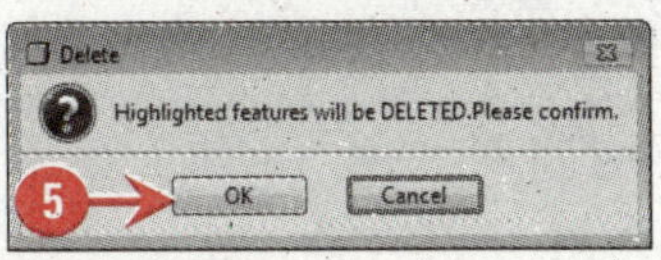

Fig.Pro/E-4.150

6. The selected feature is deleted, as shown in Fig.Pro/E-4.151:

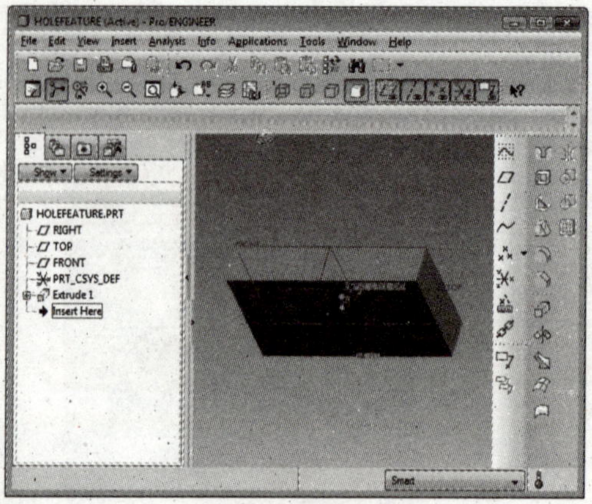

Fig.Pro/E-4.151

Note

> Click the Cancel button to abort the process.

7. *Save* the file.

This is the process to delete a feature. With this, we come to the end of this chapter. However, before closing this chapter let's summarize the topics we have covered in this chapter.

Summary

In this chapter, you have learned about:

- The process to start the Part mode
- Datums and datum plane creation
- The creation of a base feature and a protrusion
- The process to create a hole in the base feature
- The process to create a cut in a feature
- The process to round the edges of a feature
- The process to create chamfer feature on a feature
- The process to create a shell
- The process to create a revolved feature
- The process to create a sweep
- The process to blend two features
- The process to create a rib to provide support to other features
- The processes to modify a feature
- The process to delete a feature

Chapter 5

Exploring Pro/ENGINEER Wildfire 4.0 Assembly Mode

In this Section

- Understanding Assembly Approaches
- Integrating Components in an Assembly
- Modifying Assembly Components
- Creating Exploded View of an Assembly

The Assembly mode is a state in Pro/ENGINEER, which a product model must pass through to complete the design process. The Assembly mode allows a user to assemble Part files in an Assembly file. These Part files are created either within the Assembly mode itself or within the Part mode. The Part mode is discussed in Chapter 4, *Exploring Pro/ENGINEER Wildfire 4.0 Part Mode*. The Assembly mode uses either the **Top-down** or the **Bottom-up** approach to assemble components within an Assembly file. In the **Top-down** approach, components of a model are created within the Assembly mode. On the other hand, the **Bottom-up** approach uses existing files created within the Part mode in the Assembly file. The **Top-down** and **Bottom-up** approaches are discussed in detail later in this chapter.

In this chapter, you learn about the fundamentals of the Assembly mode in Pro/ENGINEER Wildfire 4.0. This chapter discusses the approaches used to integrate Part files in the Assembly mode. This chapter also discusses how to position one Part file in relation to another in an Assembly file by using various assembly constraints to help position components within an assembly. In addition, you learn how to integrate multiple Part files in an Assembly file and modify the appearance of the Assembly file. Finally, you learn how to create an exploded view, where all the components assembled inside an Assembly file are moved from their original location so that they can be easily viewed. An exploded view helps you to view all the components placed inside an Assembly file.

You should know about the approaches supported by Pro/ENGINEER Wildfire 4.0 to assemble Part files in an Assembly file before you start assembling them.

Now, let's begin with a discussion on the approaches available in Pro/ENGINEER Wildfire 4.0 to integrate the components with the Assembly mode.

Understanding Assembly Approaches

You learned how to create Part files in the Part mode in **Chapter 4, Exploring Pro/ENGINEER Wildfire 4.0 Part Mode**. Part files are often created in the Part mode where the entities contained in them can be viewed easily to help users in designing features, such as hole, cut and shell, of a Part file. However, you also can create Part files within the Assembly mode. Part files are assembled in the Assembly mode on the basis of the design mode used to create these files. Pro/ENGINEER Wildfire 4.0 provides the following approaches to integrate the components within the Assembly mode:

- Top-down Approach
- Bottom-up Approach

Let's discuss these approaches one by one.

Top-Down Approach

The **Top-down** approach is a method of integrating components, such as Part files created within the Assembly mode. These Part files are assembled by using assembly constraints, which will be discussed later in the chapter. You must know how to create a Part file in the Assembly mode before you start assembling Part files.

Perform the following steps to create a Part file in the Assembly mode:

1. *Select* the **File→New** option from the **Menu Bar**, as shown in Fig.Pro/E-5.1:

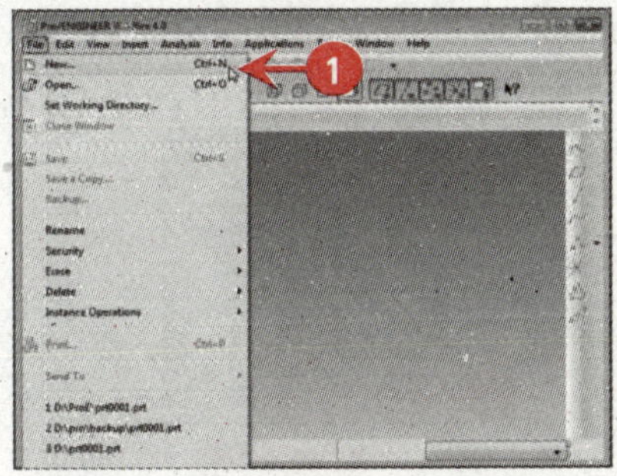

Fig.Pro/E-5.1

The **New** dialog box appears and displays the various modes available in Pro/ENGINEER Wildfire 4.0 to create a model, as shown in Fig.Pro/E-5.2:

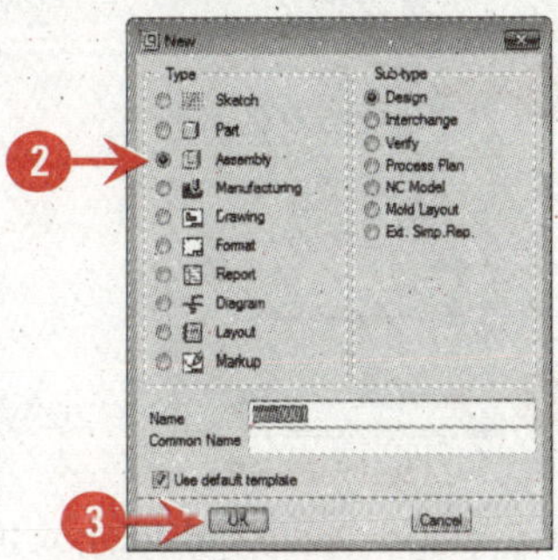

Fig.Pro/E-5.2

2. *Select* the **Assembly** radio button from the **Type** area in the **New** dialog box (Fig.Pro/E-5.2). The **Design** radio button is selected by default from the **Sub-type** area in the **New** dialog box. A default name (asm0001) of the Assembly file is also shown in the **Name** text box (Fig.Pro/E-5.2).
3. *Click* the **OK** button in the **New** dialog box (Fig.Pro/E-5.2). An Assembly file containing three default datum planes: ASM_RIGHT, ASM_TOP, and ASM_FRONT are created, as shown in Fig.Pro/E-5.3:

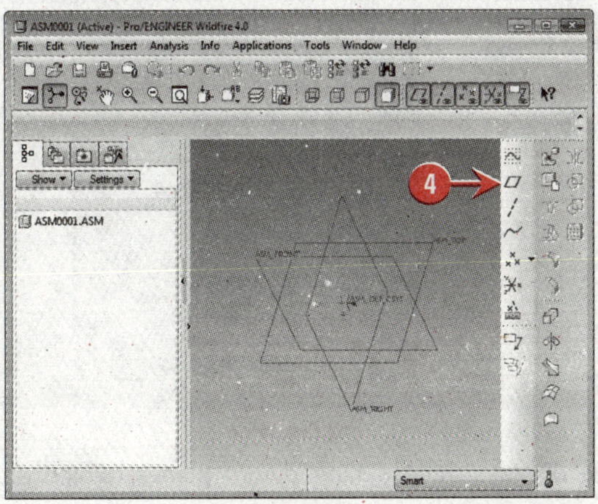

Fig.Pro/E-5.3

4. *Click* the **Create** button ( ) from the **Tool chest** located at the right side of the **Drawing Area** (Fig.Pro/E-5.3). The **Component Create** dialog box appears and displays various components that you can use in

the Assembly mode. You can specify a name of the Part file in the Name text box. In our case, the Part file is PRT0004, as shown in Fig.Pro/E-5.4:

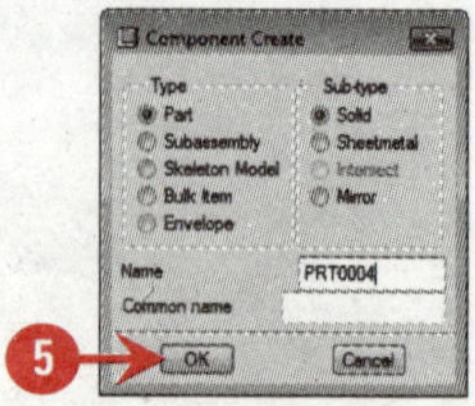

Fig.Pro/E-5.4

5. *Click* the OK button in the Component Create dialog box (Fig.Pro/E-5.4). The Creation Options dialog box appears and displays various options for creating a Part file, as shown in Fig.Pro/E-5.5:

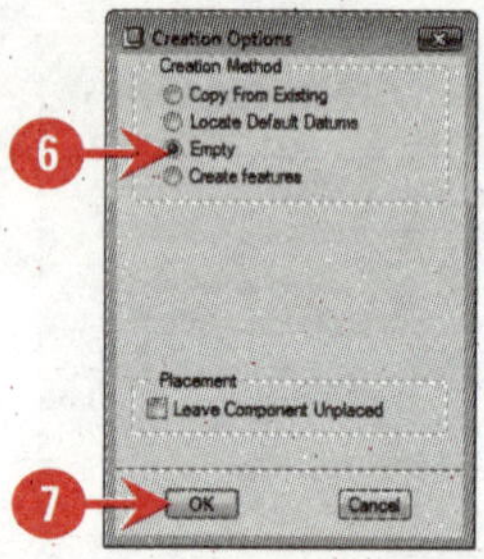

Fig.Pro/E-5.5

6. *Select* the Empty radio button in the Creation Options dialog box (Fig.Pro/E-5.5) to create a Part file with no initial geometry.
7. Next, *click* the OK button in the Creation Options dialog box (Fig.Pro/E-5.5). The Part file (PRT0004.PRT) is created in the Assembly file named ASM0001.ASM, as shown in Fig.Pro/E-5.6:

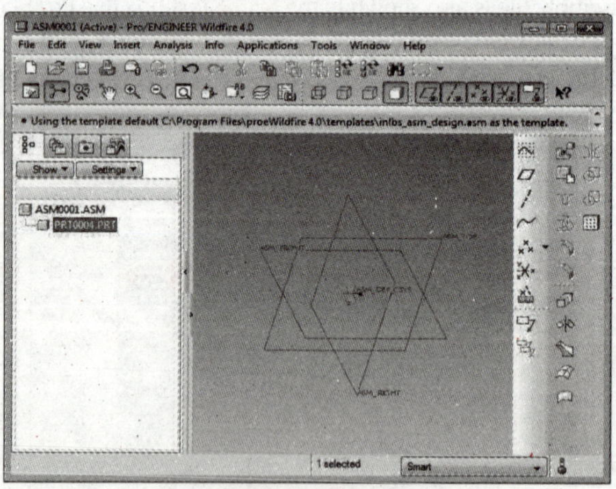

Fig.Pro/E-5.6

The PRT0004.PRT Part file displays three default datum planes, which are required to create any base feature.

8. Create a base feature using the Extrude feature (Fig.Pro/E-5.7) as discussed in the Chapter 4, Exploring the Pro/ENGINEER Wildfire 4.0 Part Mode, under the section "Creating an Extruded Feature":

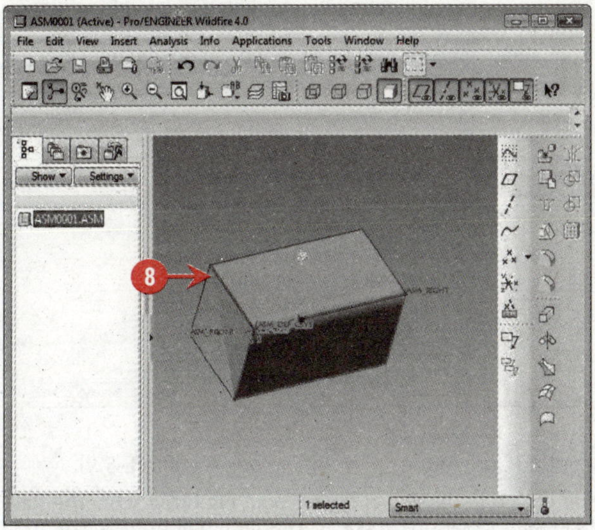

Fig.Pro/E-5.7

Now, let's discuss the Bottom-up approach.

Bottom-up Approach

The Bottom-up approach is a method of assembling existing Part files created within the Part mode in an Assembly file. These Part files are also integrated within the Assembly file through assembly constraints. You must know how to insert an existing Part file in an Assembly file before you start assembling them.

Perform the following steps to insert an existing Part file in an Assembly file:

1. *Create* an Assembly file (.asm) by following the steps given in the preceding section (Top-down Approach). An Assembly file containing three default datum planes is created, as shown in Fig.Pro/E-5.8:

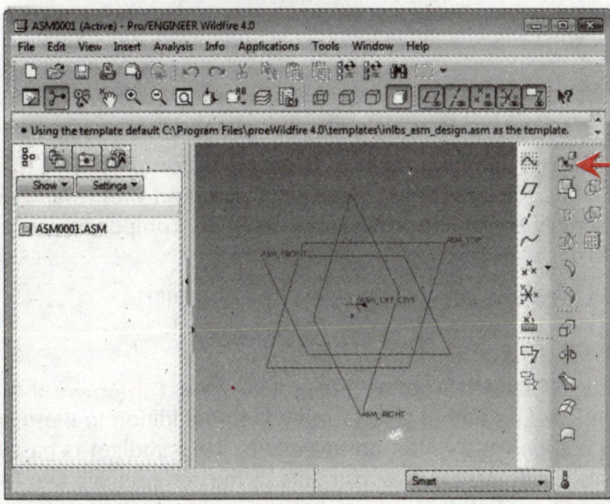

Fig.Pro/E-5.8

2. *Click* the Assemble button ( ) from the Tool chest located at the right side of the Drawing Area (Fig.Pro/E-5.8). The Open dialog box appears, where you can browse the local system for existing Part files, as shown in Fig.Pro/E-5.9:

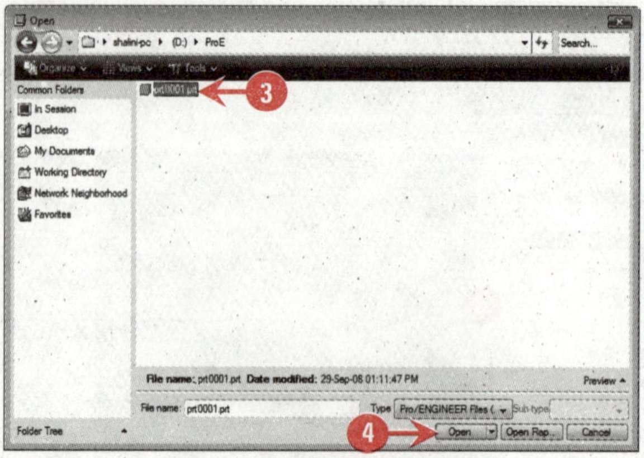

Fig.Pro/E-5.9

3. *Select* the Part file (prt0001.prt) in the **Open** dialog box (Fig.Pro/E-5.9).
4. Next, *click* the **Open** button in the **Open** dialog box (Fig.Pro/E-5.9). The Part file (prt0001.prt) is inserted in the Assembly file named ASM0001.ASM, as shown in Fig.Pro/E-5.10:

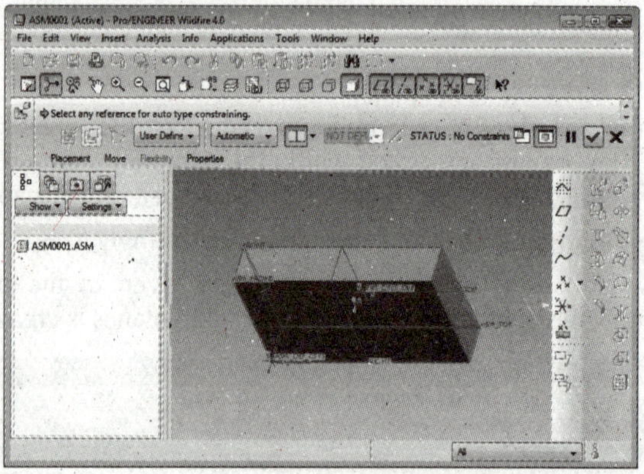

Fig.Pro/E-5.10

Part files placed in an Assembly file by using the Assemble button should be placed in relation to the other Part files in the Assembly file. This helps in assembling these Part files in an Assembly file. Pro/ENGINEER Wildfire 4.0 provides assembly constraints to integrate assembly components by placing these components in relation to one another in an Assembly.

Next, let's discuss how to integrate components in an Assembly file.

Integrating Components in an Assembly

As discussed earlier, Pro/ENGINEER Wildfire 4.0 provides the **Top-down** and **Bottom-up** approaches for integrating or assembling Part files within an Assembly file. In addition to these approaches, there is another paradigm for integrating components within an assembly. The paradigm is based on the implementation of assembling components by using assembly constraints. If the components are assembled by using assembly constraints, the assembly created is called a parametric assembly. However, if the components are packaged and are not assembled by using the assembly constraints, the assembly created is called a non-parametric assembly.

Pro/ENGINEER Wildfire 4.0 provides the following methods to create parametric and non-parametric assemblies:

- ❑ Using Assembly constraints
- ❑ Packaging components

Let's discuss the methods one by one.

Using Assembly Constraints

After placing Part files in an Assembly file, it is important to specify how you want to arrange these Part files in relation to one another. To define the position of one Part file in relation to

other, you must use assembly constraints available in the **Constraint-type** drop-down list of the **Component Placement** dashboard, as shown in Fig.Pro/E-5.11:

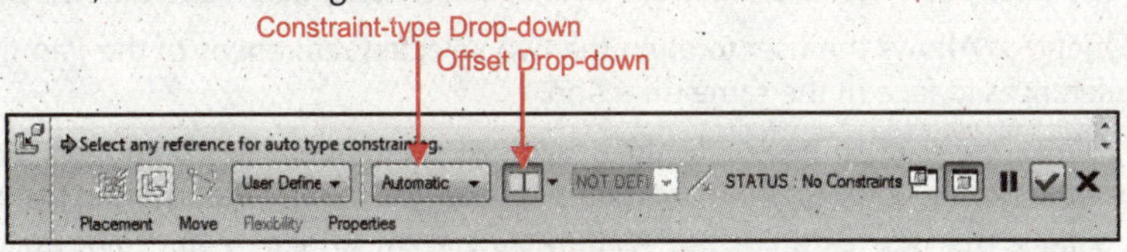

Fig.Pro/E-5.11

The **Component Placement** dashboard is located below the Toolchest at the top of the Drawing Area. It appears automatically when a Part file is placed in an Assembly file. Pro/ENGINEER Wildfire 4.0 provides various assembly constraints in the **Constraint-type** drop-down list, of the Component Placement dashboard, as shown in Fig.Pro/E-5.12:

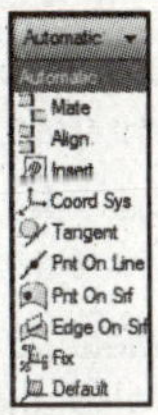

Fig.Pro/E-5.12

Fig.Pro/E-5.12 shows the following assembly constraints available in the **Constraint-type** drop-down list of the **Component Placement** dashboard:

- ❑ **Automatic:** Determines automatically the appropriate constraint to apply on the selected reference. Suppose you want to align two Part files and have selected the axes of both the files. In this case, Pro/ENGINEER Wildfire 4.0 automatically determines that you want to apply the **Align** constraint and applies the constraint to both the Part files. The **Automatic** constraint is the default constraint in the **Component Placement** dashboard.

  - **Mate:** Allows a user to position two selected references (planes or surfaces) from each Part file in such way that they face each other. The **Mate** constraint can position the two references as coincident, orient, or offset with the help of the **Automatic** drop-down list available in the **Component Placement** dashboard (Fig.Pro/E-5.11). The **Mate** constraint is used in the following three combinations with the following Offset drop-down list options.
  - **Mate Offset:** Allows the user to keep the two selected planes from each Part file at some distance from each other. The value of the **Offset** option determines the distance between the two planes of the Part files.
  - **Mate Coincident:** Allows the user to position the two selected faces coplanar to each other. Coplanar faces are those faces that lie in the same plane.
  - **Mate Oriented:** Allows the user to position the two selected planes or faces of the Part files and simultaneously orient the other planes or faces to face in the same direction.

- **Align:** Allows the user to align the two selected planes, axes, or points from each Part file to face in the same direction. An Align constraint can make two selected references coincident, orient, or offset with the help of the Offset drop-down list available in the Component Placement dashboard (Fig.Pro/E-5.11). The Align constraint is used in the following three combinations with the following Offset drop-down list options:
- **Align Offset:** Allows the user to align the two selected planes or faces of the Part files at some distance between them. The Offset value determines the distance between the two references.
- **Align Coincident:** Allows the user to align the two selected references in such a way that they lie in the same plane. Users can use the Align Coincident constraint to make the two planes coplanar.
- **Align Oriented:** Allows the user to align the two selected references of the Part files while orienting these references to face in the same direction.
- **Insert:** Allows the user to assemble the revolved component (a component with curved edges), such as a rod, into a hole. An Insert constraint makes the axes of the inserted surfaces coaxial.

❑ **Coord Sys:** Allows the user to align the coordinate system of a Part file with the coordinate system available in the Assembly file to place the Part file firmly in the Assembly file.

- **Tangent:** Allows the user to place a Part file tangentially to the other Part file. A Tangent constraint also places the Part files in the same plane, thereby making them coplanar.
- **Pnt On Line:** Allows the user to align the selected point on the first component (Part file) with the selected edge, axis, and datum curve on the second component (Part file).
- **Pnt On Srf:** Allows the user to align the selected point on the first component (Part file) with the selected surface on the second component (Part file).
- **Edge On Srf:** Allows the user to align the selected edge of the first component (Part file) with the selected surface on the second component (Part file).
- **Default:** Allows the user to align the default coordinate system of a Part file with the default coordinate system of the assembly. This also aligns the datum planes of the Part file with the datum planes of the assembly.

❑ **Fix:** Allows the user to fix the current position of a moved component (Part file).

Now, let's learn how to integrate the components by using assembly constraints within an assembly file. To do so, perform the following steps:

1. *Select* the File→New option from the Menu Bar, as shown in Fig.Pro/E-5.13:

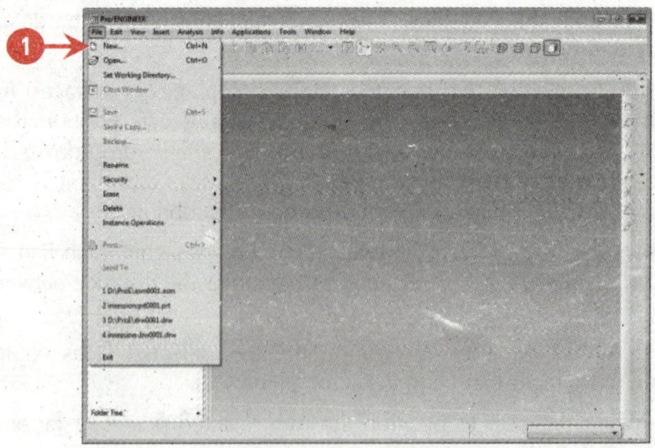

Fig.Pro/E-5.13

The **New** dialog box appears and displays the various modes available in Pro/ENGINEER Wildfire 4.0 to create a model, as shown in Fig.Pro/E-5.14:.

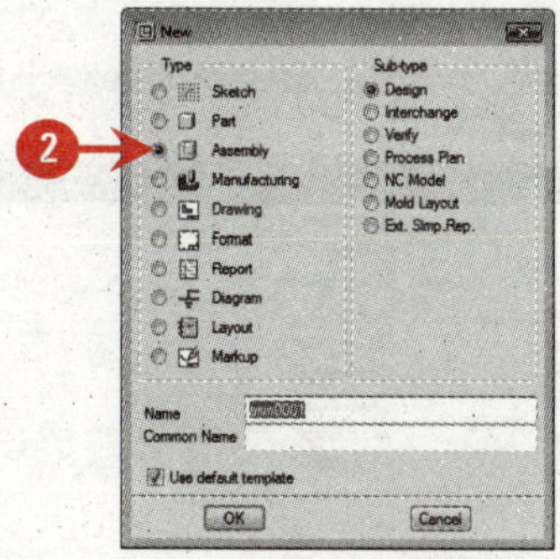

Fig.Pro/E-5.14

2. *Select* the **Assembly** radio button from the **Type** area in the **New** dialog box (Fig.Pro/E-5.14). The default name of the Assembly file (asm0001) is shown in the **Name** text box.

3. *Specify* the name of the Assembly file in the **Name** text box as ASSEMBLYDEMO and *click* the **OK** button, as shown in Fig.Pro/E-5.15:

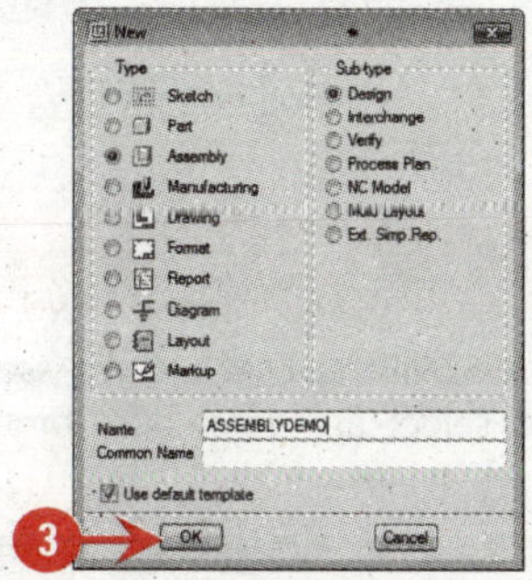

Fig.Pro/E-5.15

An Assembly file containing three default datum planes is created in the current session of Pro/ENGINEER Wildfire 4.0, as shown in Fig.Pro/E-5.16:

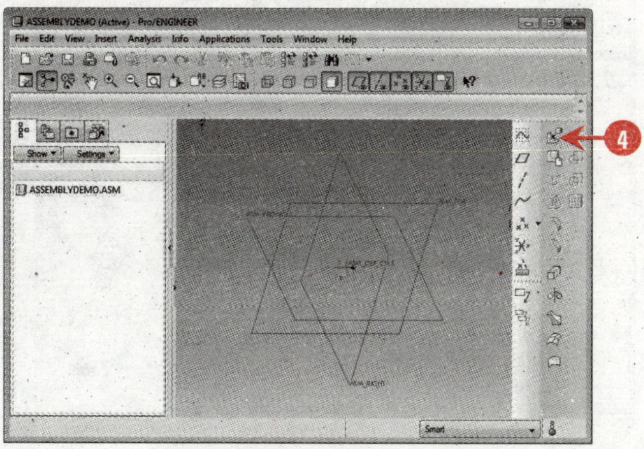

Fig.Pro/E-5.16

4. *Click* the **Assemble** button ( ) from the **Tool chest** located at the right side of the **Drawing Area** (Fig.Pro/E-5.16). The **Open** dialog box appears. Here, you can select the Part files that you need to assemble, as shown in Fig.Pro/E-5.17:

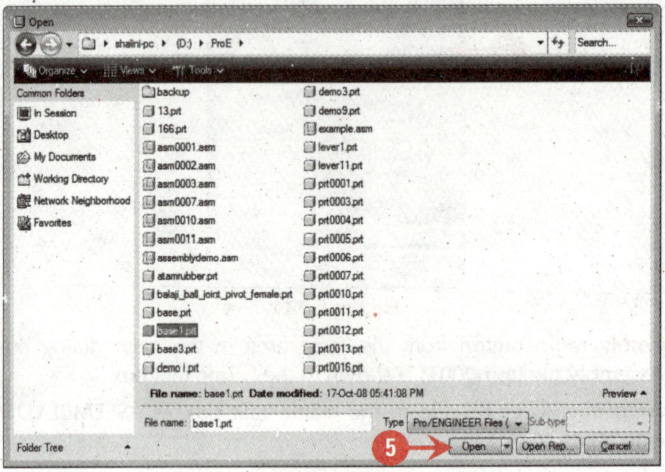

Fig.Pro/E-5.17

5. *Select* the base1.prt component in the **Open** dialog box and *click* the **Open** button (Fig.Pro/E-5.17).

Note

> The base1.prt file already exists.

The base1.prt file is inserted into the ASSEMBLYDEMO.ASM Assembly file. The **Component Placement** dashboard appears and prompts you to select any reference for the Automatic constraint, as shown in Fig.Pro/E-5.18:

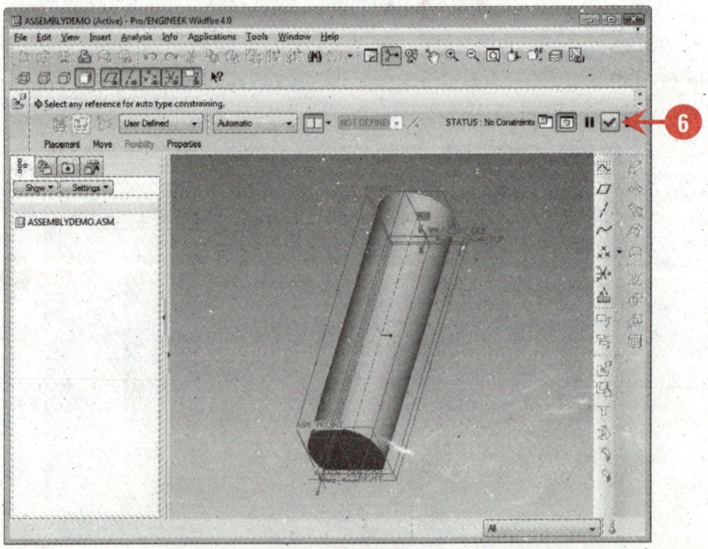

Fig.Pro/E-5.18

The **Automatic** constraint is the default constraint in the **Component Placement** dashboard, which automatically determines the appropriate constraint to apply on the selected reference.

6. *Accept* the default constraint (**Automatic**) from the **Constraint type** drop-down list in the **Component Placement** dashboard by *clicking* the **Build feature** button (Fig.Pro/E-5.18). The BASE1.PRT component is successfully inserted in the Assembly file, as shown in Fig.Pro/E-5.19:

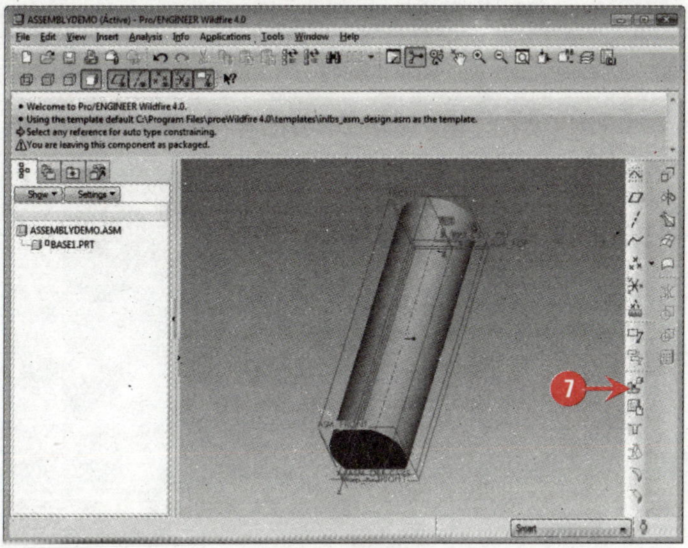

Fig.Pro/E-5.19

7. Now, let's insert a second component to the Assembly file. For this, *click* the **Assemble button** ( ) from the Tool chest located at the right side of the drawing area.

The **Open** dialog box appears (Fig.Pro/E-5.20).

8. *Select* the part file (base.prt) (Fig.Pro/E-5.20).
9. *Click* the **Open** button, as shown in Fig.Pro/E-5.20:

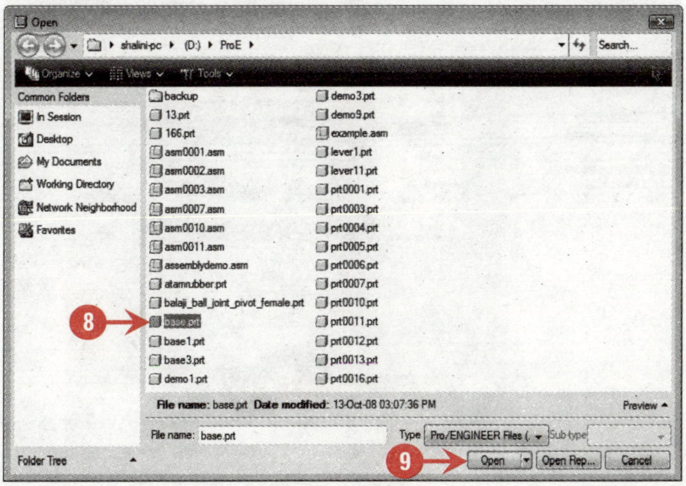

Fig.Pro/E-5.20

The second component is inserted in the Assembly file and the Component Placement dashboard appears. The constraints available in the Component Placement dashboard are used to specify how the components (base1.prt and base.prt in our case) are placed in relation to one another in the Assembly file.

10. *Select* the Insert option from the Constraint type drop-down list in the Component Placement dashboard, as shown in Fig.Pro/E-5.21:

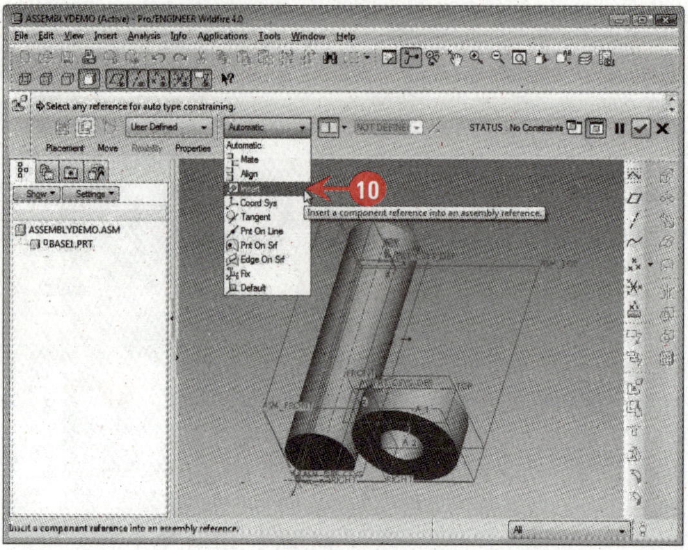

Fig.Pro/E-5.21

The Insert constraint is applied on the BASE.PRT component, and a prompt appears in the message area to let you select a surface on the BASE1.PRT component so that you can insert the BASE.PRT component in it (Fig.Pro/E-5.21).

11. *Select* the surface on the BASE1.PRT component to apply the Insert constraint, as shown in Fig.Pro/E-5.22:

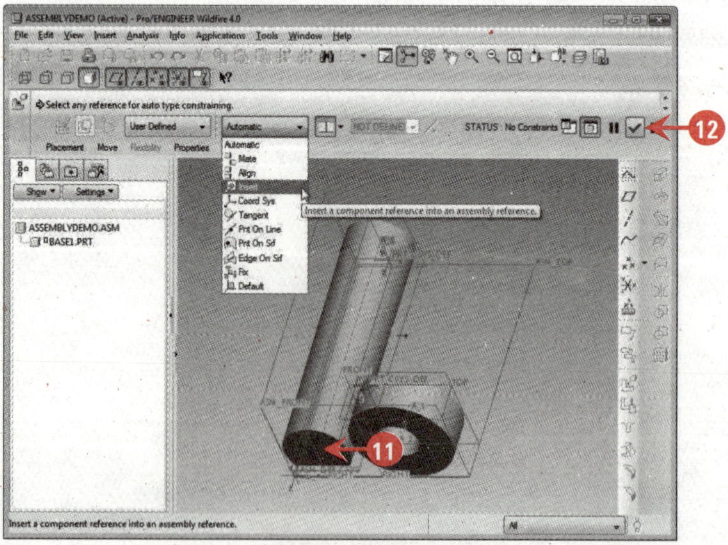

Fig.Pro/E-5.22

12. *Click* the **Build feature** button (✔) to assemble both the components (BASE1.PRT and BASE.PRT) by using the **Insert** constraint (Fig.Pro/E-5.22).

The assembled components are shown in Fig.Pro/E-5.23:

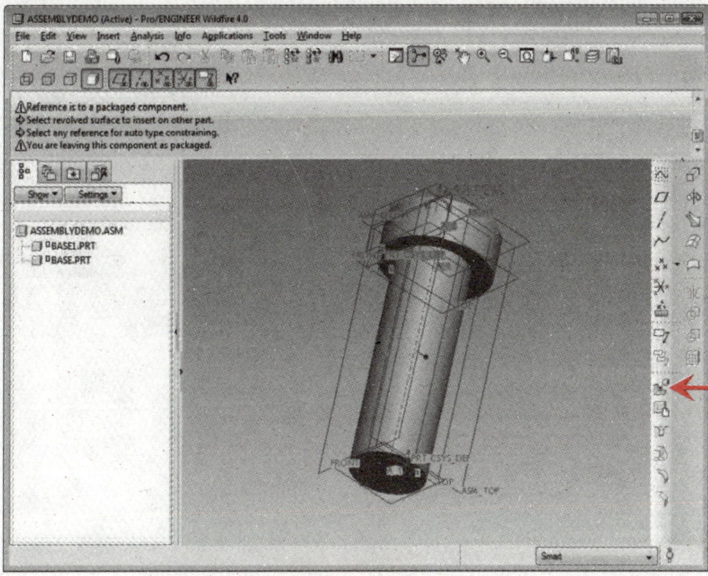

Fig.Pro/E-5.23

13. Now, let's insert the third component (**BASE.PRT**) in the ASSEMBLYDEMO.ASM Assembly file. To do this, *click* the **Assemble** button ( )(Fig.Pro/E-5.23).

The **Open** dialog box appears (Fig.Pro/E-5.24).

14. *Select* the **base.prt** file (Fig.Pro/E 5.24).

15. *Click* the **Open** button, as shown in Fig.Pro/E-5.24:

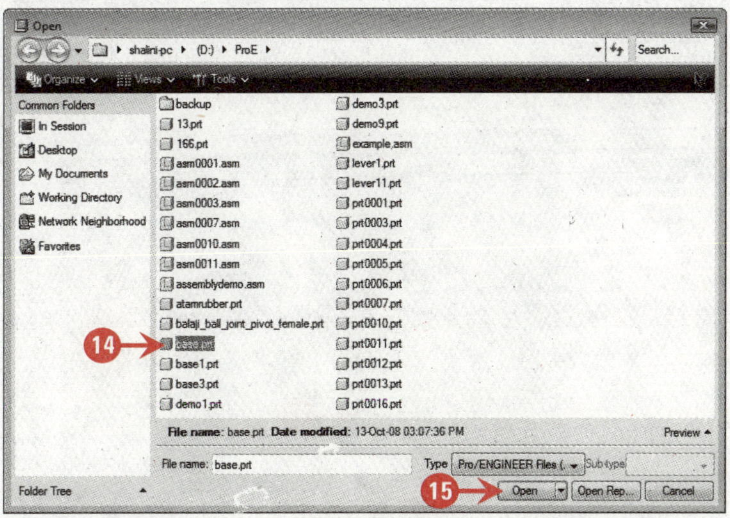

Fig.Pro/E-5.24

The **BASE.PRT** component is inserted in the Assembly file and the **Component Placement** dashboard appears, as shown in Fig.Pro/E-5.25:

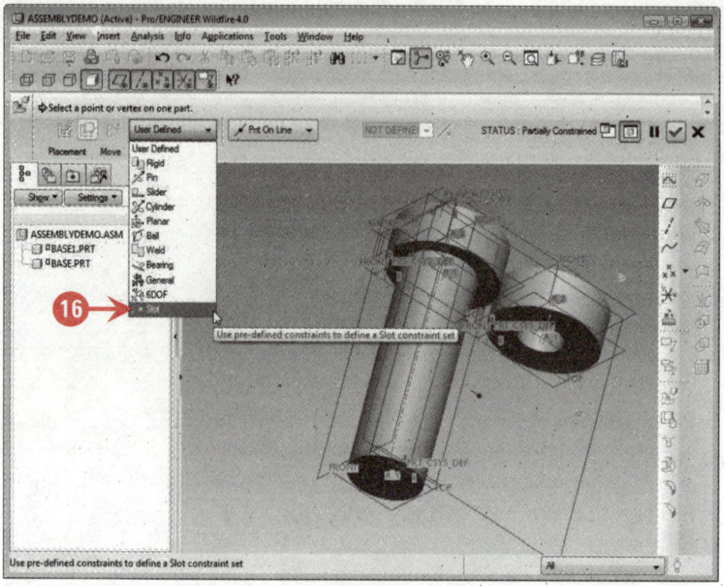

Fig.Pro/E-5.25

16. *Select* a constraint from the Automatic **constraint type** drop-down list. In our case, we have selected the **Pnt On Line** constraint. In addition, select the **Slot** option from the **User Defined** drop-down list (Fig.Pro/E-5.25). A prompt to select a point on one component appears.
17. *Select* a point on the third component (BASE.PRT) in the Assembly file. The **Pnt On Line** constraint is applied on the component. A red line is displayed to select an axis on the first component (BASE1.PRT) in the Assembly file, as shown in Fig.Pro/E-5.26:

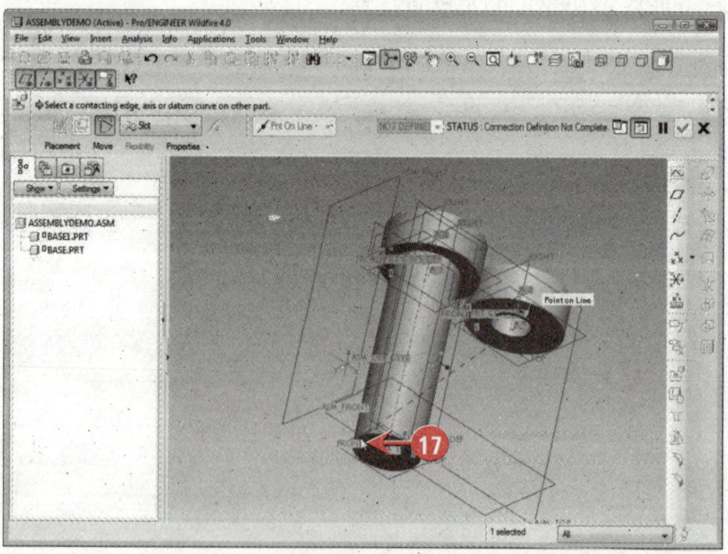

Fig.Pro/E-5.26

18. *Click* the **Build feature button** (✔) to assemble both the components by using the Pnt On Line constraint, as shown in Fig.Pro/E-5.27:

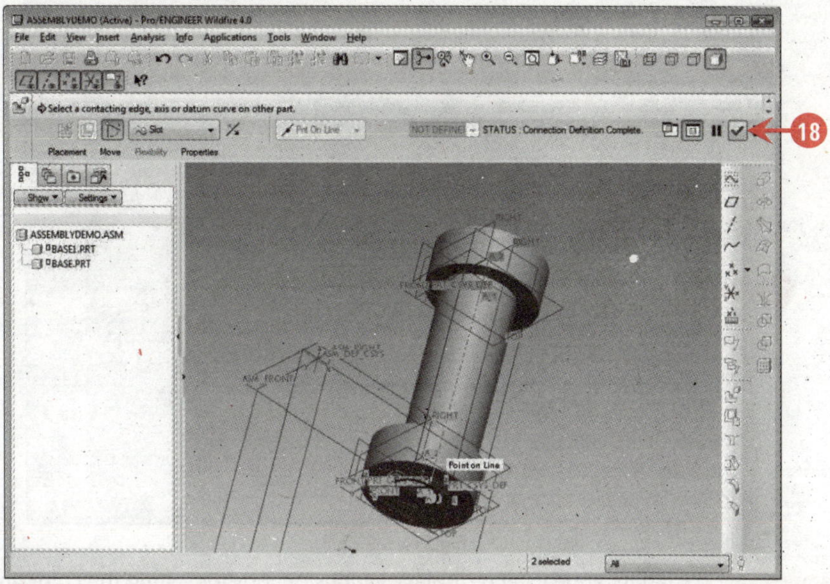

Fig.Pro/E-5.27

The selected components are assembled by using the **Pnt On Line** constraint, as shown in Fig.Pro/E-5.28:

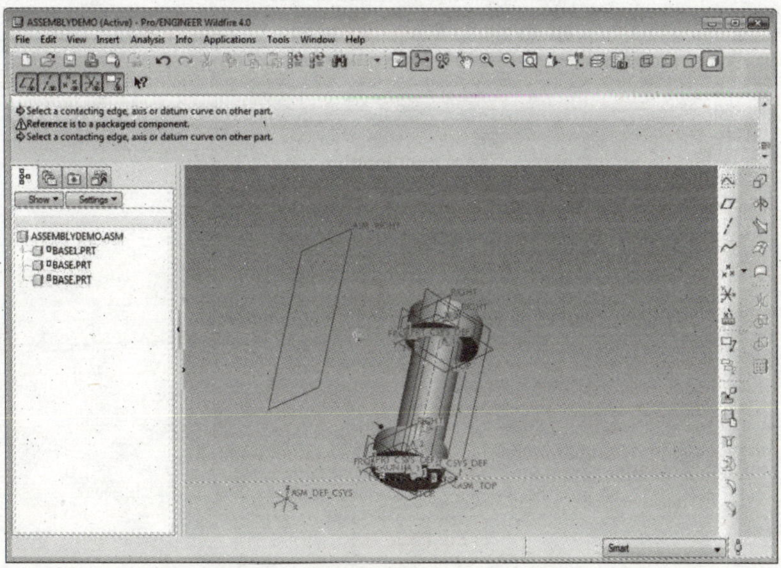

Fig.Pro/E-5.28

Pro/ENGINEER Wildfire 4.0 allows you to set the color of the components inside an assembly. Now, let's learn how to set the color of the identical components assembled in the assembly (ASSEMBLYDEMO.ASM). To learn the process to set color on assembly components, we use the ASSEMBLYDEMO.ASM file. Following steps help you set colors on assembly components:

1. *Select* the **View→Color and Appearance** option from the **Menu Bar**, as shown in Fig.Pro/E-5.29:

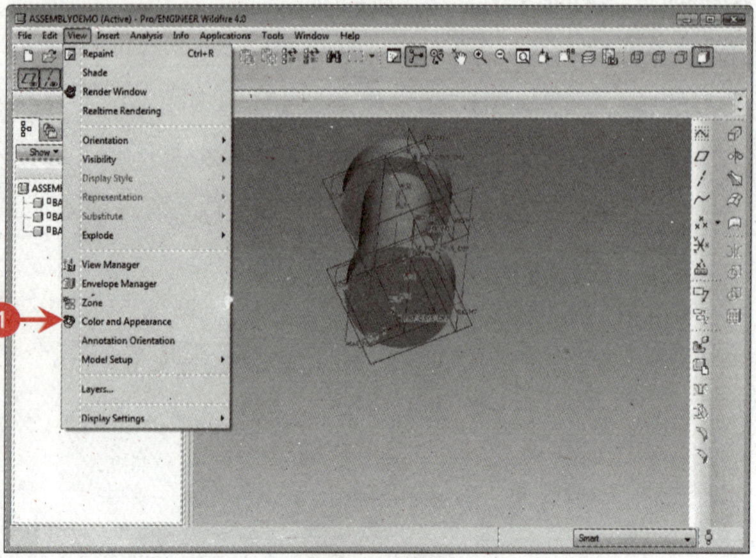

Fig.Pro/E-5.29

The **Appearance Editor** window appears and displays various options for setting the color and appearance of assembly components, as shown in Fig.Pro/E-5.30:

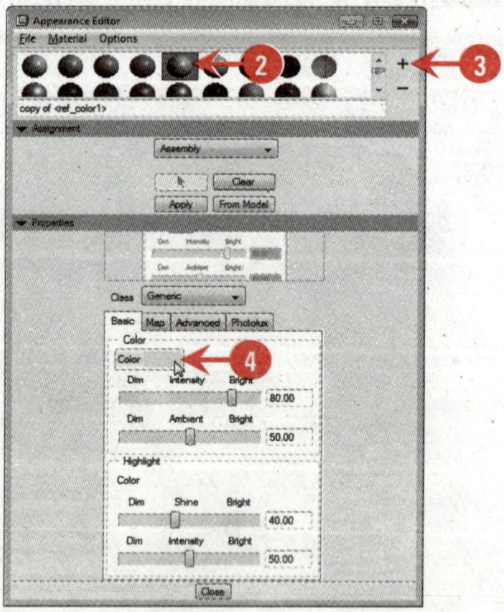

Fig.Pro/E-5.30

2. *Select* an appearance from the pre-defined set of appearances at the top of the Appearance **Editor window** (Fig.Pro/E-5.30).
3. Next, *click* the (+) button to create a copy of the selected color (Fig.Pro/E-5.30).
4. Now, *click* the **Color button** under the Basic tab of Properties palette in the **Appearance Editor window** (Fig.Pro/E-5.30). The **Color Editor** dialog box appears, as shown in Fig.Pro/E-5.31:

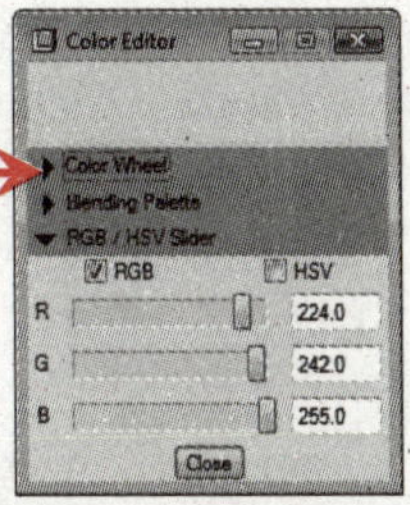

Fig.Pro/E-5.31

5. *Select* the **Color Wheel** option (Fig.Pro/E-5.31) to display the **Color Wheel** palette from where you can select a color an assembly component, as shown in Fig.Pro/E-5.32:

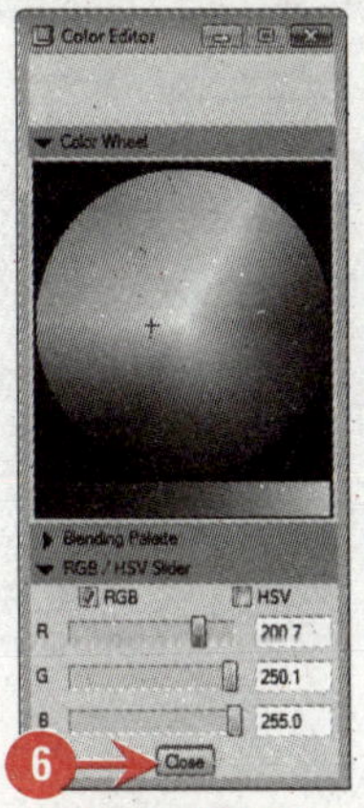

Fig.Pro/E-5.32

6. *Select* the desired color and *click* the **Close** button on the **Color Editor** window (Fig.Pro/E-5.32).
7. Next, *select* the **Components** option from the drop-down list available in the Assignment palette in the **Appearance Editor** window, as shown in Fig.Pro/E-5.33:

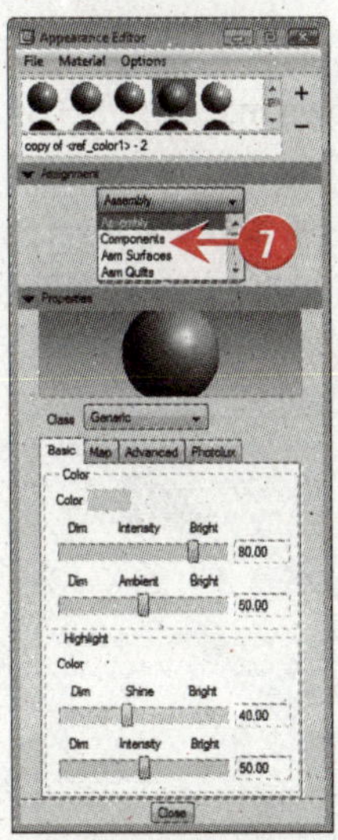

Fig.Pro/E-5.33

The Select dialog box appears and prompts the user to select the desired assembly component to apply the color selected from the Color Wheel palette, as shown in Fig.Pro/E-5.34:

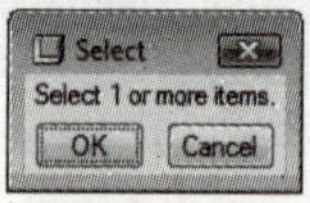

Fig.Pro/E-5.34

8. *Select* any of the BASE.PRT component from the Drawing Area to set the color for the component, as shown in Fig.Pro/E-5.35:

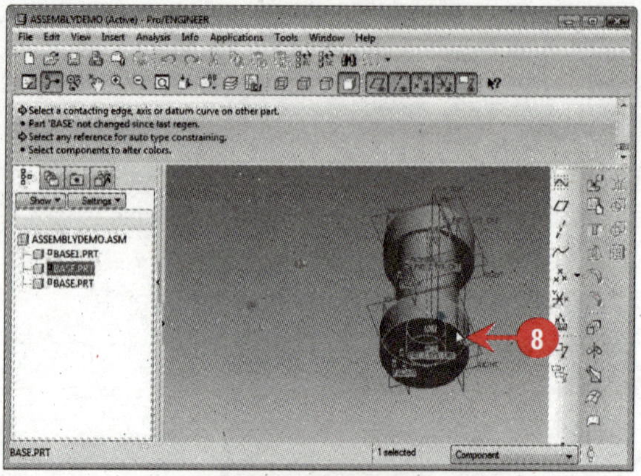

Fig.Pro/E-5.35

The selected component is highlighted with a blue outline (Fig.Pro/E-5.35).

9. Next, *click* the OK button to close the Select dialog box, as shown in Fig.Pro/E-5.36:

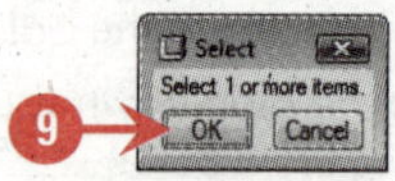

Fig.Pro/E-5.36

10. Now, to apply the color on the selected component, *click* the Apply button in the Assignment palette in the Appearance Editor window, as shown in Fig.Pro/E-5.37:

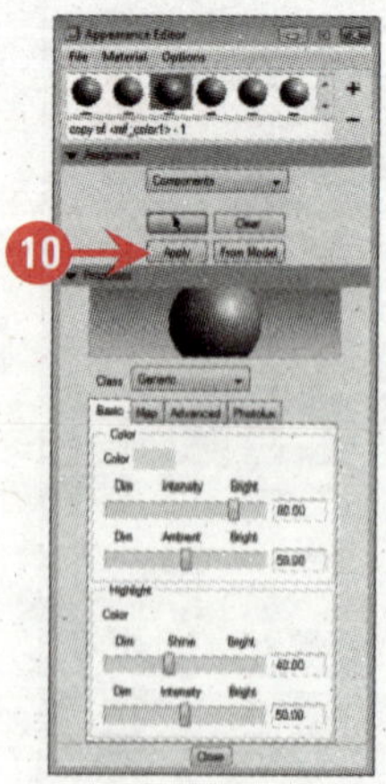

Fig.Pro/E-5.37

The specified color is applied to the selected component (BASE.PRT), as shown in Fig.Pro/E-5.38:

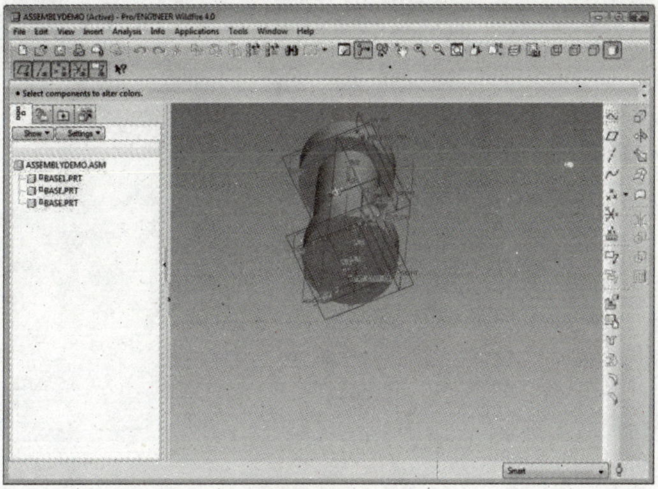

Fig.Pro/E-5.38

11. Now, you need to apply color to the other identical component (BASE.PRT). To do this, *click* the **Select** button ( ) from the **Assignment** palette in the **Appearance Editor** window, as shown in Fig.Pro/E-5.39:

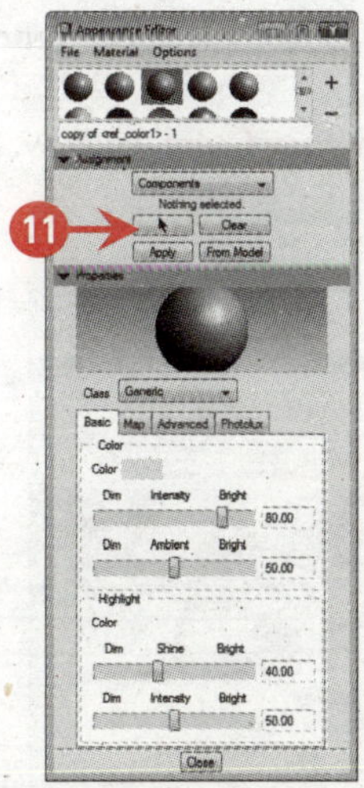

Fig.Pro/E-5.39

The **Select** dialog box reappears, prompting you to select the identical component (BASE.PRT) that you want to color, as shown in Fig.Pro/E-5.40:

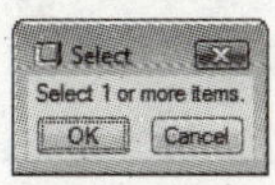

Fig.Pro/E-5.40

12. *Select* the identical BASE.PRT component from the **Drawing Area** to set the specified color for the component, as shown in Fig.Pro/E-5.41:

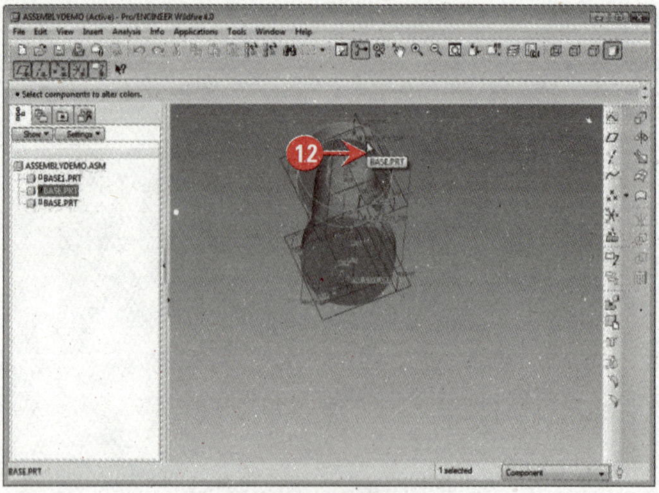

Fig.Pro/E-5.41

13. Next, *click* the **OK** button in the **Select** dialog box, as shown in Fig.Pro/E-5.42:

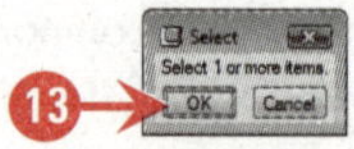

Fig.Pro/E-5.42

This closes the **Select** dialog box. Now, to complete the process of applying the color to the selected component, move to the next step.

14. *Click* the **Apply** button from the **Appearance Editor** window, as shown in Fig.Pro/E-5.43:

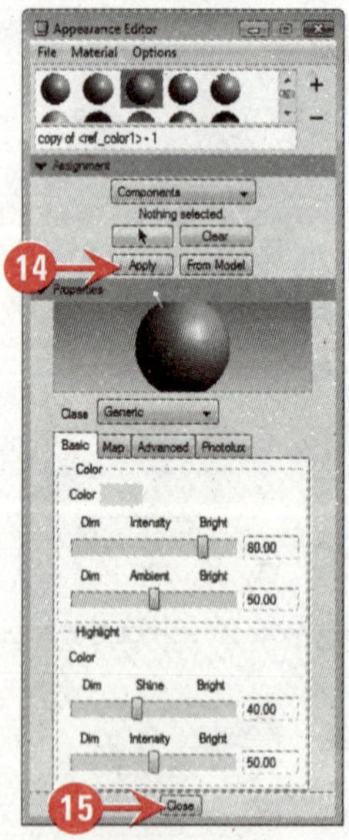

Fig.Pro/E-5.43

15. *Click* the **Close** button in the Appearance Editor window to close it (Fig.Pro/E-5.43).

The specified color is applied to the selected component, as shown in Fig.Pro/E-5.44:

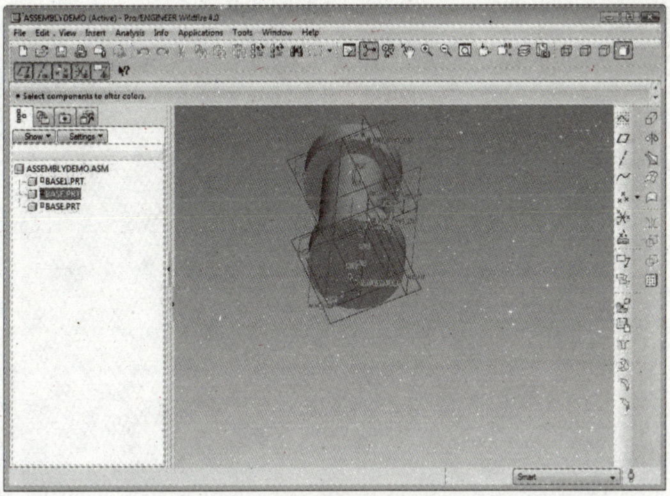

Fig.Pro/E-5.44

16. After changing the appearance of the components in the ASSEMBLYDEMO.ASM file, let's save the changes made the file. For this, *select* the **File→Save** option from the **Menu Bar**, as shown in Fig.Pro/E-5.45:

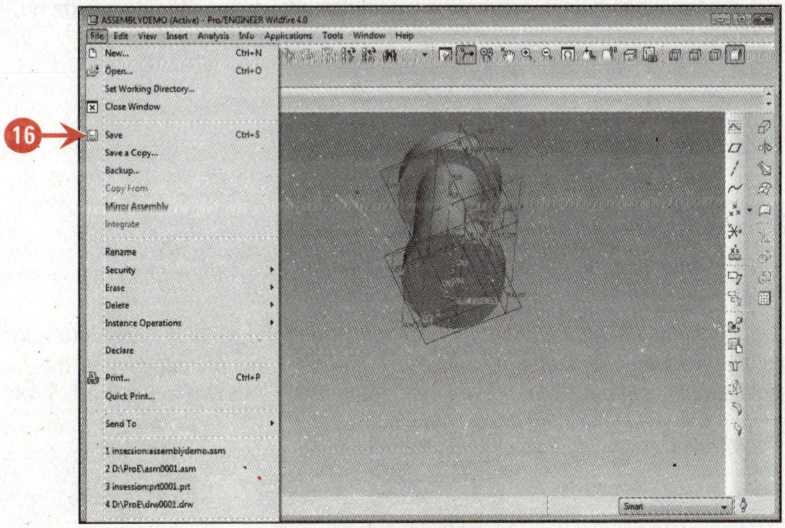

Fig.Pro/E-5.45

Next, we discuss how to package components in an assembly.

Packaging Components

Packaging is a mechanism of assembling components when their exact positions in the Assembly file are not known. In such cases, the components cannot be assembled by using assembly constraints and can only be packaged.

Perform the following steps to package a component in an Assembly file:

1. *Create* an Assembly file and *select* the **Insert→Component→Package** option from the **Menu Bar**, as shown in Fig.Pro/E-5.46:

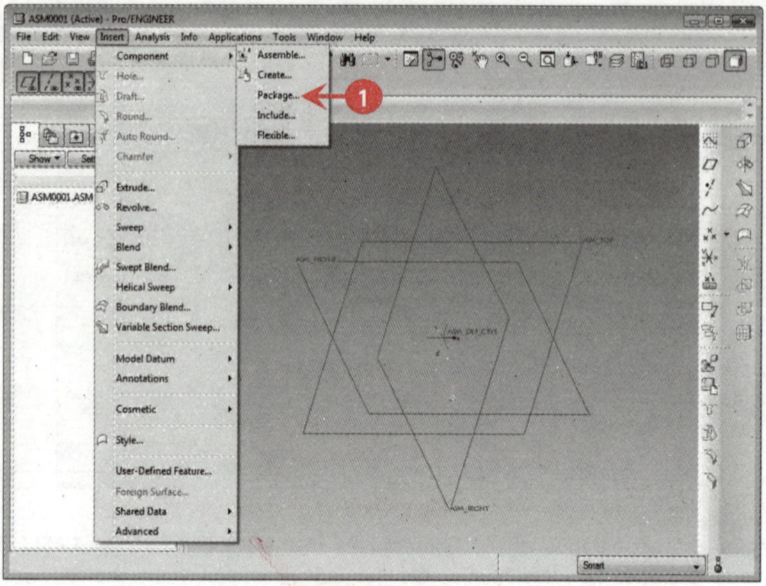

Fig.Pro/E-5.46

The **Menu Manager** menu appears and displays various options in the **PACKAGE** menu, as shown in Fig.Pro/E-5.47:

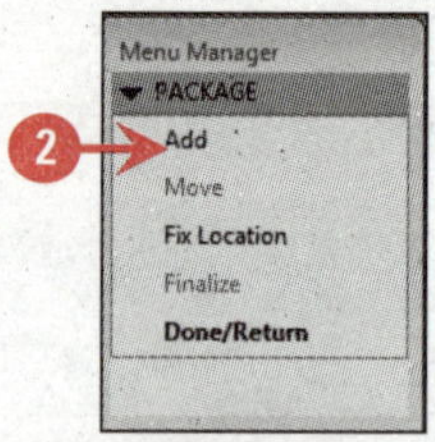

Fig.Pro/E-5.47

2. *Select* the **Add** option under the **PACKAGE** menu from the **Menu Manager** menu to add a component that you want to package (Fig.Pro/E-5.47). The **GET MODEL** option appears in the **Menu Manager** menu and displays various sub-options to add the component, as shown in Fig.Pro/E-5.48:

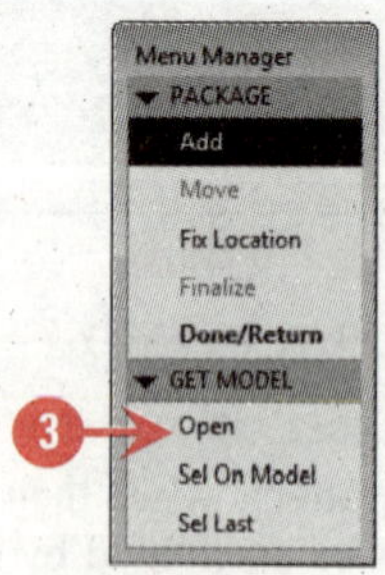

Fig.Pro/E-5.48

3. *Select* the **Open** sub-option under the **GET MODEL** option in the **Menu Manager** menu (Fig.Pro/E-5.48). The **Open** dialog box appears and displays the existing Part files available for packaging in the Assembly file (**ASM0001.ASM** in our case), as shown in Fig.Pro/E-5.49:

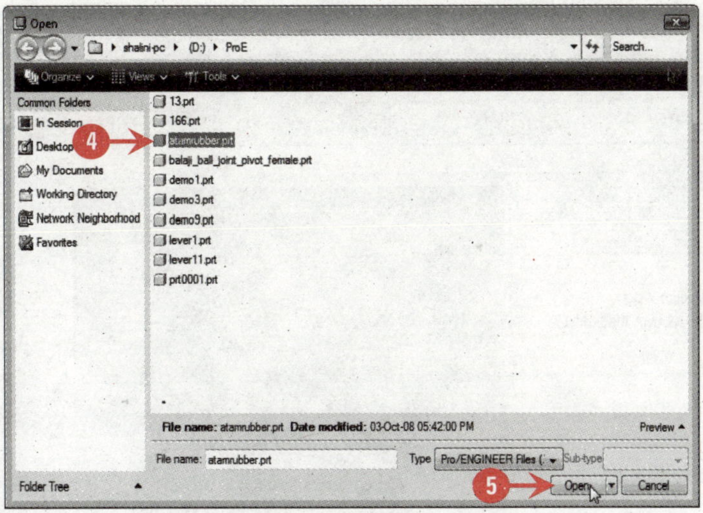

Fig.Pro/E-5.49

4. *Select* the Part file that you want to package from the Open dialog box. In our case, we have selected the **atamrubber.prt** Part file (Fig.Pro/E-5.49).

5. Next, *click* the **Open** button (Fig.Pro/E-5.49).

The selected Part file is placed in the ASM0001.ASM Assembly file and the **Move** dialog box appears. You can use the **Move** dialog box to move a packaged component to a new location (Fig.Pro/E-5.50).

6. *Select* a radio button from the **Motion Type** area in the **Move** dialog box to specify the type of motion you want to apply to the component. By default the **Orient Mode** radio button is selected and we continue with it (Fig.Pro/E-5.50).

7. Now, *click* the **OK** button, as shown in Fig.Pro/E-5.50:

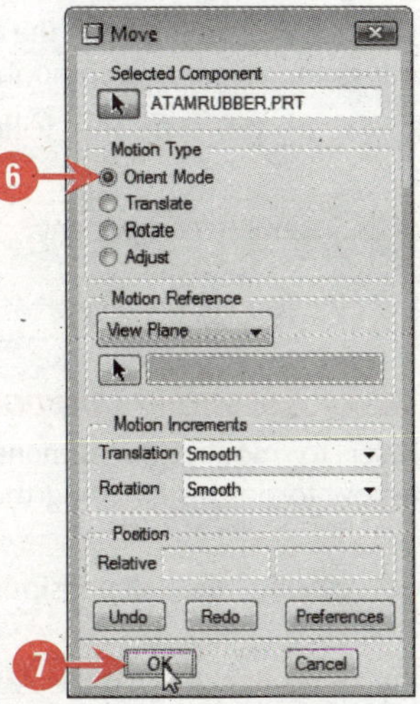

Fig.Pro/E-5.50

The selected component (ATAMRUBBER.PRT) is packaged successfully in the Assembly file, as shown in Fig.Pro/E-5.51:

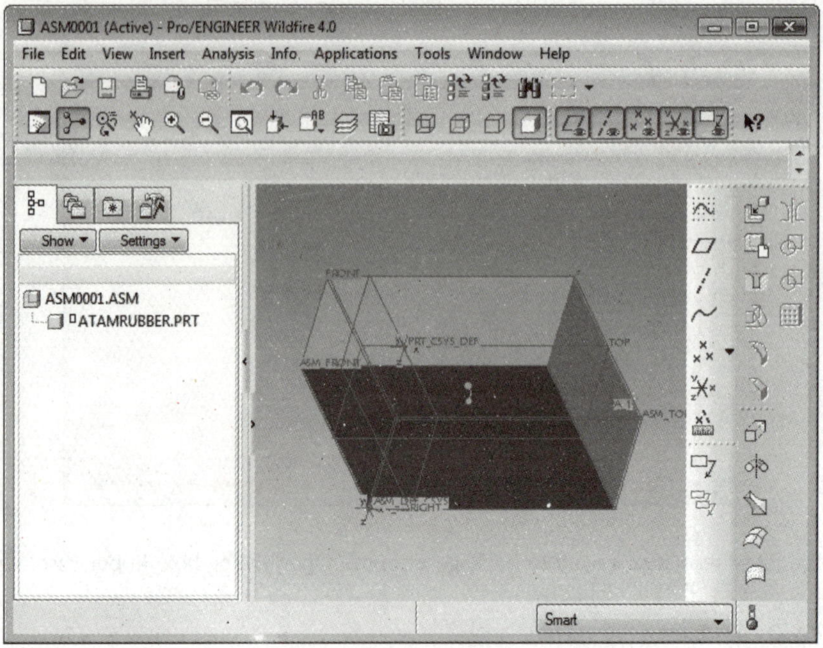

Fig.Pro/E-5.51

Now, let's discuss how to modify assembly components within an assembly.

Modifying Assembly Components

Users often face a situation where they want to edit the features of assembly components in the Assembly mode. This may happen when the result from a Part file does not reflected in the Assembly mode. A user can perform this task easily in Pro/ENGINEER Wildfire 4.0 by redefining the features of the assembled components in the Assembly mode. However, you can also edit the assembly components in the Part mode by modifying the dimensions of the features of the assembled components.

Pro/ENGINEER Wildfire 4.0 allows you to perform the following manipulations on assembly components:

- Modify the dimensions of a feature of an assembly component
- Redefine the features of an assembly component

Let's discuss these one by one.

Modifying Dimensions of a Feature of an Assembly Component

Pro/ENGINEER Wildfire 4.0 allows a user to modify the dimensions of the features of an assembly component in the Part mode. Let's learn how to do this by using the ASSEMBLYDEMO.ASM Assembly file created earlier in the chapter.

Now, perform the following steps to modify the dimensions of a component feature in the ASSEMBLYDEMO.ASM Assembly file:

1. *Select* the **File→Open** option from the **Menu Bar** to open the Assembly file in which you want to make modifications, as shown in Fig.Pro/E-5.52:

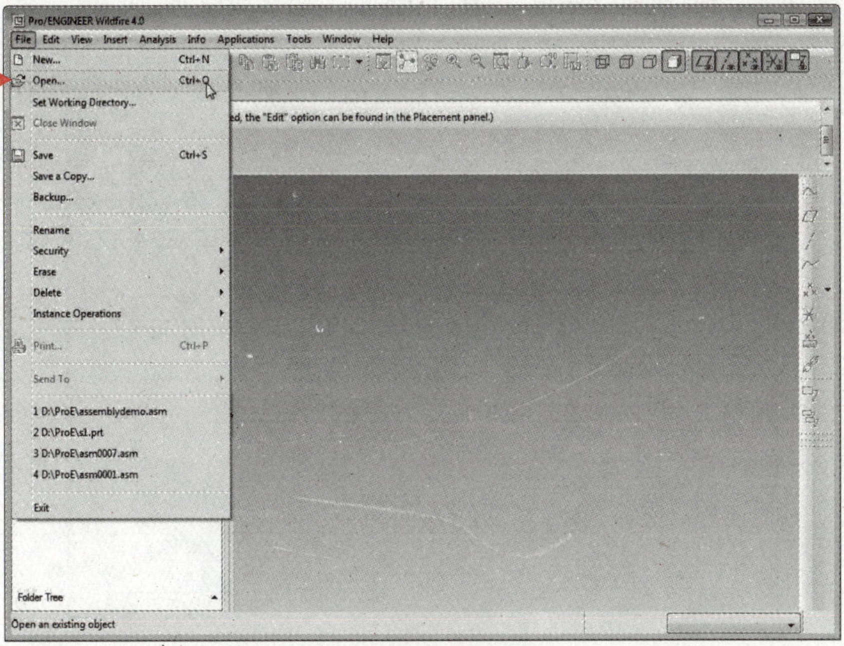

Fig.Pro/E-5.52

The **Open** dialog box appears and displays a list of Pro/ENGINEER Wildfire 4.0 files, as shown in Fig.Pro/E-5.53:

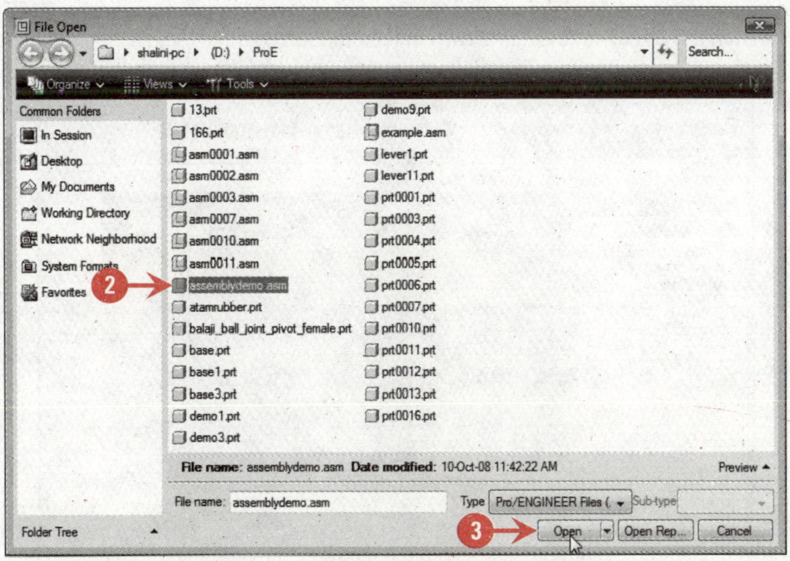

Fig.Pro/E-5.53

2. *Select* the **assemblydemo.asm** Assembly file (Fig.Pro/E-5.53).
3. *Click* the **Open** button (Fig.Pro/E-5.53). The selected Assembly file opens in the current session of Pro/ENGINEER Wildfire 4.0, as shown in Fig.Pro/E-5.54:

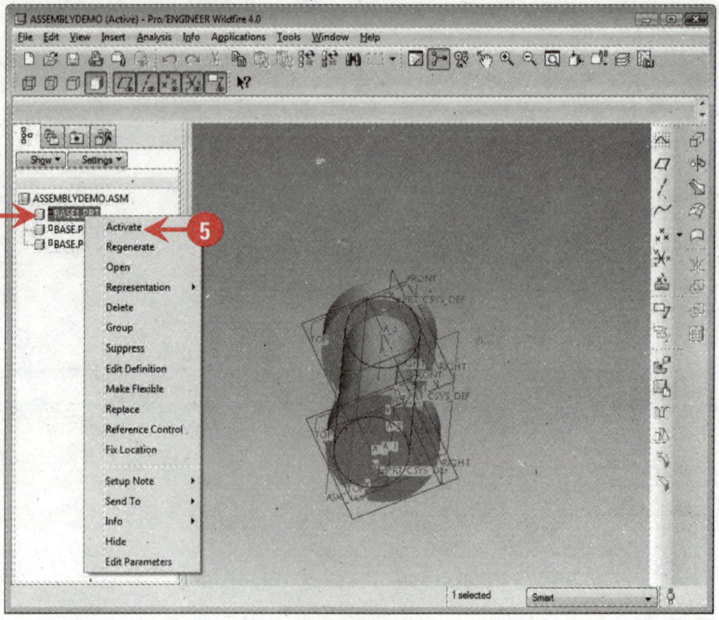

Fig.Pro/E-5.54

4. *Right-click* the desired component (BASE1.PRT) from the **Model Tree** located at the left side of the **Drawing Area.** A context menu appears (Fig.Pro/E-5.54).
5. *Select* the **Activate** option from the context menu (Fig.Pro/E-5.54). A small green button appears on the selected component in the **Model Tree** and the components other than those of the **BASE1.PRT** component become faded in the **Drawing Area**, as shown in Fig.Pro/E-5.55:

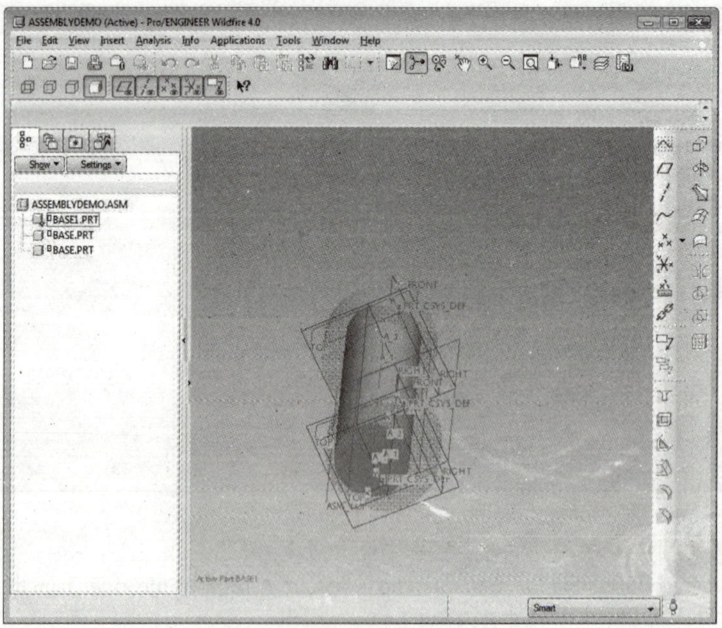

Fig.Pro/E-5.55

6. *Right-click* the selected component (BASE1.PRT) and select the **Open** option from the context menu, as shown in Fig.Pro/E-5.56:

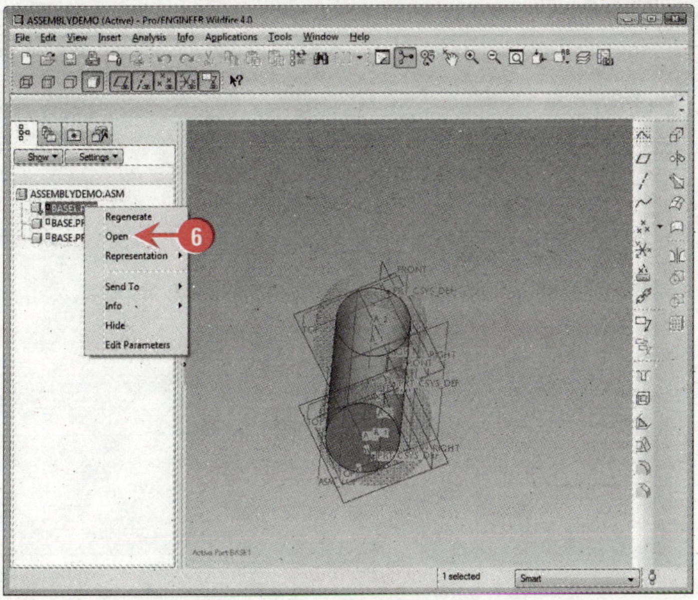

Fig.Pro/E-5.56

The selected component opens as a Part file in a separate window, as shown in Fig.Pro/E-5.57:

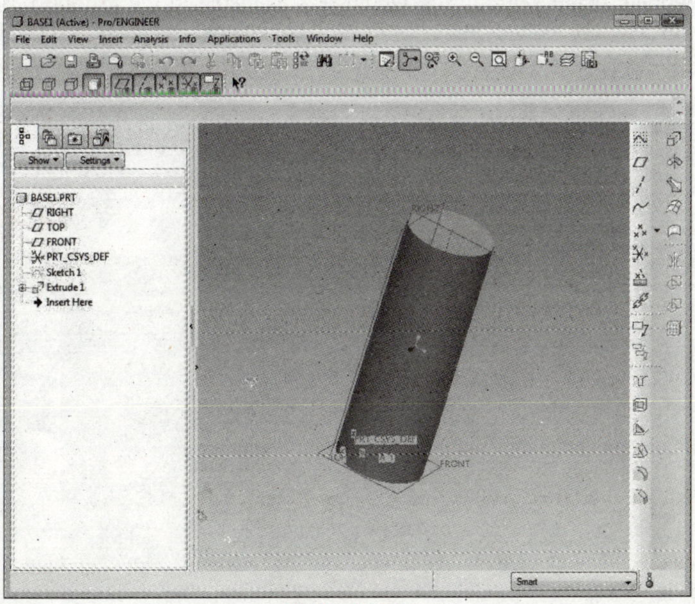

Fig.Pro/E-5.57

7. *Right-click* the desired feature (Extrude) in the **Model Tree** and *select* the **Edit Definition** option from the context menu, as shown in Fig.Pro/E-5.58:

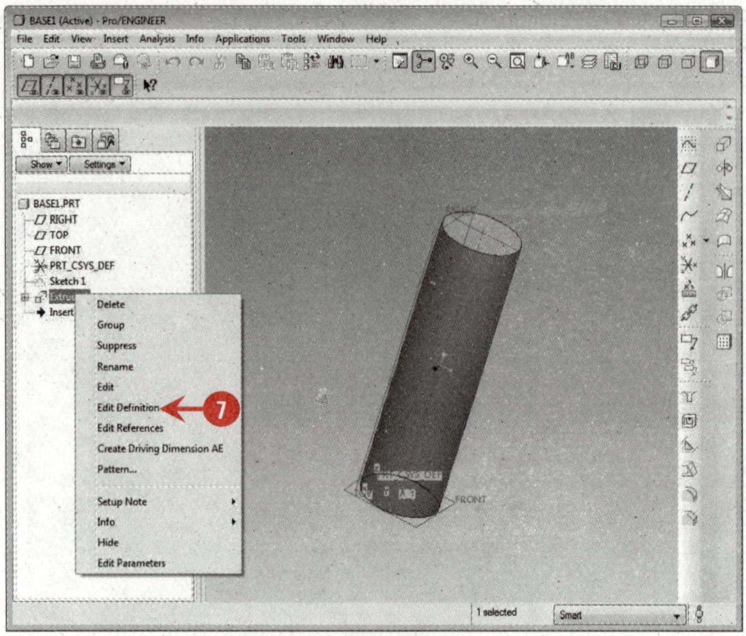

Fig.Pro/E-5.58

The **Extrude** dashboard appears (Fig.Pro/E-5.59). Here, you can modify the dimensions of the **Extrude** feature in the BASE1.PRT component by changing its existing depth value.

8. *Enter* an appropriate depth value in the Depth box available in the **Extrude** dashboard to modify the dimensions of the **Extrude** feature. In our case, we have entered the value **400.00** (Fig.Pro/E-5.59).
9. *Press* the **ENTER** key to reflect the new depth value in the **Extrude** dashboard.

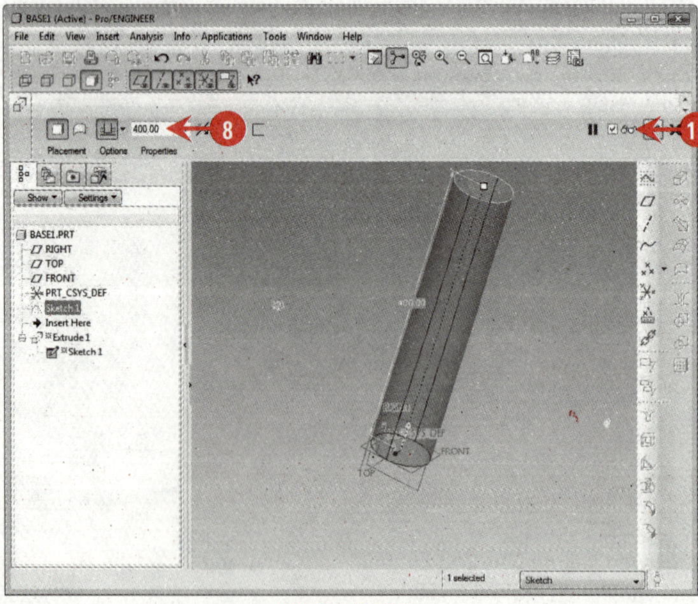

Fig.Pro/E-5.59

The depth value entered in the **Extrude** dashboard modifies the previous depth value of the **Extrude** feature applied on the assembled component.

10. *Click* the **Build feature** button (✔) in the **Extrude** dashboard to apply and save the changes made in the **Extrude** feature of the component (Fig.Pro/E-5.59).
11. Next, *select* the **Edit→Regenerate** option from the **Menu Bar**, as shown in Fig.Pro/E-5.60:

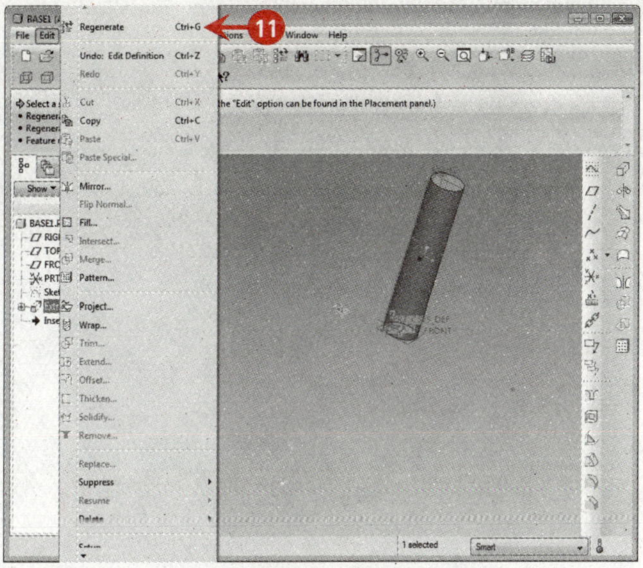

Fig.Pro/E-5.60

The component (BASE1.PRT) is regenerated (that is, the modifications made in the dimensions of the **Extrude** feature are permanently saved in memory), as shown in Fig.Pro/E-5.61:

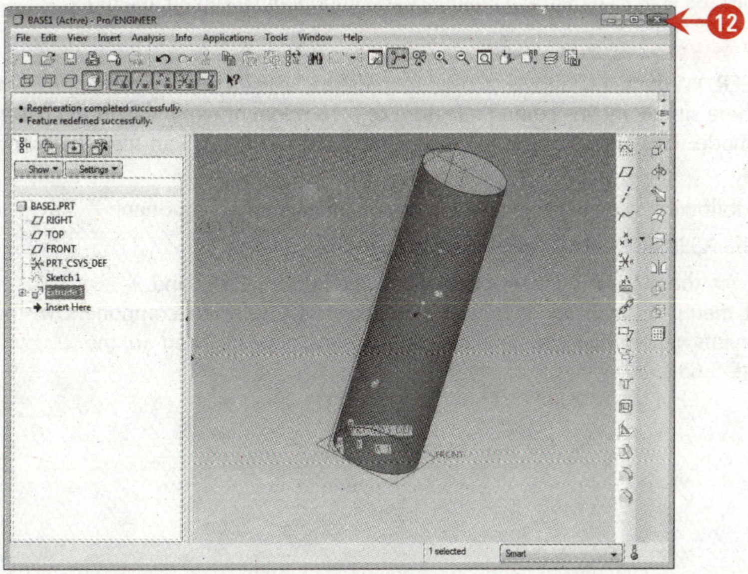

Fig.Pro/E-5.61

12. *Close* the **Part file** window (Fig.Pro/E-5.61). The modifications made in the dimensions of the component are incorporated in the **Assembly** file (ASSEMBLYDEMO.ASM), as shown in Fig.Pro/E-5.62:

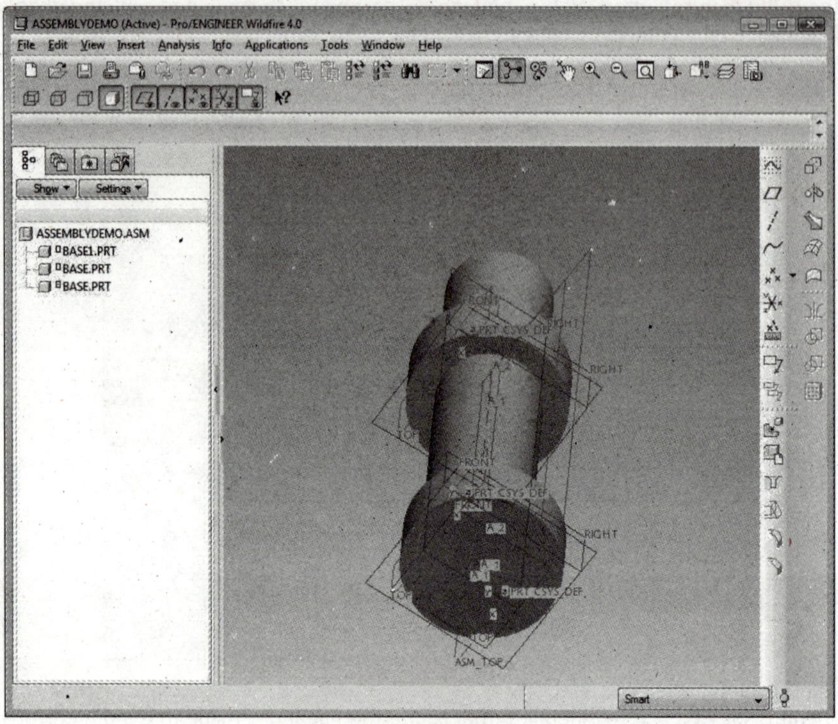

Fig.Pro/E-5.62

Now, let's discuss how to redefine a feature of a component within an assembly.

Redefining a Feature of an Assembly Component

Pro/ENGINEER Wildfire 4.0 allows a user to redefine a feature of a component in the Assembly mode. Redefining here simply means editing a feature of a component while it is in the Assembly mode rather than in the Part mode. This allows a user to get the expected result in a Part file when the Part file is imported to an assembly.

Perform the following steps to redefine a feature of an assembly component:

1. *Open* the ASSEMBLYDEMO.ASM Assembly file.
2. *Right-click* the **BASE.PRT** component from the **Model Tree**, and *select* the **Activate** option from the context menu. A small green button appears on the selected component in the **Model Tree** and the components other than the selected component appear faded in the **Drawing Area**, as shown in Fig.Pro/E-5.63:

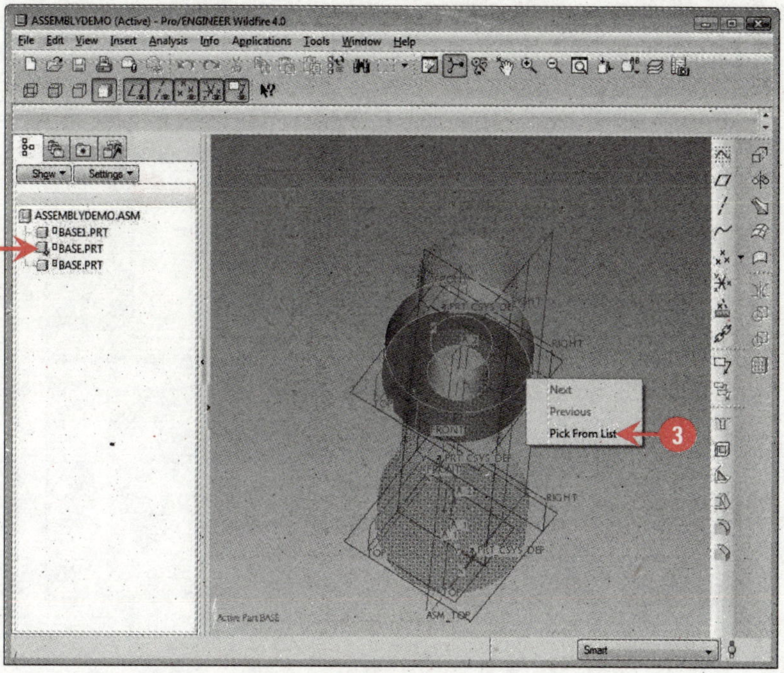

Fig.Pro/E-5.63

3. *Right-click the* BASE1.PRT component and *select* the Pick From List option from the context menu (Fig.Pro/E-5.63). The Pick From List dialog box appears, as shown in Fig.Pro/E-5.64:

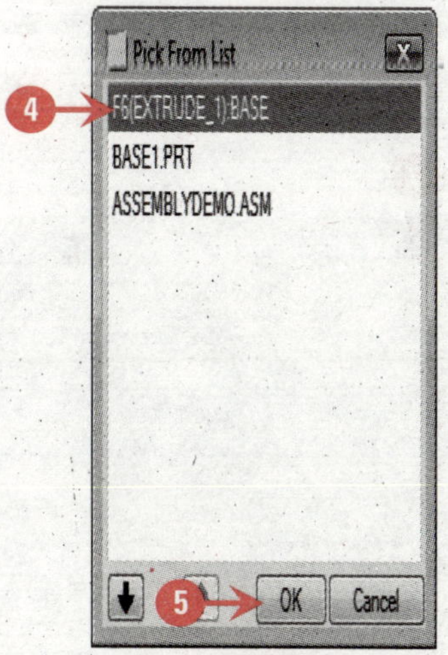

Fig.Pro/E-5.64

The Pick From List dialog box is used to redefine a feature of a selected component.

4. *Select* the Extrude feature of the selected component (Fig.Pro/E-5.64), F6(EXTRUDE_1): BASE in our case.
5. *Click* the OK button (Fig.Pro/E-5.64). The component whose Extrude feature is selected appears faded, as shown in Fig.Pro/E-5.65:

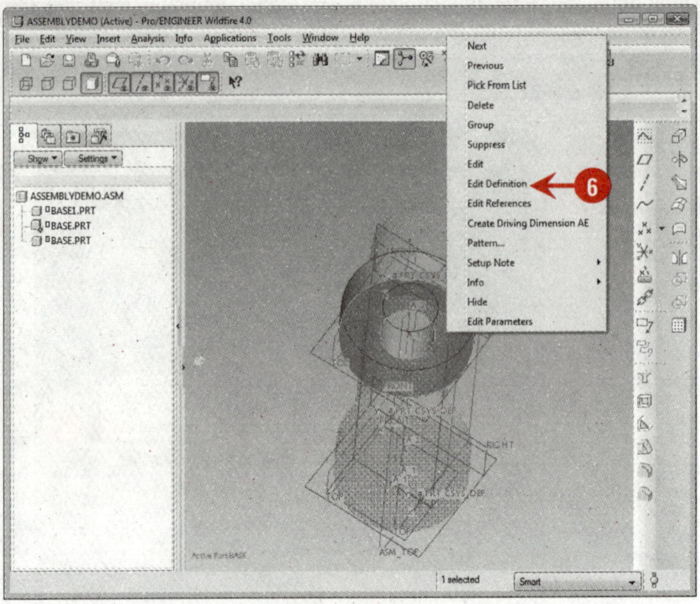

Fig.Pro/E-5.65

6. *Right-click* the **Drawing Area** and select the **Edit Definition** option from the context menu (Fig.Pro/E-5.65).
7. Next, *specify* a depth value as per your requirement in the **Depth** box of the **Extrude** dashboard, as shown in Fig.Pro/E-5.66:

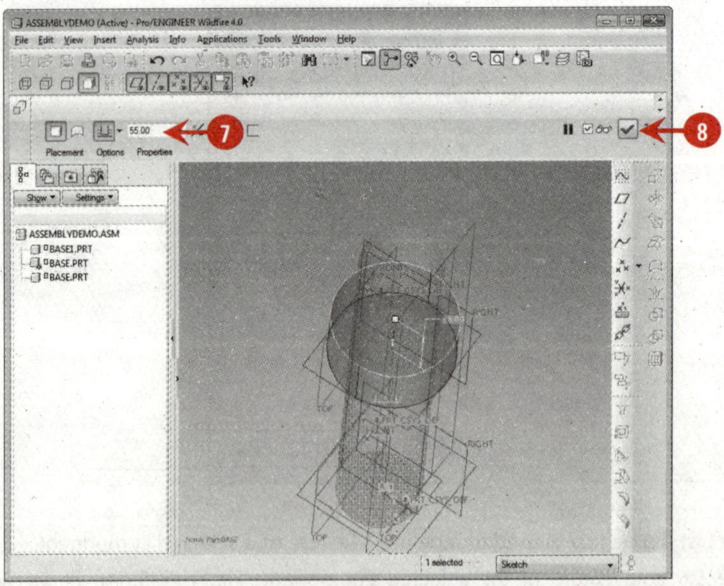

Fig.Pro/E-5.66

The depth value that you specify in the **Extrude** dashboard modifies the previous depth value of the selected component.

8. *Click* the Build feature button in the **Extrude** dashboard to apply and save the changes made in **Extrude** feature of the selected component (Fig.Pro/E-5.66). The modifications made in the **Extrude** feature of the **BASE.PRT** component are incorporated within the **ASSEMBLYDEMO.ASM** Assembly file, as shown in Fig.Pro/E-5.67:

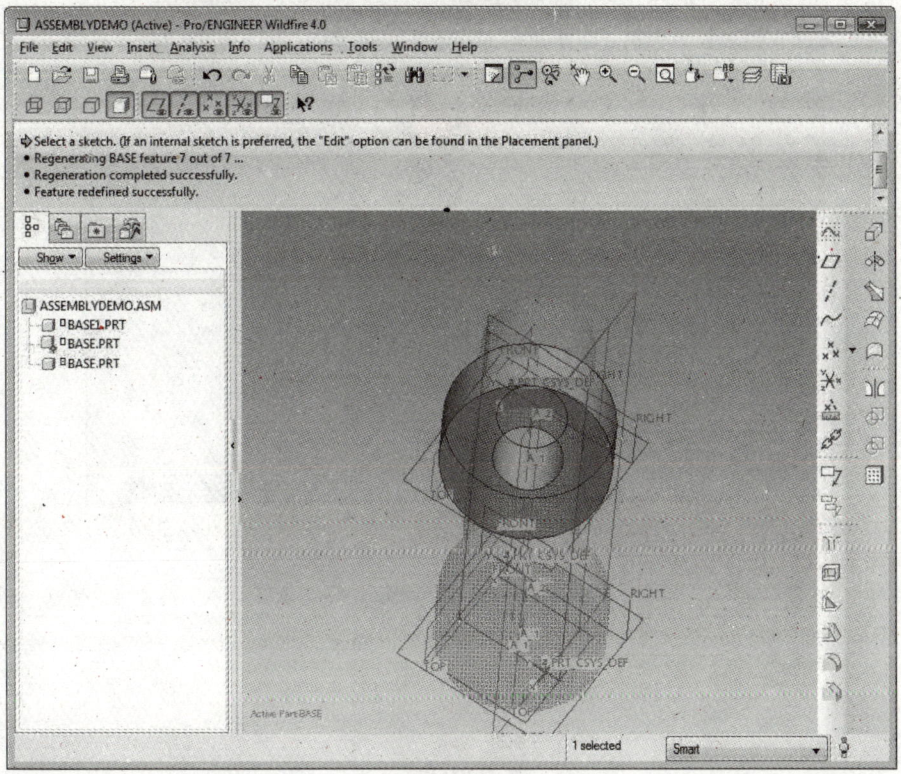

Fig.Pro/E-5.67

We learned about modifying assembly components in the previous section. Now, let's discuss how to create an exploded view of the Assembly.

Creating Exploded View of an Assembly

Users often find it difficult to view all the components placed inside an Assembly file, which, in turn, can lead a user to misinterpret the information intended to be conveyed. For example, a wrong value for a dimension can be considered by a user. Pro/ENGINEER Wildfire 4.0 provides a solution to such problems by creating an exploded view of an assembly. The exploded view is a state in which all the components assembled inside an Assembly file are moved from their actual location so that they can be easily viewed. Let's take a look at the steps required to create an exploded view of the ASSEMBLYDEMO.ASM Assembly file created earlier in the chapter.

Perform the following steps to create an exploded view of the ASSEMBLYDEMO.ASM Assembly file:

1. *Select* the **View→Explode→Edit Position** option from the **Menu Bar**, as shown in Fig.Pro/E-5.68:

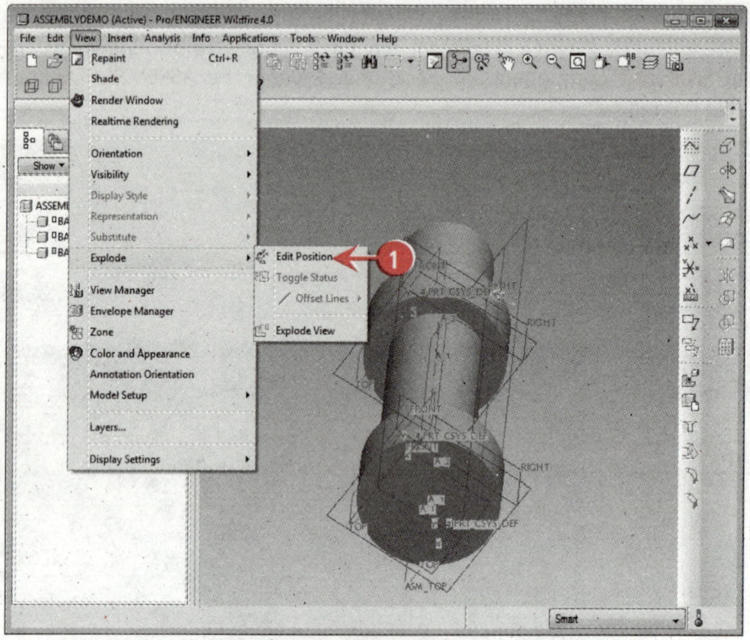

Fig.Pro/E-5.68

The Explode Position dialog box appears, as shown in Fig.Pro/E-5.69:

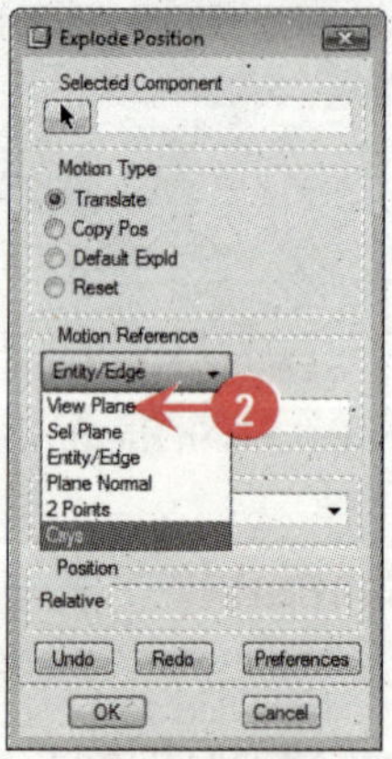

Fig.Pro/E-5.69

The Select dialog box also appears along with the Explode Position dialog box, as shown in Fig.Pro/E-5.70:

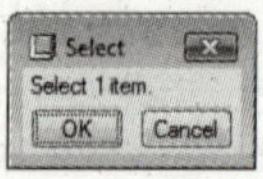

Fig.Pro/E-5.70

2. *Select* the **View Plane** option by clicking the Entity/Edge drop-down list in the **Motion Reference** area of the **Explode Position** dialog box (Fig.Pro/E-5.69).
3. Now, *select* any component and, keeping the mouse-button pressed, *drag* the component to the desired location. After reaching the location, release the mouse-button, as shown in Fig.Pro/E-5.71:

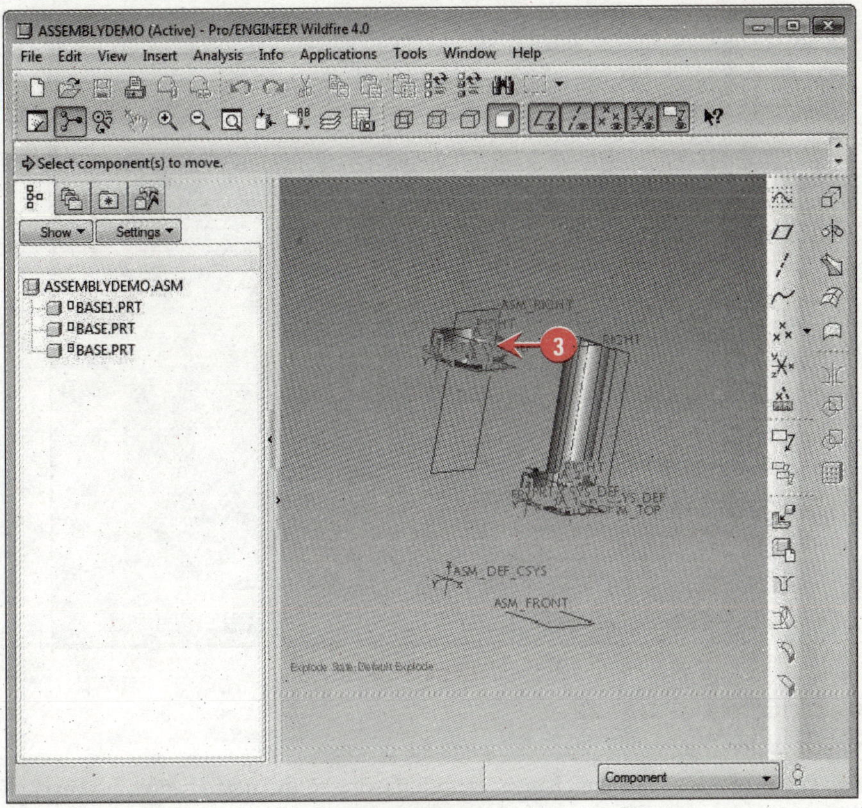

Fig.Pro/E-5.71

When you release the mouse button, the Menu Manager menu appears, as shown in Fig.Pro/E-5.72:

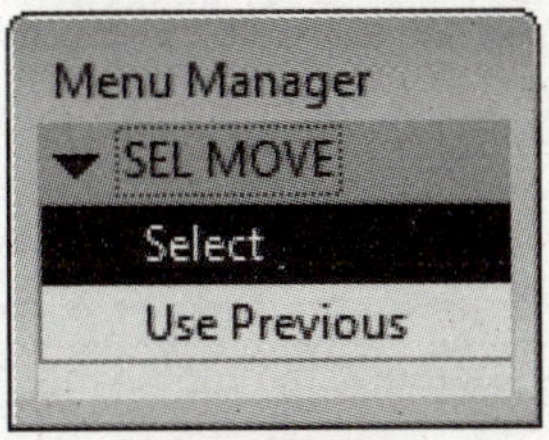

Fig.Pro/E-5.72

The Menu Manager menu helps you select another component to move. Following two options are present in the Menu Manager menu:

- **Select:** Allows you to select and move another component.
- **Use Previous:** Allows you to select and move the previously moved component.

By default the Select option is selected and we continue with this option.

Repeat the step 3 to move the other components in the ASSEMBLYDEMO.ASM Assembly file to display their exploded view (Fig.Pro/E-5.71). In our case, we have moved the another component, as shown in Fig.Pro/E-5.73:

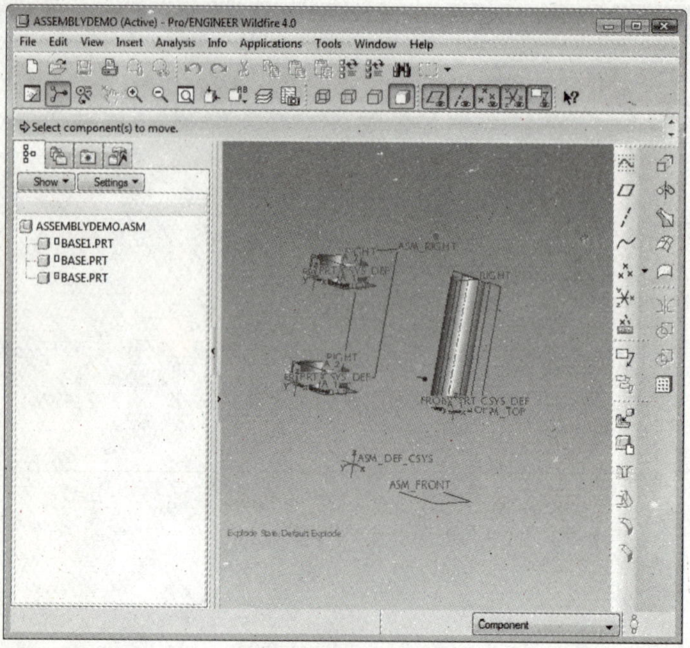

Fig.Pro/E-5.73

4. *Click* the **OK** button (Fig.Pro/E-5.70).
5. *Click* the **OK** button (Fig.Pro/E-5.69) to complete the process.

With this, we come to the end of the chapter. Before moving to the next chapter, however, let's recap the main points covered in this chapter through a brief summary.

Summary

In this chapter, you learned about:

- Assembly approaches including Top-down and Bottom-up approach to assemble components in an assembly
- Integrating assembly components with the help of assembly constraints and by packaging assembly components
- Manipulating components of an assembly by modifying the dimensions of their features and by redefining their features.
- Creating an exploded view of an assembly to move the assembly components from their actual location so that they can be easily viewed in an assembly.

Chapter 6

Exploring Pro/ENGINEER Wildfire 4.0 Drawing Mode

In this Section

- Starting the Drawing Mode
- Generating Drawing Views
- Modifying the Drawing Views
- Inducing Dimensions and Tolerance
- Creating Notes

Pro/ENGINEER Wildfire 4.0 allows users to create drawing views, which contain complete and detailed information about the geometry of a component (Part) to manufacture it easily. The drawing views are created to convey all necessary information about a component for proper & easy manufacturing of the component. Pro/ENGINEER Wildfire 4.0 provides the **Drawing** mode to generate the drawing views of the Parts and the Assemblies created in a Pro/ENGINEER Wildfire 4.0 session.

In this chapter, you learn the fundamentals of the **Drawing** mode in Pro/ENGINEER Wildfire 4.0. To generate the drawing view, you need to first start the **Drawing** mode. Therefore, this chapter begins with the discussion on starting the **Drawing** mode. The chapter then proceeds further to discuss the various drawing views. You also learn to modify the drawing views and induce dimensions and tolerances in the drawing views. Towards the end of the chapter, you learn about modifying dimensions and creating notes in the drawing views.

Starting the Drawing Mode

The **Drawing** mode is used to generate the drawing view from the Parts and Assemblies created in a Pro/ENGINEER Wildfire 4.0 session. To generate the drawing view, first, you need to start the **Drawing** mode. Let's take a look at the steps required to start the **Drawing** mode.

Perform the following steps to start the **Drawing** mode:

1. *Select* the **File →New** option from the **Menu Bar**, as shown in the Fig.Pro/E-6.1:

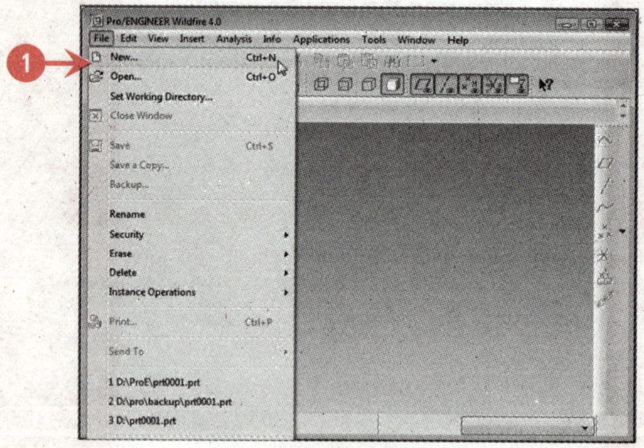

Fig.Pro/E-6.1

The **New** dialog box appears and displays the various modes available in Pro/ENGINEER Wildfire 4.0 (Fig.Pro/E-6.2).

2. *Select* the **Drawing** radio button from the **Type** Area in the **New** dialog box, as shown in Fig.Pro/E-6.2:

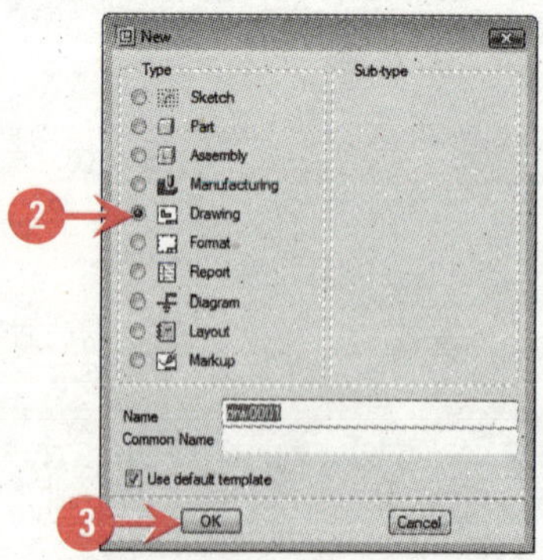

Fig.Pro/E-6.2

A default name of the Drawing file is shown in the **Name** text box. We continue with the default name. However, you can change the default name.

3. *Click* the **OK** button in the **New** dialog box (Fig.Pro/E-6.2). The **New Drawing** dialog box appears and displays various options to specify a template, which will be applied to a new Drawing file in a Pro/ENGINEER Wildfire 4.0 session, as shown in Fig.Pro/E-6.3:

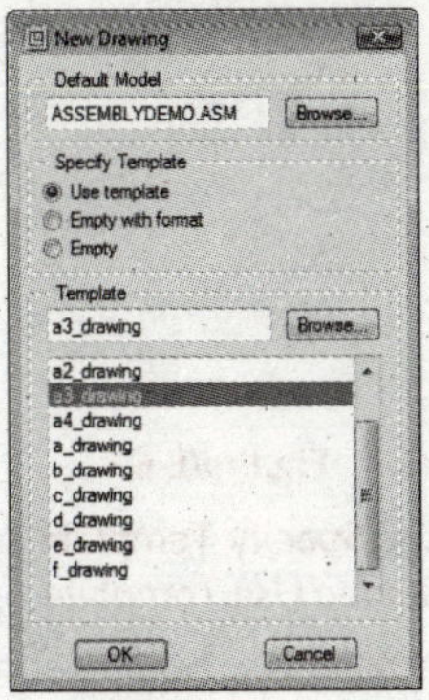

Fig.Pro/E-6.3

The **Default Model** Area in the **New Drawing** dialog box allows the user to specify the name of the model for which views are to be generated.

Note

> By default a model name appears in the Default Model area. To change the default model name, click the Browse button and select the model.

In our case, the ASSEMBLYDEMO.ASM is displayed by default in the Name edit box in the **Default Model** Area because it is opened in the current session of the Pro/ENGINEER Wildfire 4.0. In our case, we will use the ASSEMBLYDEMO.ASM file that has already been created in the chapter 5, *Exploring Pro/ENGINEER Wildfire 4.0 Assembly Mode*.

4. *Select* one of the following radio buttons from the **Specify Template** Area in the **New Drawing** dialog box:

- **Use template:** Allows users to select any pre-defined drawing template for the new Drawing file (Fig.Pro/E-6.3). It is selected by default in the **Specify Template** Area.
- **Empty with format:** Allows users to select a system defined format for the new Drawing file, as shown in Fig.Pro/E-6.4:

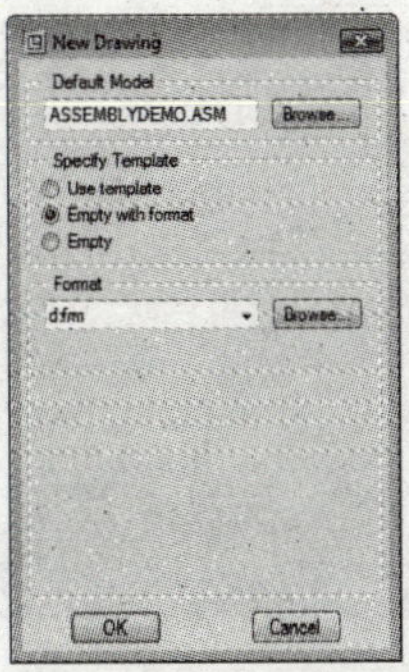

Fig.Pro/E-6.4

- **Empty:** Allows user to create a new Drawing file with a specified size and orientation, as shown in Fig.Pro/E-6.5:

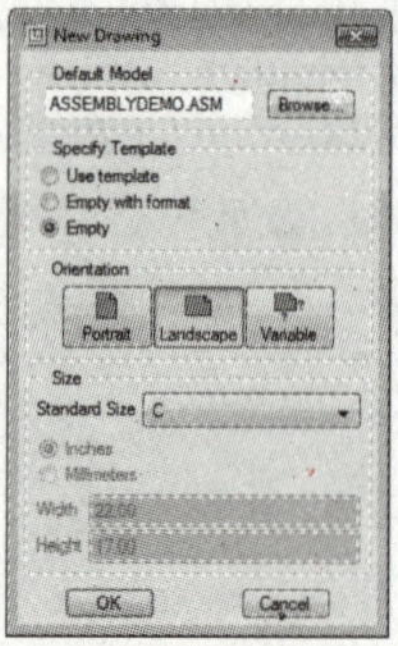

Fig.Pro/E-6.5

Once the desired option is selected from the **Specify Template** Area in the **New Drawing** dialog box, the next step is to apply it. In our case, we have *selected* **Use Template** option from the **Specify Template** Area.

5. *Click* the **OK** button in the **New Drawing** dialog box, as shown in Fig.Pro/E-6.6:

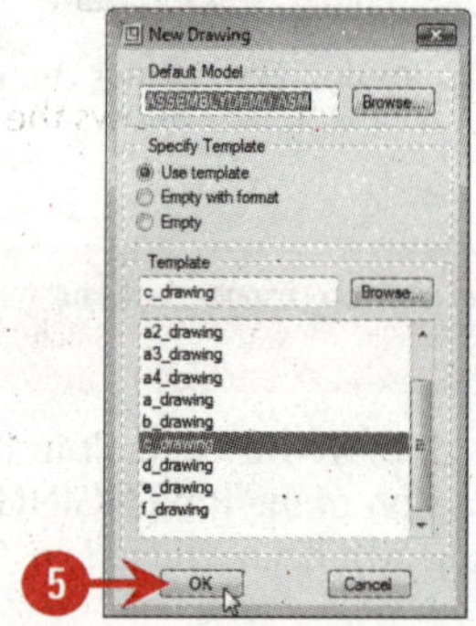

Fig.Pro/E-6.6

A Drawing file with a size and orientation specified in the pre-defined template selected is created, as shown in Fig.Pro/E-6.7:

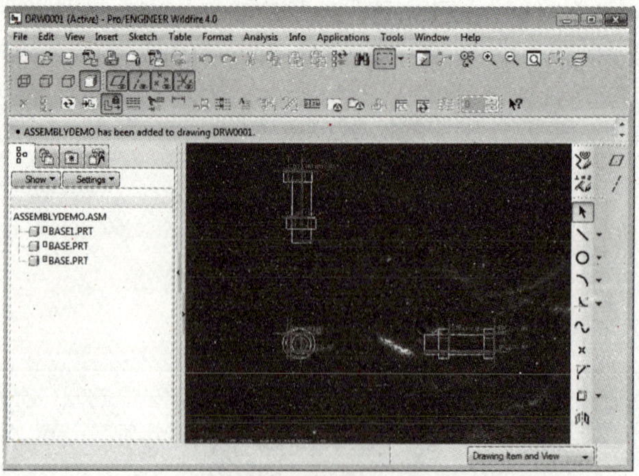

Fig.Pro/E-6.7

Now that the Drawing file has been created, you can now generate drawing views. Now, let's discuss about generating drawing views in the following section.

Generating Drawing Views

Drawing views convey the necessary information about a component for proper and easy manufacturing of the component. There are several different types of drawing views that you can generate in a Drawing file. However, it is important to generate the **General** view before placing any other view in the Drawing file. Pro/ENGINEER Wildfire 4.0 provides the following drawing views:

- General view
- Projection view
- Detailed view
- Auxiliary view
- Revolved view

Now, let's discuss each of these drawing views.

General view

A **General** view is the first view that must be placed in the Drawing file. The **General** view, by default, is placed in its default orientation. However, you can reorient it using the default datum planes or the predefined named views. Let's take a look at the steps required to generate a **General** view.

Perform the following steps to generate a **General** view:

1. *Create* a new Drawing file named **GENERALVIEW.DRW** with the default model named **ASSEMBLY.DEMO.ASM.**
2. In the Drawing file *Select* the **Insert→Drawing View→General** option from the **Menu Bar**, as shown in Fig.Pro/E-6.8:

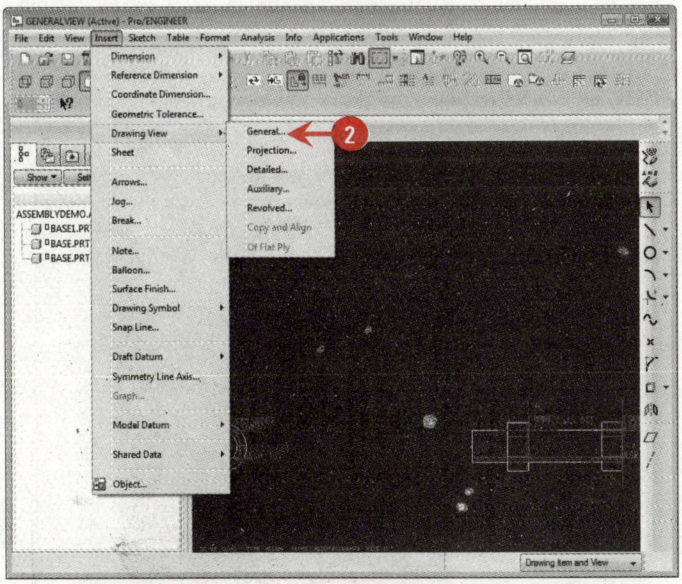

Fig.Pro/E-6.8

The **Select Combined State** dialog box appears and displays two options in the **Combined state names** collector, as shown in Fig.Pro/E-6.9:

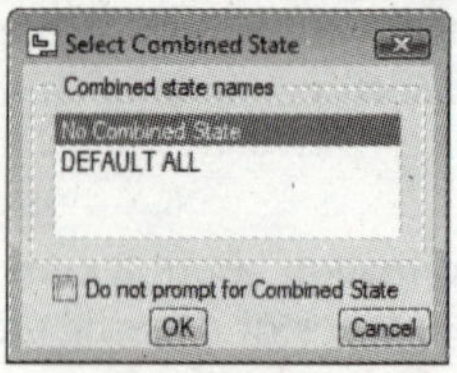

Fig.Pro/E-6.9

The **Select Combined State** dialog box provides two options:

- **No Combined State:** If you select the No Combined State option, then an unexploded view of the drawing view is placed in the Drawing file.
- **DEFAULT ALL:** If you select the DEFAULT ALL option, then an exploded view of the drawing view is placed in the Drawing file.

In our case, we want an unexploded view to be placed in the Drawing file. So, we select the **No Combined State** option.

Note

> The Select Combined State dialog box appears only when the selected model is an Assembly.

3. *Select* the **No Combined State** option from the **Combined state names collector** and *click* the **OK** button, as shown in Fig.Pro/E-6.10:

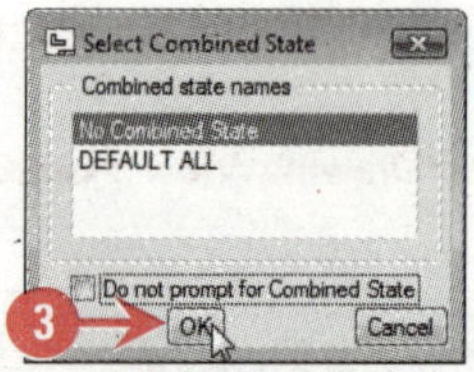

Fig.Pro/E-6.10

A prompt to select a CENTER POINT to place the drawing view appears (Fig.Pro/E-6.11).

4. *Select* a centre point on the **Drawing Area**, as shown in Fig.Pro/E-6.11:

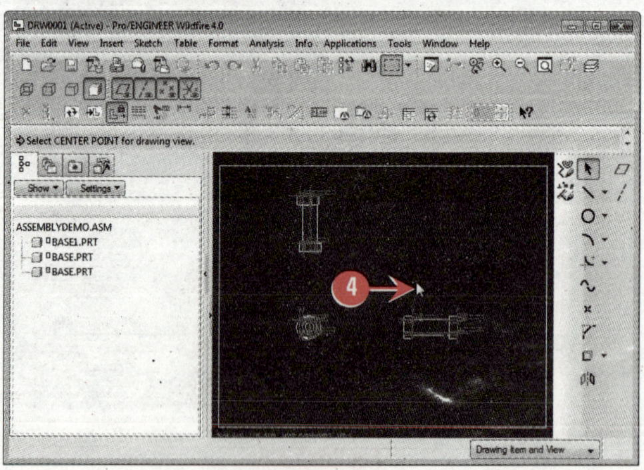

Fig.Pro/E-6.11

Once you select the **CENTER POINT**, the **Drawing View** dialog box also appears (Fig.Pro/E-6.12).

5. *Select* the Standard Orientation from the Model view names list box in the Drawing View dialog box as shown in Fig.Pro/E-6.12:

6. *Click* the Apply button, as shown in Fig.Pro/E-6.12:

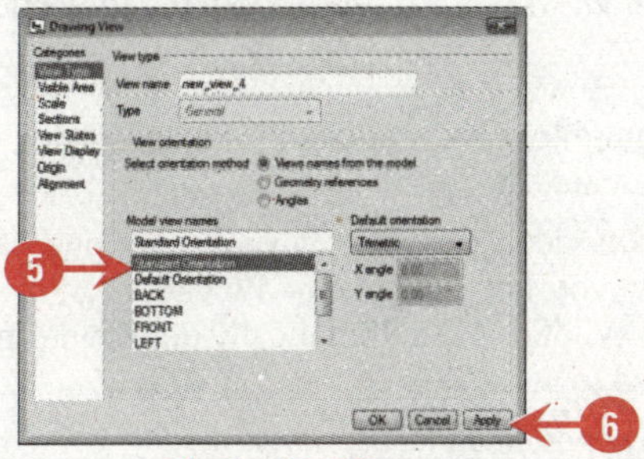

Fig.Pro/E-6.12

When you click the Apply button (Fig.Pro/E-6.12), the cancel button changes to the Close button (Fig.Pro/E-6.13).

In the Drawing View dialog box, the General option is selected by default in the Type drop down list. The Views names from the model radio button is selected by default, in the View orientation Area to display the various orientations to choose. Once, the Standard Orientation option is selected from Model view names list box, you need to click the Close button to apply the selected orientation on the General view.

7. Now, *click* the Close button to complete the process of generating the General view, as shown in Fig.Pro/E-6.13:

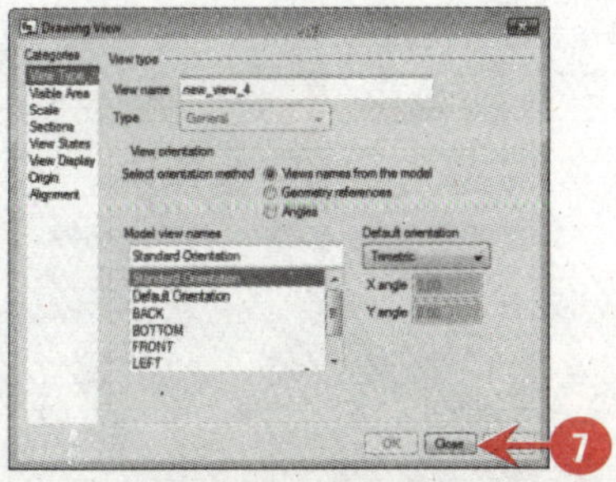

Fig.Pro/E-6.13

The General view of the selected model (ASSEMBLYDEMO.ASM) is generated, as shown in Fig.Pro/E-6.14:

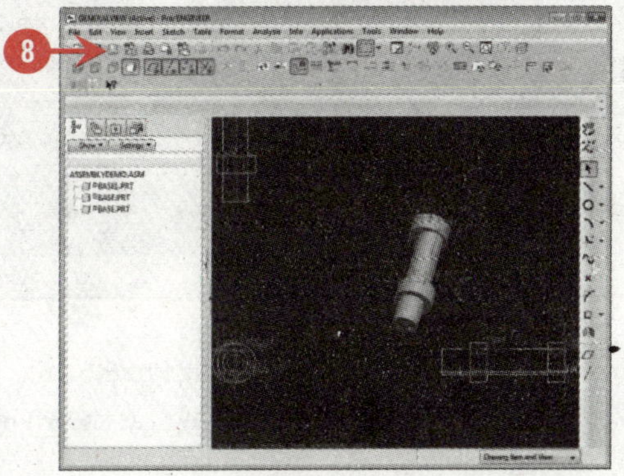

Fig.Pro/E-6.14

8. Now, save the file in the working directory by *clicking* the **Save** button (Fig.Pro/E-6.14) in the **File** toolbar.

Once the General view is generated, you can place the Projection view on it. The General view mostly acts as the parent view for the other views. Let's discuss about the Projection view in the following section.

Projection View

A **Projection** view is an orthographic view generated from an existing drawing view. In an orthographic view, all objects appear at the same scale.

Before generating the **Projection** view, you must have at least one parent view present on the Drawing file. To generate a **Projection** view, you need to specify a location on the **Drawing Area** with respect to another drawing view. Pro/ENGINEER Wildfire 4.0 automatically determines how to project the generated view. Let's take a look at the steps required to generate a **Projection** view.

Perform the following steps to generate a **Projection** view:

1. *Start* a new Drawing file named **PROJECTIONVIEW.DRW** with the default model named **ASSEMBLYDEMO.ASM.** Since, at least one parent view should be present before generating a **Projection** view; we will generate a **General** view first.
2. *Generate* the **General** view. A **Projection** view can now be placed in the Drawing file.
3. *Select* the **Insert →Drawing View→Projection** option from the **Menu Bar**, as shown in Fig.Pro/E-6.15:

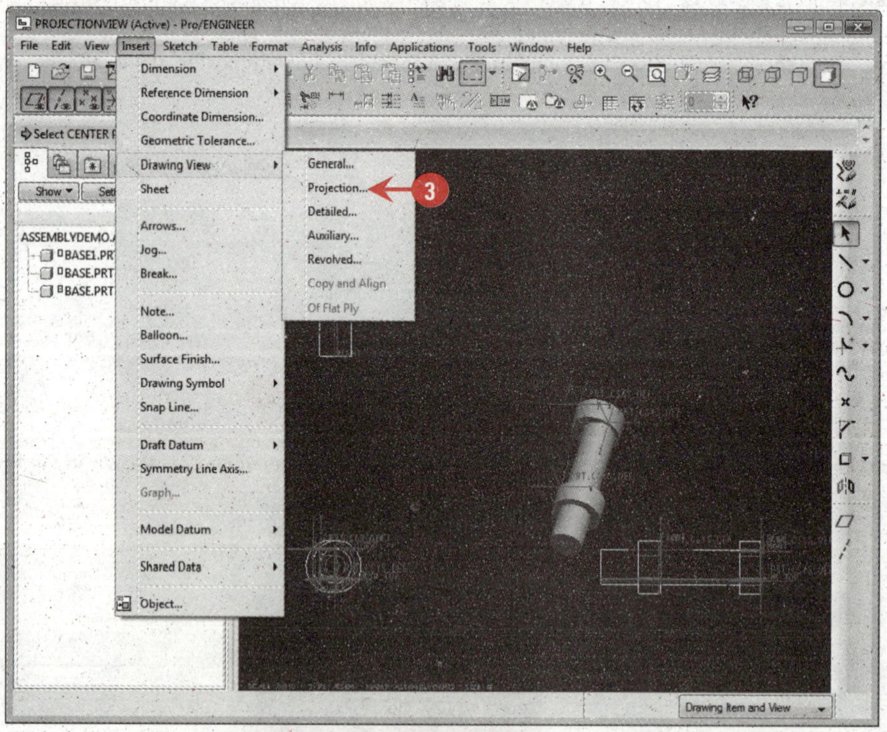

Fig.Pro/E-6.15

A box with a yellow color outline appears in the Drawing file (Fig.Pro/E-6.16).

4. *Move* the box to a location where you need to place the view, as shown in Fig.Pro/E-6.16:

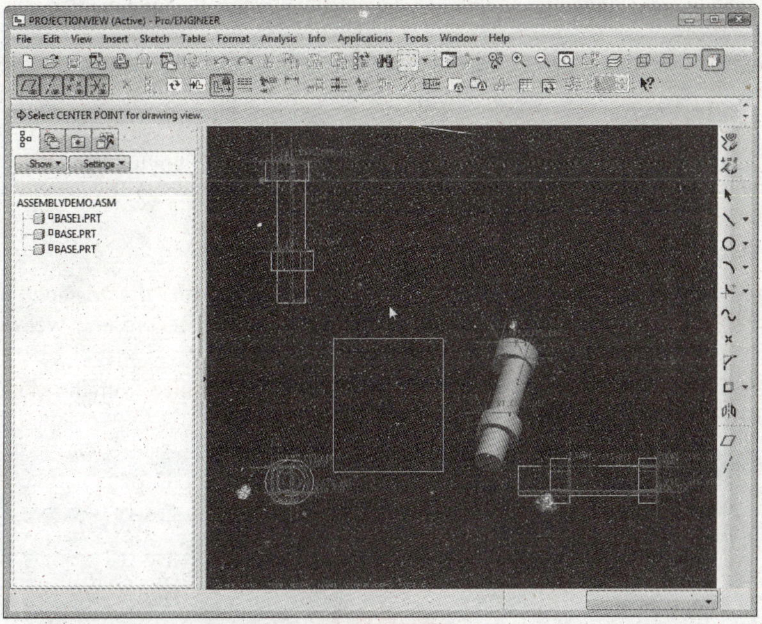

Fig.Pro/E-6.16

When you click on the desired location, the **Projection** view of the selected model (ASSEMBLYDEMO.ASM) is generated, as shown in Fig.Pro/E-6.17:

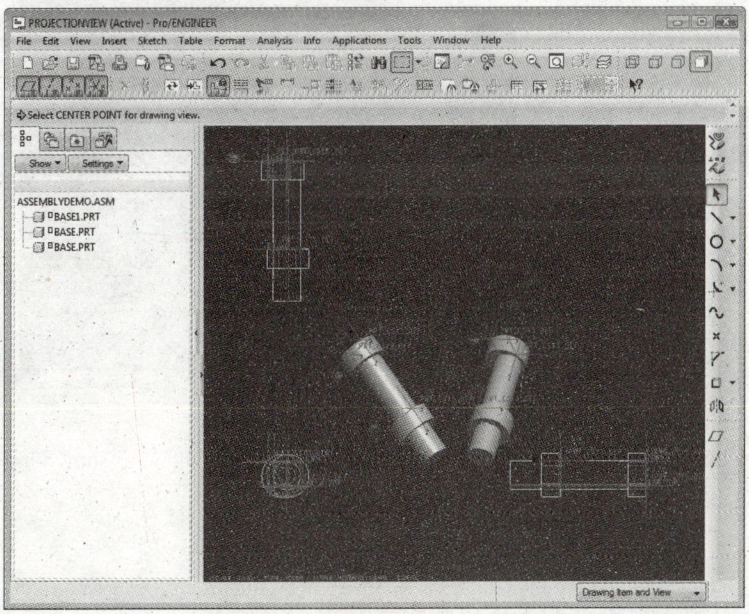

Fig.Pro/E-6.17

5. Now, save the file in the working directory. Once the Projection view has been generated, you can place other views on it.

Let's discuss about the Detailed view in the following section.

Detailed View

A **Detailed** view is a view that displays a particular portion of an existing view in a larger viewing scale to help the user clearly and easily see the geometry and dimensions in the view. Dimensions are discussed in detail in the section, *Inducing Dimensions and Tolerance* later in the chapter.

Let's look at the steps required to generate a **Detailed** view. Before generating the **Detailed** view, you must have at least one parent view present on the Drawing file. In our case, we will generate a **Detailed** view on the existing **General** view present in the Drawing file.

Perform the following steps to generate a **Detailed** view:

1. *Start* a new Drawing file named **DETAILEDVIEW.DRW** with the default model named as **ASSEMBLYDEMO.ASM.** Since, at least one parent view should be present; we will first generate a **General** view, before generating a **Detailed** view.
2. *Generate* the **General** view. *Select* the **Insert→Drawing View→Detailed** from the **Menu Bar**, as shown in Fig.Pro/E-6.18:

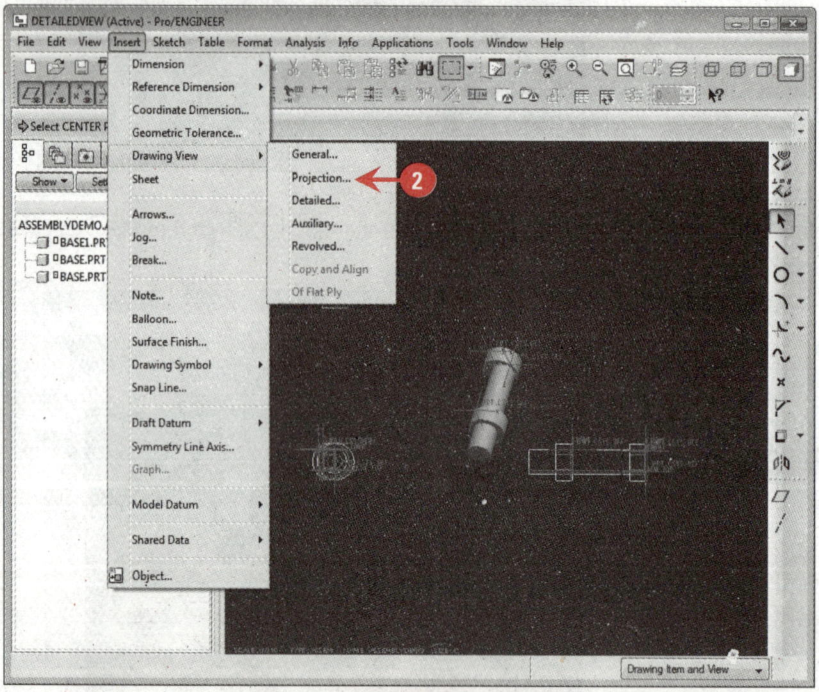

Fig.Pro/E-6.18

The **Select** dialog box appears to select a point for detail on an existing view, as shown in Fig.Pro/E-6.19:

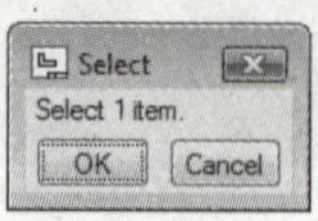

Fig.Pro/E-6.19

When you select a point on an existing view, the **Select** dialog box automatically gets closed.

You are prompted to sketch a spline around the point selected of an existing view that you want to enlarge.

3. *Draw* the spline in such a way that it covers the point selected, as shown in Fig.Pro/E-6.20:

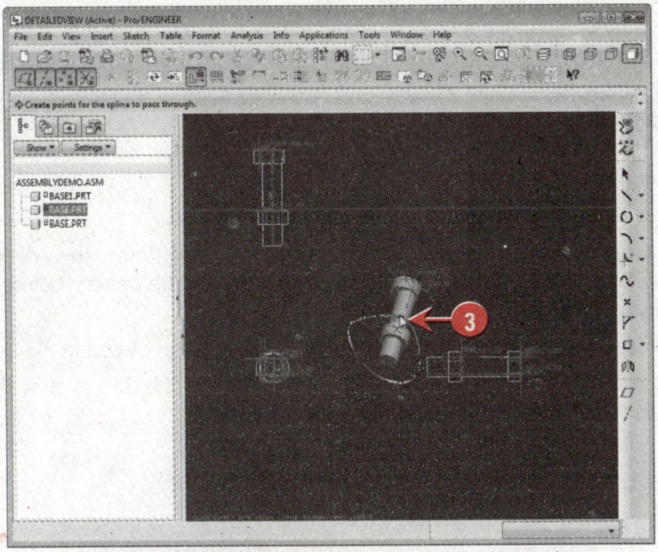

Fig.Pro/E-6.20

Note

Spline must not be drawn using the Sketcher toolbar located at the right side of the Drawing Area. Click the Drawing Area to start drawing the spline.

4. *Click* the middle mouse button to complete the spline and move the cursor to the location where you need to place the view, as shown in Fig.Pro/E-6.21:

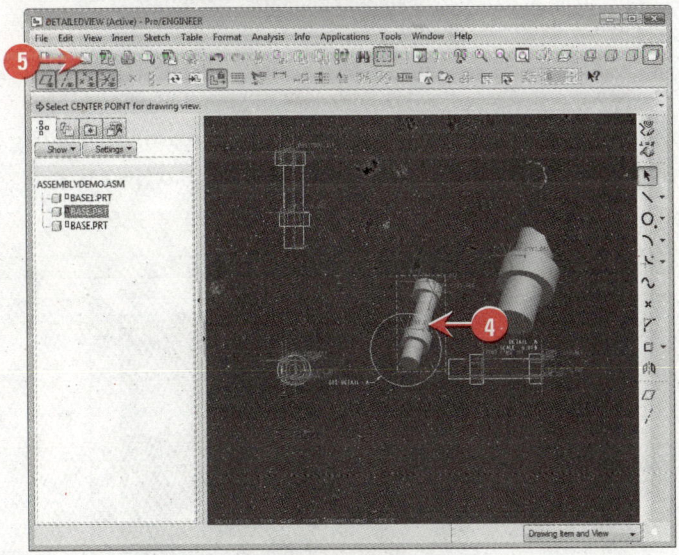

Fig.Pro/E-6.21

5. *Save* the file in the working directory.

Let's discuss about the Auxiliary view in the following section.

Auxiliary View

An Auxiliary view is similar to the Projection view. The only difference being that the view is generated at right angles to a selected surface, datum plane, or a specified edge of an existing view. Before generating the Auxiliary view, at least one parent view must be present on the Drawing file. In our case, we are considering the Projection view as the parent view. Let's generate an Auxiliary view on the existing Projection view present in the Drawing file.

Perform the following steps to generate an Auxiliary view:

1. *Start* a new Drawing file named AUXILIARYVIEW.DRW with the default model named as ASSEMBLYDEMO.ASM. Since, at least one parent view should be present before generating an Auxiliary view; we will first generate a Projection view.
2. *Generate* the Projection view. Now, an Auxiliary view can now be placed in the Drawing file.
3. *Select* the Insert→Drawing View→Auxiliary option from the Menu Bar, as shown in Fig.Pro/E-6.22:

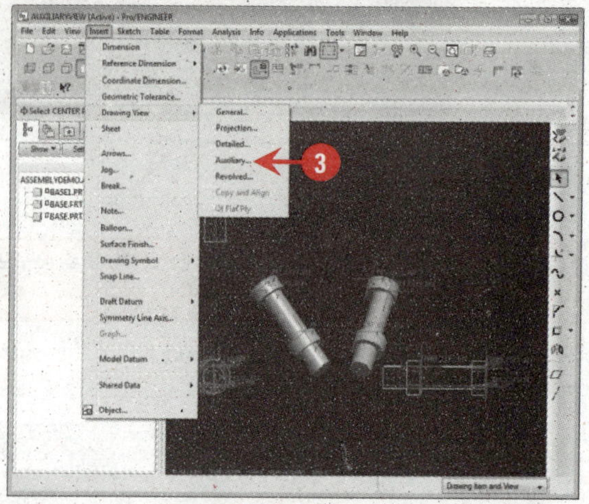

Fig.Pro/E-6.22

The Select dialog box appears to select an item (edge) on an existing view (Fig.Pro/E-6.19).

4. *Select* the desired edge on the Projection view, as shown in Fig.Pro/E-6.23:

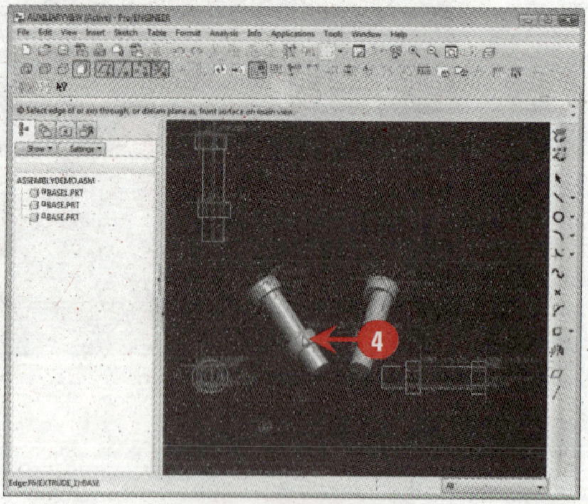

Fig.Pro/E-6.23

A box with a yellow color outline appears in the **Drawing Area**.

5. *Move* the box to the location where you want to place the view, as shown in Fig.Pro/E-6.24:

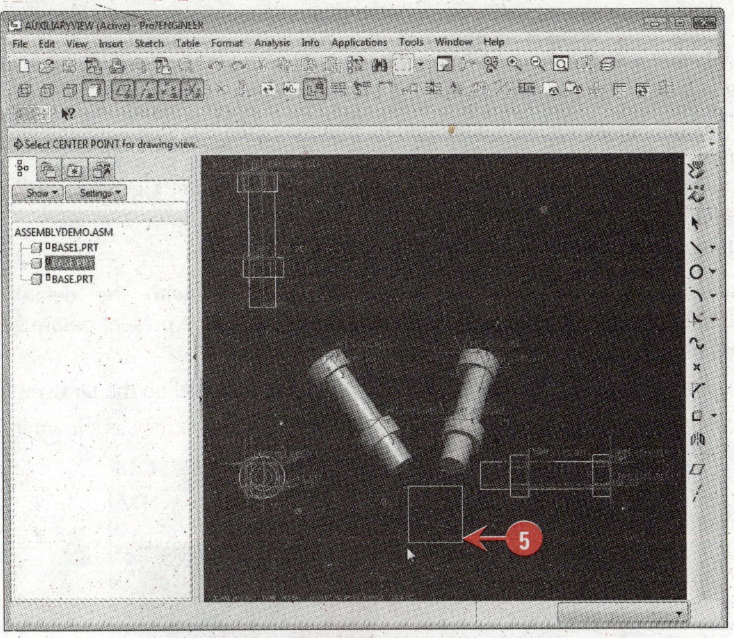

Fig.Pro/E-6.24

When you click on the desired location, the **Auxiliary** view of the selected view is generated, as shown in Fig.Pro/E-6.25:

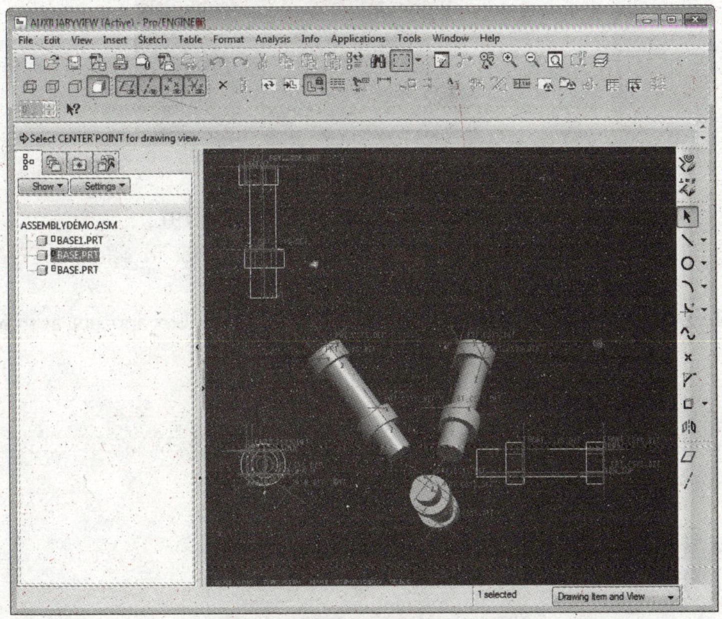

Fig.Pro/E-6.25

6. *Save* the file in the current directory.

Now, in the following section, let's discuss the Revolved view.

Revolved View

A **Revolved** view is that section of an existing view which has been revolved to 90 degrees around a cutting plane. A cutting plane is a plane that shows the cutting pattern of a section. An area cross-section of the existing view cut by the cutting plane is known as section.

At least one parent view must be present on the Drawing file before generating the **Revolved** view. In our case, we are considering the Projection view as the parent view. Let's generate a **Revolved** view on the existing **Projection** view present in the Drawing file.

Perform the following steps to generate a **Revolved** view:

1. *Start* a new Drawing file named **REVOLVEDVIEW.DRW** with the default model named as **ASSEMBLYDEMO.ASM.** Since, at least one parent view should be present before generating a **Revolved** view; we will first generate a **Projection** view.
2. *Generate* the **Projection** view. A **Revolved** view can now be placed on the Drawing file.
3. *Select* the **Insert→Drawing View→Revolved** option from the **Menu Bar**, as shown in Fig.Pro/E-6.26:

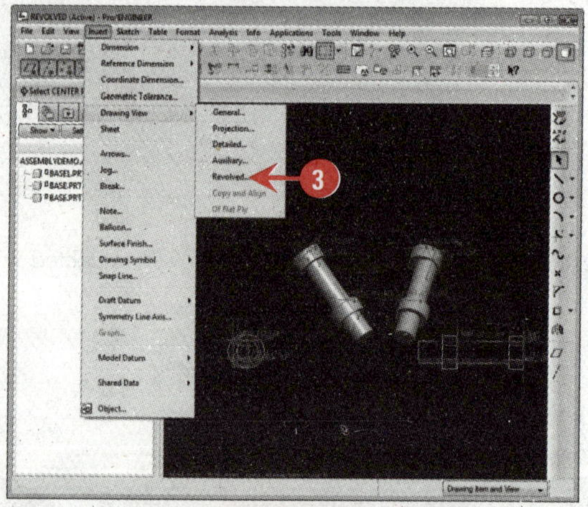

Fig.Pro/E-6.26

The **Select** dialog box appears to select a **Projection** view (Fig.Pro/E-6.19).

Once a **Projection** view on the **Drawing Area** is selected, a prompt to select a **CENTER POINT** is also displayed.

4. *Click* the center point in the drawing area. The **Drawing View** dialog box appears, as shown in Fig.Pro/E-6.27:

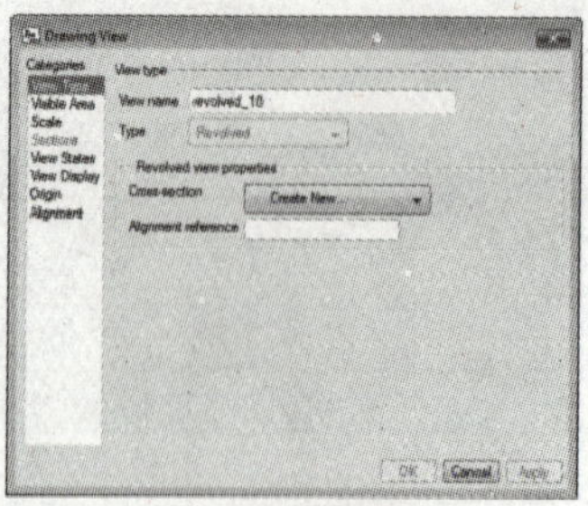

Fig.Pro/E-6.27

Apart from **Drawing View** dialog box, the **Menu Manager** menu also appears (Fig.Pro/E-6.28).

5. *Select* **Planar→Single→Done** from the **XSEC CREATE** sub-menu, as shown in Fig.Pro/E-6.28:

Fig.Pro/E-6.28

A text box appears in the message area that prompts you to specify a name for the cross-section (Fig.Pro/E-6.29).

6. *Specify* the name of the cross-section in the **Message Input** window (Fig.Pro/E-6.29). We specified crosssection as the name, as shown in Fig.Pro/E-6.29:

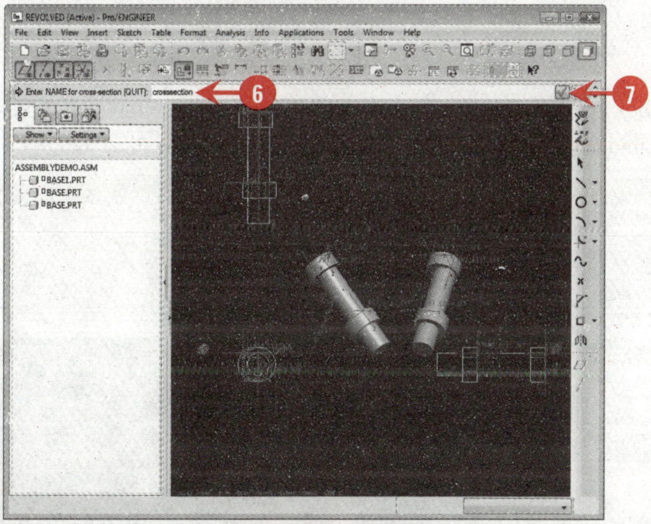

Fig.Pro/E-6.29

7. *Click* the **Build feature** button (☑) on the **Message Input** window (Fig.Pro/E-6.29). The Menu Manager menu with the **SETUP PLANE** sub-menu appears (Fig.Pro/E-6.30) to either select or create a datum plane on the selected view.

8. *Select* the **Make Datum** option from the **SETUP PLANE** sub-menu to create a datum, as show in Fig.Pro/E-6.30:

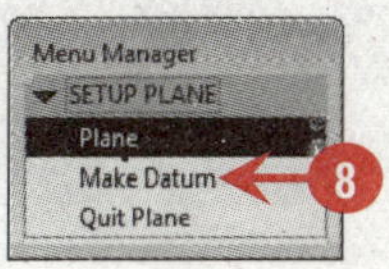

Fig.Pro/E-6.30

The **DATUM PLANE** sub-menu appears under the **Make Datum** menu (Fig.Pro/E-6.31).

9. *Click* the **Normal** option from the **DATUM PLANE** to create a datum plane normal or perpendicular to the selected view, as shown in Fig.Pro/E-6.31:

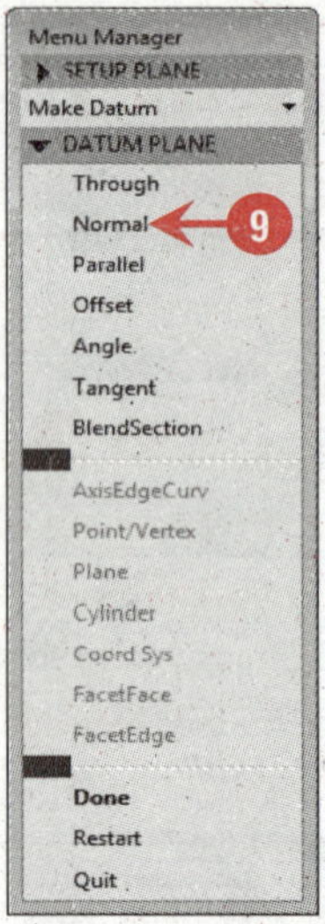

Fig.Pro/E-6.31

The **Select** dialog box appears to select a point (edge) on a selected view (Fig.Pro/E-6.19).

10. *Select* the desired edge on the **Projection** view, as shown in Fig.Pro/E-6.32:

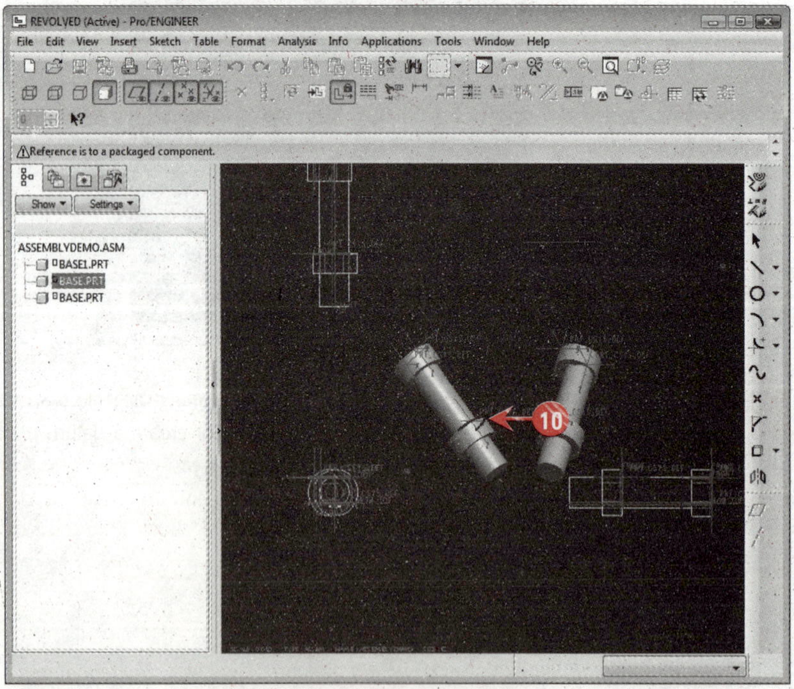

Fig.Pro/E-6.32

11. *Select* the **Done** option in the **Menu Manager** menu to finish the creation of the datum plane, as shown in Fig.Pro/E-6.33:

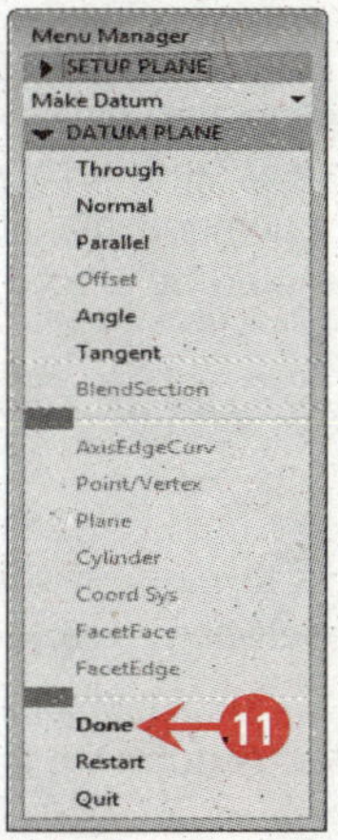

Fig.Pro/E-6.33

The DATUM PLANE sub-menu disappears.

12. *Click* the **Apply** button in the **Drawing View** dialog box, as shown in Fig.Pro/E-6.34:

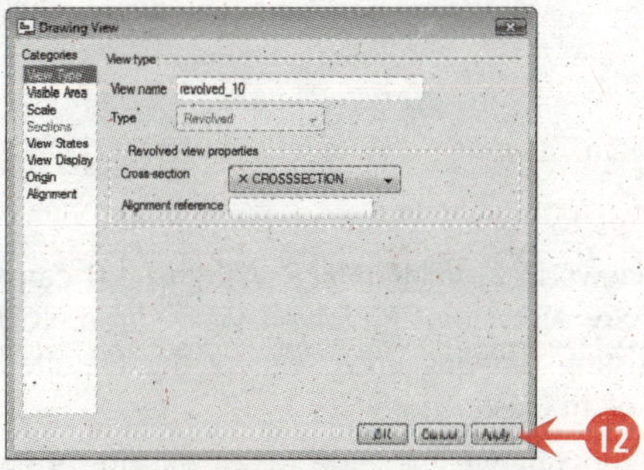

Fig.Pro/E-6.34

13. *Click* the **Close** button to exit the **Drawing View** dialog box. The **Revolved** view of the selected model (ASSEMBLYDEMO.ASM) is generated, as shown in Fig.Pro/E-6.35:

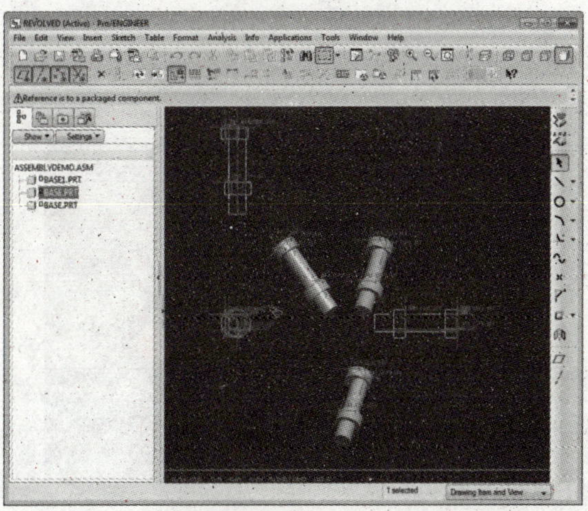

Fig.Pro/E-6.35

14. *Save* the file in the working directory.

Till now, we discussed the drawing views supported by Pro/ENGINEER Wildfire 4.0 such as General view, Projection view, Detailed view, Auxiliary view, and Revolved view. These drawing views are required to convey all the necessary information about a component so that the component can be manufactured easily. A General view must be the first view that should be placed in the Drawing file so that other drawing views can be generated on top of it.

Next, let's discuss about modifying these drawing views.

Modifying the Drawing Views

Pro/ENGINEER Wildfire 4.0 allows the users to easily modify existing drawing views. Most of the options available for modifying drawing views are selected from the context menu, which is displayed when you right-click a drawing view.

Pro/ENGINEER Wildfire 4.0 allows the users to perform the following modifications on the drawing views:

- Changing the type of a view
- Changing the scale of a view
- Reorienting the view
- Moving a view
- Erasing a view
- Deleting a view

Now, let's discuss each of them.

Changing the Type of a View

There are several drawing views in Pro/ENGINEER Wildfire 4.0 can be , such as General view, Projection view, Detailed view, Auxiliary view, and Revolved view. Pro/ENGINEER Wildfire 4.0 allows the user to change the type of any existing drawing view, except the Detailed view. Let's take a look at the steps required to change the type of an existing view.

Perform the following steps to change the type of an existing view. In our case, we have selected the Projection view to change its type.

1. *Open* the **PROJECTIONVIEW.DRW** file and *Right-click* the Projection view to invoke the context menu, as shown in Fig.Pro/E-6.36:

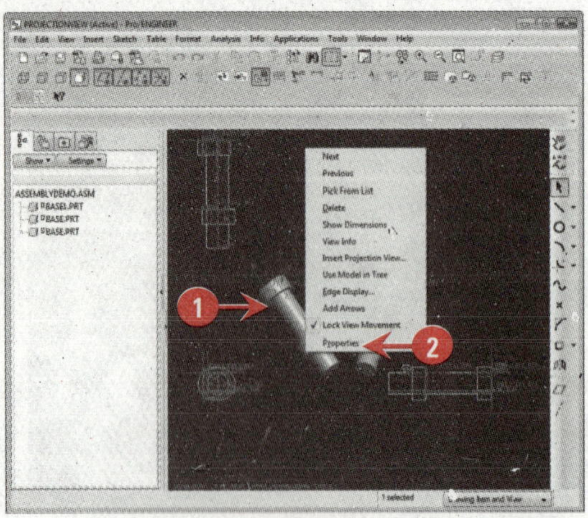

Fig.Pro/E-6.36

2. *Select* the **Properties** option from the context menu (Fig.Pro/E-6.36). The **Drawing View** dialog box appears and displays various options for modifying the view (Fig.Pro/E-6.37).
3. *Select* the desired view type from the **Type** drop down list in the **View Type option in the Categories** Area. In our case, we have selected the General view, as shown in Fig.Pro/E-6.37:

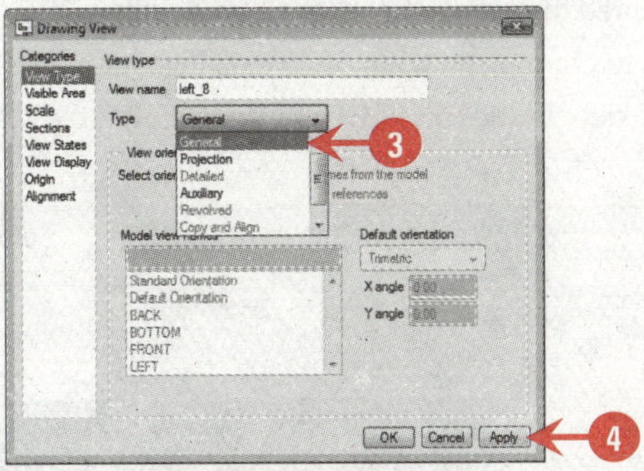

Fig.Pro/E-6.37

4. *Click* the **Apply** button to apply the selected view type to the current Projection view (Fig.Pro/E-6.37).

When you *click* the **Apply** button, the **Cancel** button changes to the **Close** button.

5. *Click* the **Close** button to exit the **Drawing View** dialog box. The selected **Projection** view is successfully changed into **General** view, as shown in Fig.Pro/E-6.38:

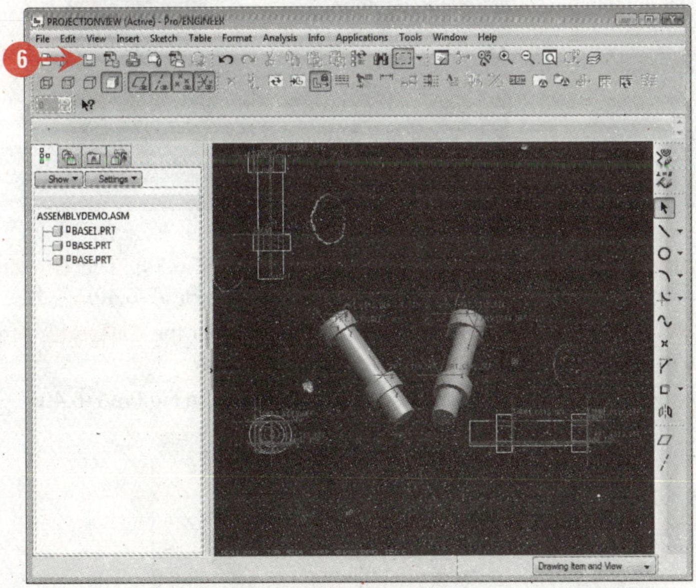

Fig.Pro/E-6.38

6. *Save* the file in the working directory.

In the next section, we discuss about changing the viewing scale of a drawing view.

Note

> Users can modify the view type of only the General, Projection, and Auxiliary view. These views are also used to change the type for the view, which has been selected to get modified.

Changing the Scale of a View

Pro/ENGINEER Wildfire 4.0 uses the drawing scale to control the scale of an existing view. In Pro/ENGINEER Wildfire 4.0, users are allowed to modify and change the scale of either an existing General view or a Detailed view only. Users can change the scale of an existing view using the Drawing View dialog box. Let's take a look at the steps required to change the scale of an existing view.

Perform the following steps to change the scale of an existing view:

1. *Right-click* the desired view in the **PROJECTIONVIEW.DRW** file. A context menu appears, as shown in Fig.Pro/E-6.39:

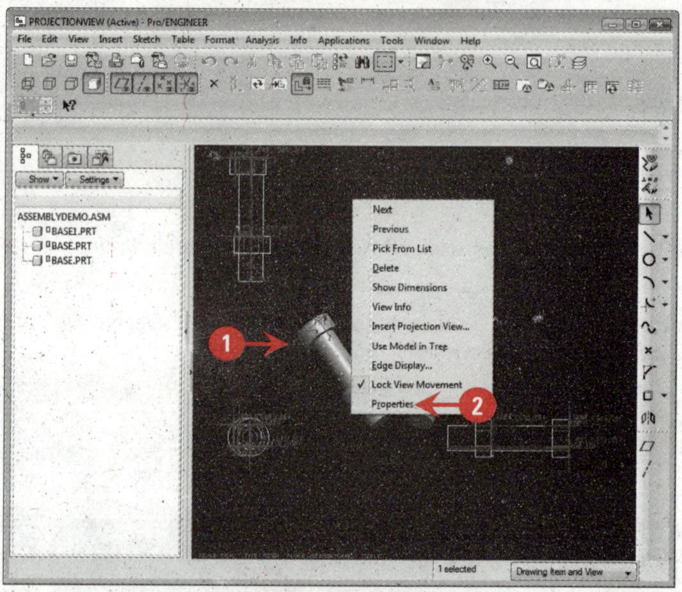

Fig.Pro/E-6.39

2. *Select* the **Properties** option from the shortcut menu (Fig.Pro/E-6.39). The **Drawing View** dialog box appears and displays various options for modifying the view (Fig.Pro/E-6.40).
3. *Select* the **Custom scale** radio button from the **Scale** menu under the **Categories** Area to specify a value (Fig.Pro/E-6.40). In our case, 0.020 is specified as the scale value.
4. *Click* the **Apply** button to confirm the specified value, as shown in Fig.Pro/E-6.40:

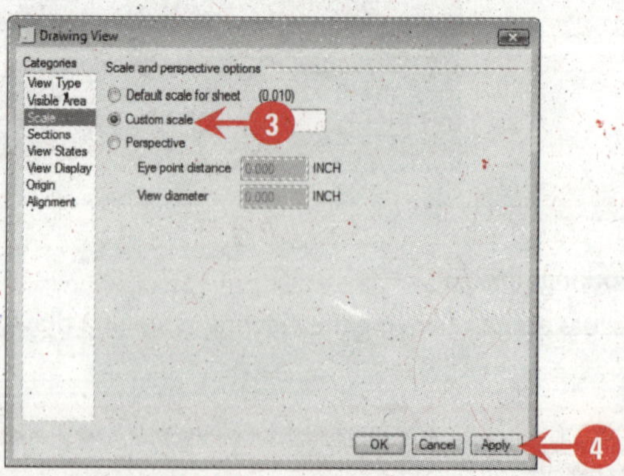

Fig.Pro/E-6.40

When you click the **Apply** button, the **Cancel** button changes to the **Close** button.

5. *Click* the **Close** button to exit the **Drawing View** dialog box (Fig.Pro/E-6.40). The view scale of the selected **General** view is successfully changed, as shown in Fig.Pro/E-6.41:

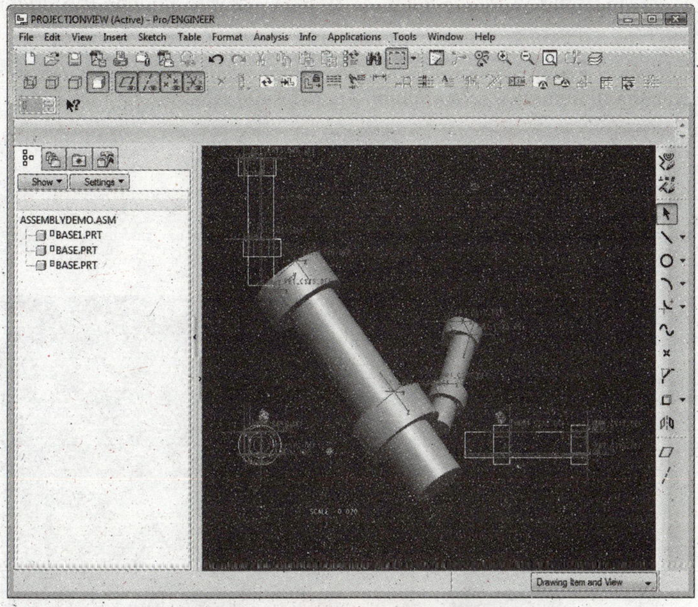

Fig.Pro/E-6.41

6. *Save* the file in the working directory.

Now, let's discuss about changing the orientation of a drawing view in the following section.

Reorienting the View

Pro/ENGINEER Wildfire 4.0 allows users to change the orientation of an existing view using the Drawing View dialog box. Let's take a look at the steps required to change the orientation of an existing view.

In our case, we have selected the General view to change its orientation. Perform the following steps to change the orientation an existing view:

1. *Right-click* the **General** view in the **PROJECTIONVIEW.DRW** file and open the **Drawing View** dialog box, as shown in Fig.Pro/E-6.42:

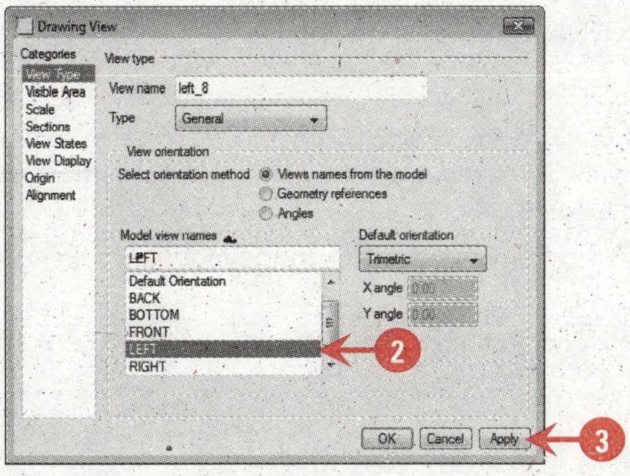

Fig.Pro/E-6.42

The **Drawing View** dialog box appears and displays various options for changing the orientation of the view. The Standard orientation is the option that is selected by default under the **Model view names** menu in the Drawing view dialog box.

2. *Select* the **LEFT** option from the **Model view names** collector to change the orientation of the selected view (Fig.Pro/E-6.42).
3. *Click* the **Apply** button (Fig.Pro/E-6.42). The orientation of the selected view is successfully changed from Standard to LEFT orientation, as shown in Fig.Pro/E-6.43:

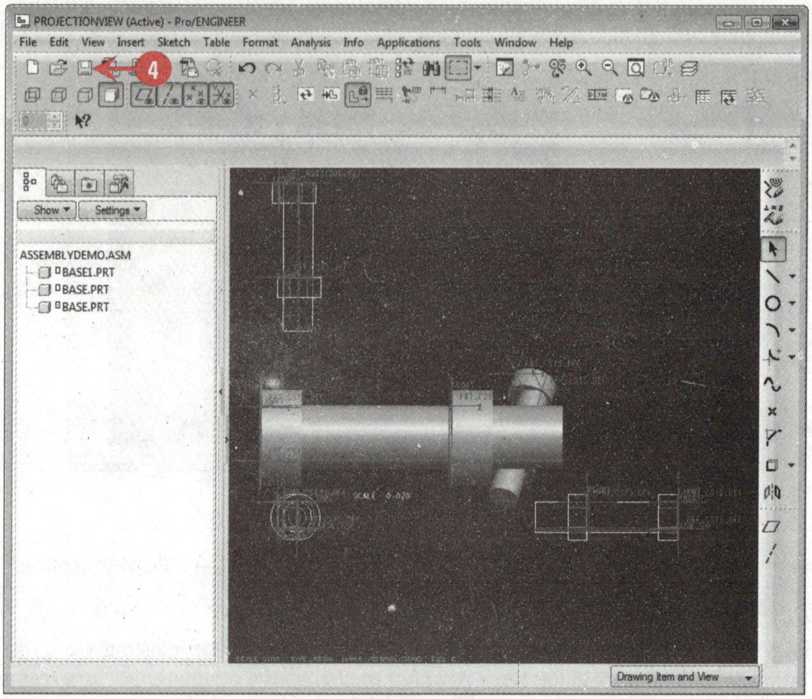

Fig.Pro/E-6.43

When you click the **Apply** button, the **Cancel** button changes to the **Close** button.

4. *Click* the **Close** button to exit the **Drawing** view dialog box.
5. *Save* the file in the working directory.

In the following section, let's now discuss about moving a drawing view from one location to another.

Moving a View

Pro/ENGINEER Wildfire 4.0 allows the user to easily move a view from one location to another in a Drawing file. However, if a parent view is moved, then the child views also move accordingly. Let's take a look at the steps required to move an existing view.

In our case, we have selected the **General** view to move a view. Perform the following steps to move an existing view:

1. *Right-click* the desired view in the **PROJECTIONVIEW.DRW** file. A context menu appears, as shown in Fig.Pro/E-6.44:

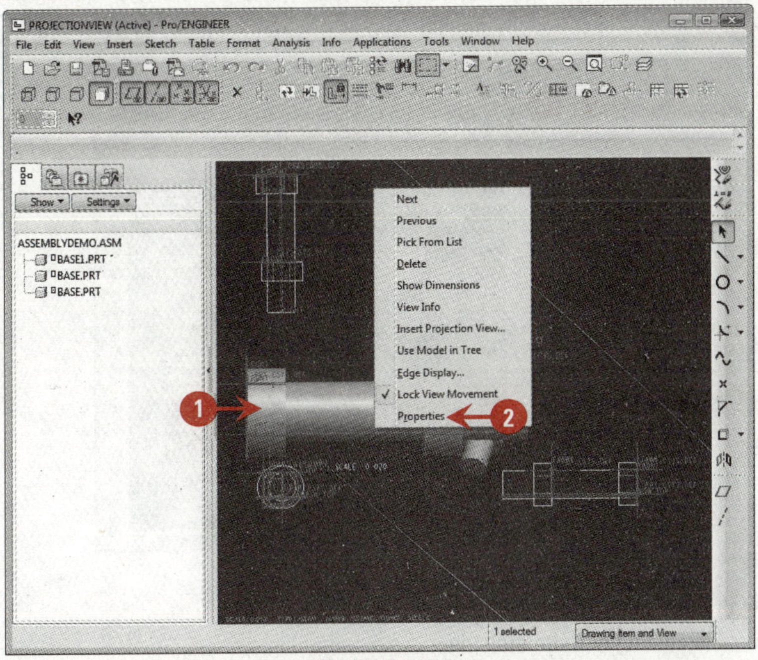

Fig.Pro/E-6.44

2. *Uncheck* the **Lock View Movement** option from the context menu (Fig.Pro/E-6.44). The selected view gets unlocked. Now the view can be moved to a desired location in the Drawing file.
3. *Drag* the selected view to a desired location, as shown in Fig.Pro/E-6.45:

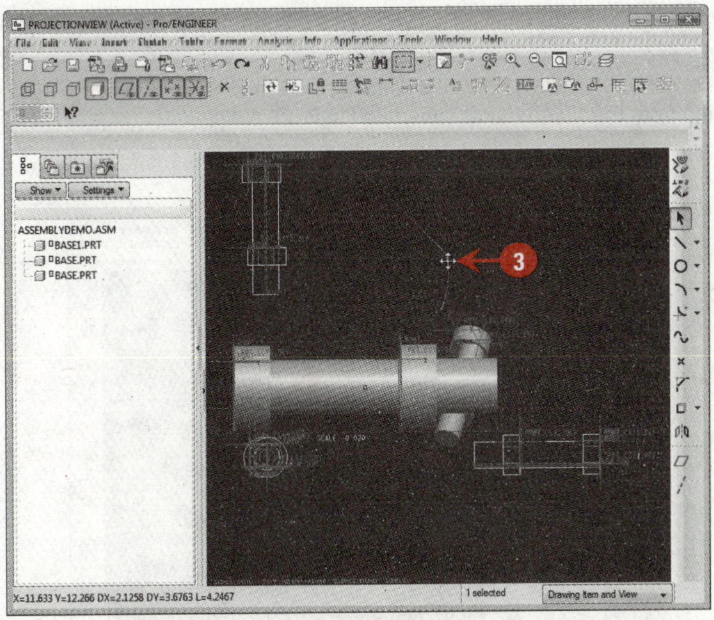

Fig.Pro/E-6.45

The selected **General** view is successfully moved from its original location, as shown in Fig.Pro/E-6.46:

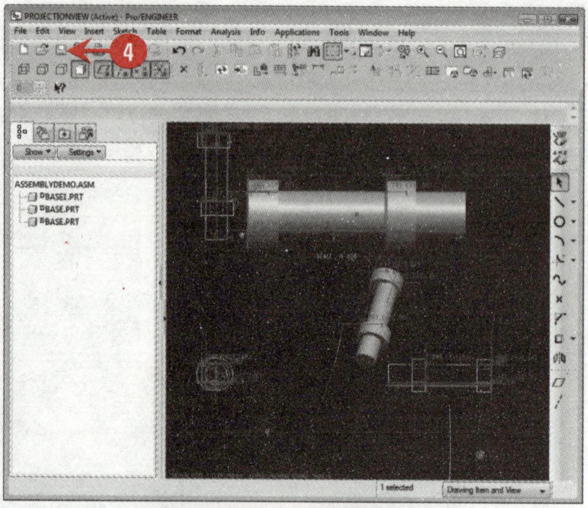

Fig.Pro/E-6.46

4. *Save* the file in the working directory.

Now, let's discuss the erasing of a drawing view in the following section.

Erasing a View

Pro/ENGINEER Wildfire 4.0 allows the user to erase a view from the Drawing file. Erasing a view temporarily removes the selected view from the Drawing file but the erased view remains in the memory of the drawing file. Let's take a look at the steps required to erase the drawing view.

In our case, we have selected the **General** view to erase. Perform the following steps to erase the selected **General** view:

1. *Open* the **PROJECTIONVIEW.DRW** file and Select the **View→Drawing Display→Drawing View Visibility** option from the **Menu Bar** as shown in Fig.Pro/E-6.47:

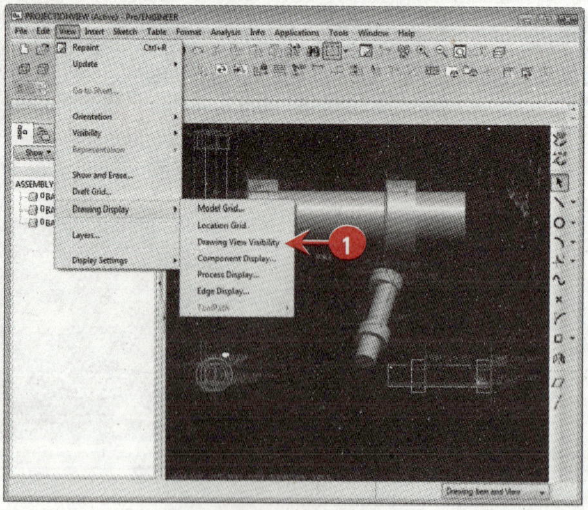

Fig.Pro/E-6.47

The **Menu Manager menu** appears, as shown in Fig.Pro/E-6.48:

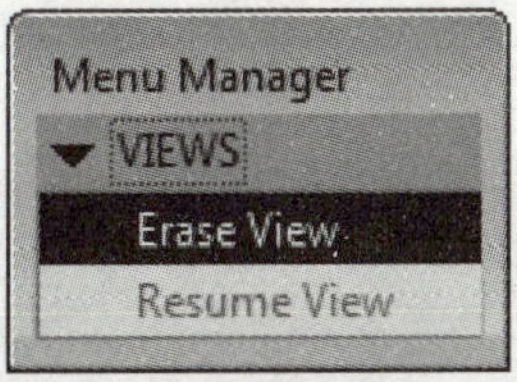

Fig.Pro/E-6.48

Apart from the **Menu Manager** menu, the **Select** dialog box also appears to select an existing view to erase, as shown in Fig.Pro/E-6.49:

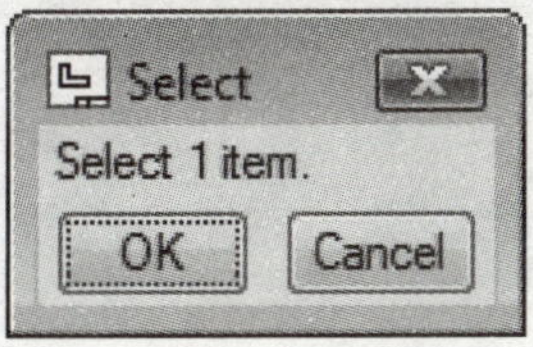

Fig.Pro/E-6.49

2. *Select* the desired view in the Drawing file to erase, as shown in Fig.Pro/E-6.50:

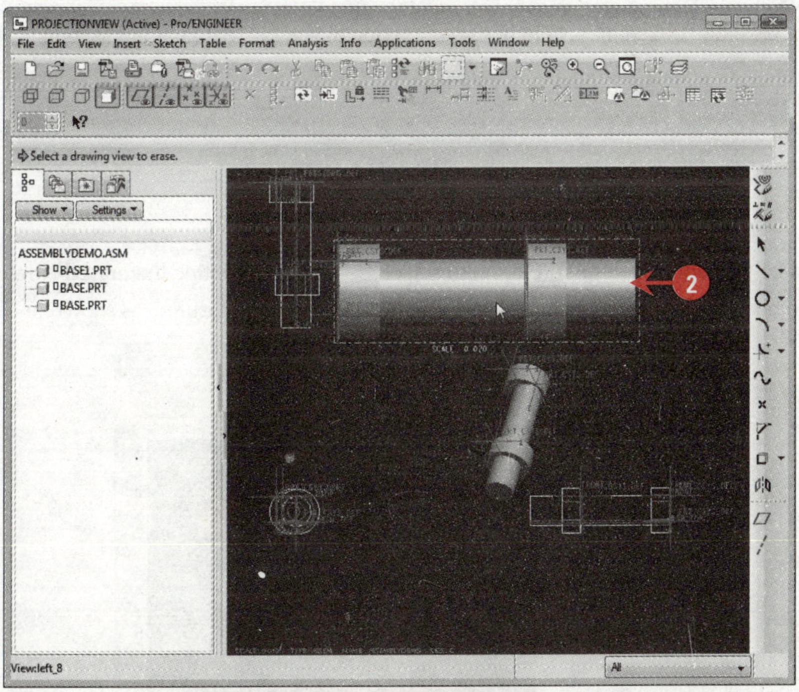

Fig.Pro/E-6.50

The selected view is erased and a box appears in place of the erased view (Fig.Pro/E-6.51).

3. *Click* the **OK** button in the **Select** dialog box to confirm the action performed (Fig.Pro/E-6.49).

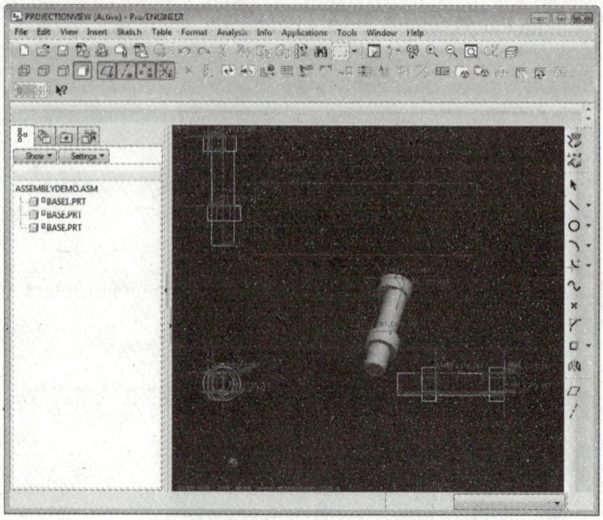

Fig.Pro/E-6.51

You can resume the view that has been temporarily removed from the Drawing file by performing the following steps:

1. *Select* the **Resume View** option from the **Menu Manager** menu, as shown in Fig.Pro/E-6.52:

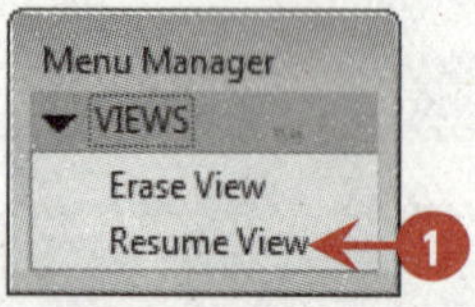

Fig.Pro/E-6.52

The **Select** dialog box appears (Fig.Pro/E-6.49) to select the box representing the erased view.

2. *Select* the box representing the erased view in the Drawing file to resume, as shown in Fig.Pro/E-6.53:

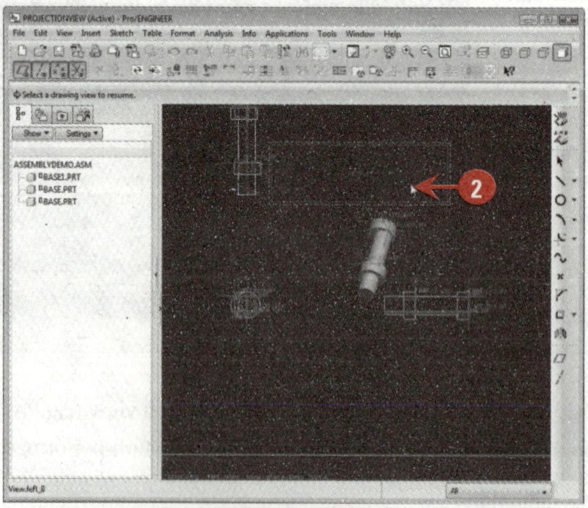

Fig.Pro/E-6.53

3. *Select* the **Done Sel** option from the **Menu Manager** menu, as shown in Fig.Pro/E-6.54:

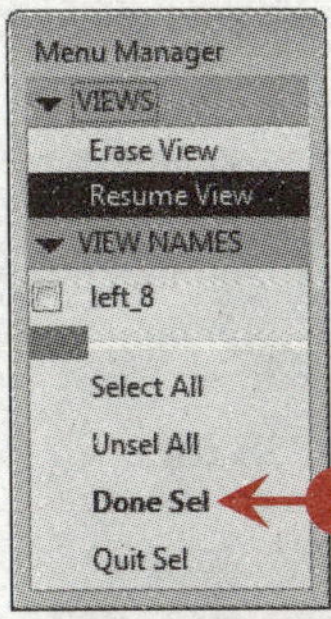

Fig.Pro/E-6.54

The erased view is successfully resumed in the Drawing file, as shown in Fig.Pro/E-6.55:

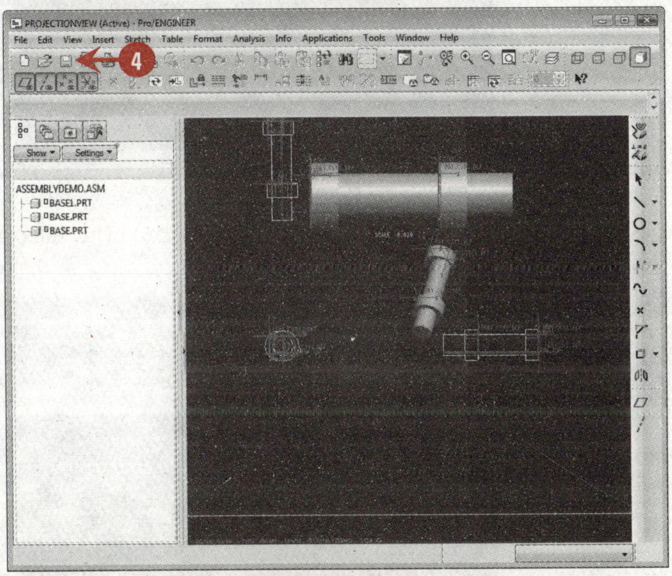

Fig.Pro/E-6.55

Note

The Menu Manager menu remains unclosed.

4. *Save* the file in the working directory.

Let's discuss about deleting a drawing view in the following section.

Deleting a View

Pro/ENGINEER Wildfire 4.0 allows the user to delete a view from the Drawing file. Deleting a view permanently removes the selected view from the memory of the drawing. Let's take a look at the steps required to delete the drawing view.

Perform the following steps to delete the selected **General** view:

1. *Open* the **PROJECTIONVIEW.DRW** file and *right-click* the desired view. A context menu appears, as shown in Fig.Pro/E-6.56:

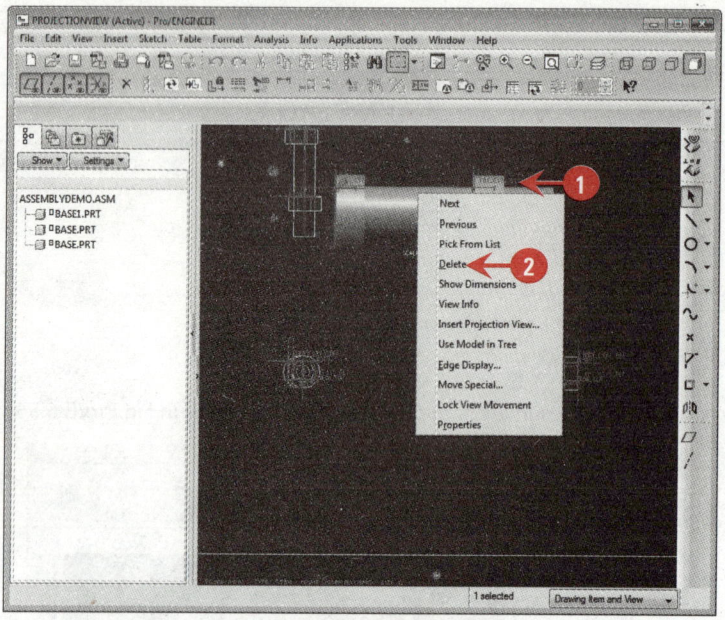

Fig.Pro/E-6.56

2. *Select* the **Delete** option from the context menu (Fig.Pro/E-6.56). The selected view is successfully deleted from the Drawing file, as shown in Fig.Pro/E-6.57:

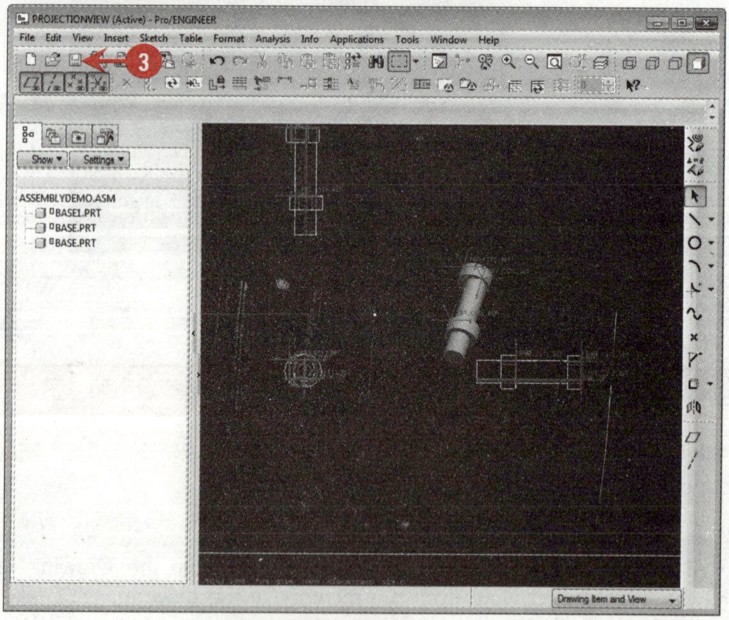

Fig.Pro/E-6.57

3. *Save* the file in the working directory.

Till now, we have discussed about modifying the existing drawing views. Let's now discuss about adding dimensions and tolerances to the drawing views

Inducing Dimensions and Tolerance

Dimensions are the minimum number of coordinates that are required to specify every point in the drawing views. Pro/ENGINEER Wildfire 4.0 allows users to generate dimensions while the drawing views are being generated. Dimensions are assigned to the features in the selected view. These dimensions are associative in nature. This associative nature ensures that any modifications made in the dimensions in the Drawing file automatically updates its Part file. Apart from adding dimensions, tolerances are also induced in the dimensions. It is difficult to manufacture a component with exact dimensions. Tolerances are values that are added to the dimensions to handle the variation occurs at the time of manufacturing.

Let's discuss inducing dimensions in the Drawing file.

In our case, we have selected the Projection view to induce dimensions. Perform the following steps to induce dimensions in the Drawing file:

1. *Create* a **Projection** view in the Drawing file, as shown in Fig.Pro/E-6.58:

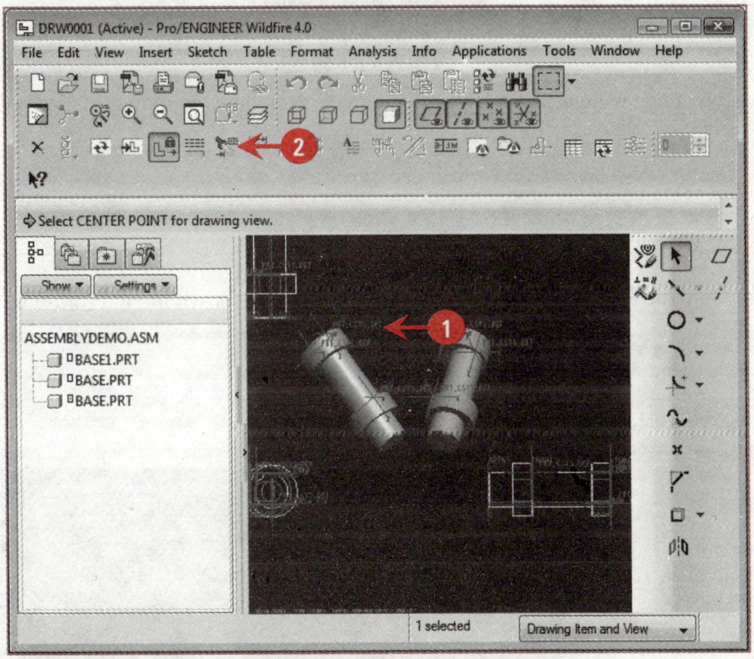

Fig.Pro/E-6.58

Note

The process to create a Projection view has been already discussed in the section, Generating Drawing Views of this chapter.

2. *Click* the **Show and Erase** button ( ) from the **Menu Bar** (Fig.Pro/E-6.58). The **Show and Erase** dialog box appears and displays various options for dimensioning the view (Fig.Pro/E-6.59).
3. *Click* the **Dimension** button in the **Type** Area in the **Show and Erase** dialog box, as shown in Fig.Pro/E-6.59:

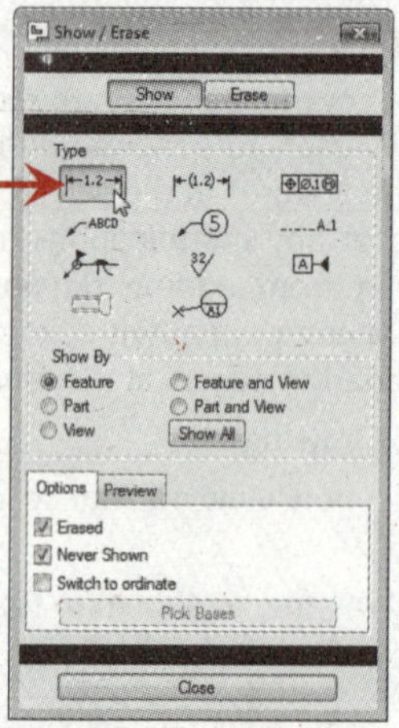

Fig.Pro/E-6.59

The **Select** dialog box appears to select a feature in a **Projection** view, as shown in Fig.Pro/E-6.60:

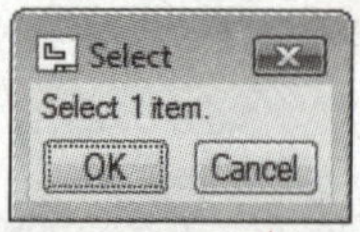

Fig.Pro/E-6.60

4. *Select* a feature in the **Projection** view to add a dimension, as shown in Fig.Pro/E-6.61:

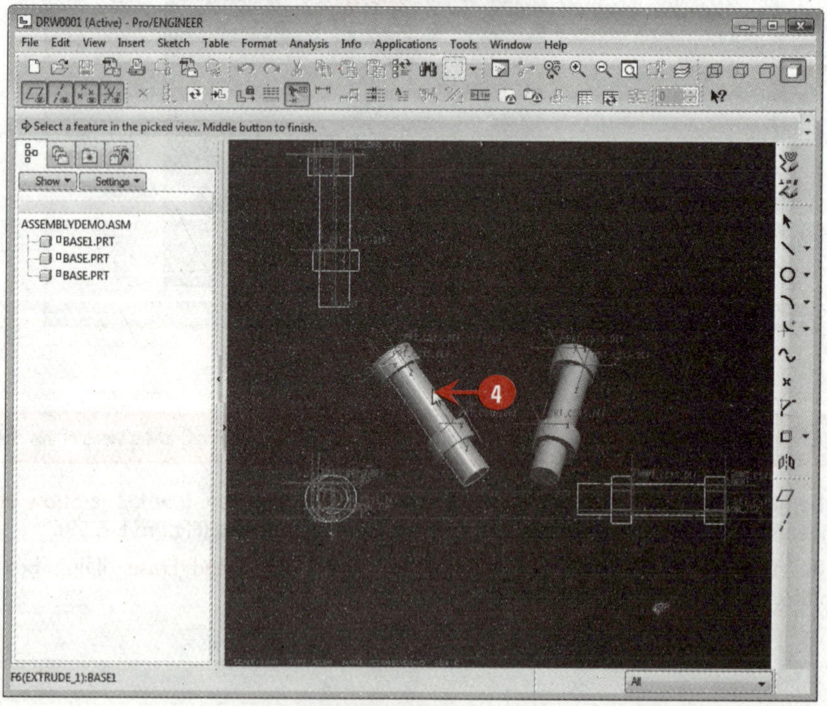

Fig.Pro/E-6.61

5. *Press* the middle mouse button and *click* the **Accept All** button in the **Show and Erase** dialog box, as shown in Fig.Pro/E-6.62:

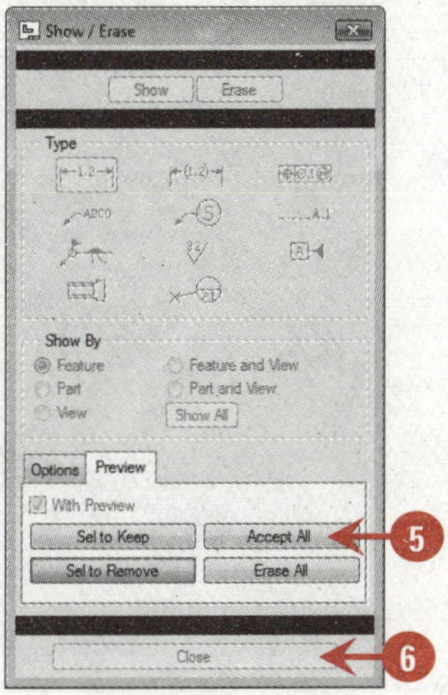

Fig.Pro/E-6.62

When you click the **Accept All**, the **Close** button becomes enable.

6. Now, *Click* the **Close** button in the **Show and Erase** dialog box (Fig.Pro/E-6.62).

All the dimensions related to the selected feature are displayed in all views, as shown in Fig.Pro/E-6.63:

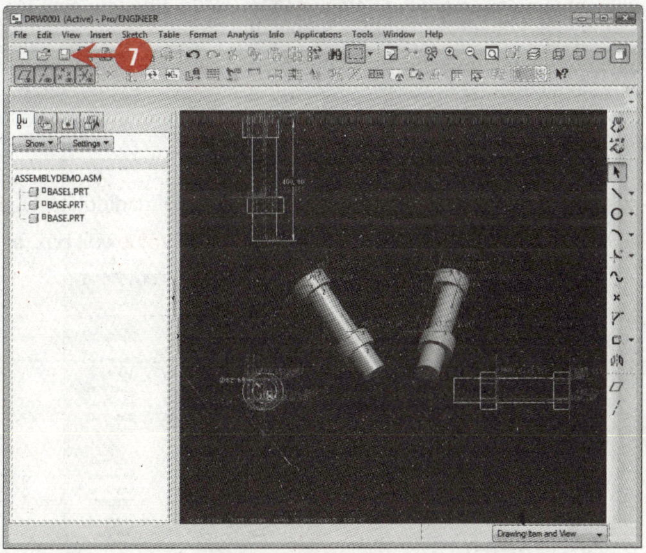

Fig.Pro/E-6.63

7. *Save* the file in the working directory.

Let's discuss the inducing of tolerances in the dimensions of an existing view.

Perform the following steps to induce tolerance in the dimensions of an existing view:

1. *Open* the **DRW0001.DRW** file and *Select* the **File→Properties** option from the **Menu Bar**, as shown in Fig.Pro/E-6.64:

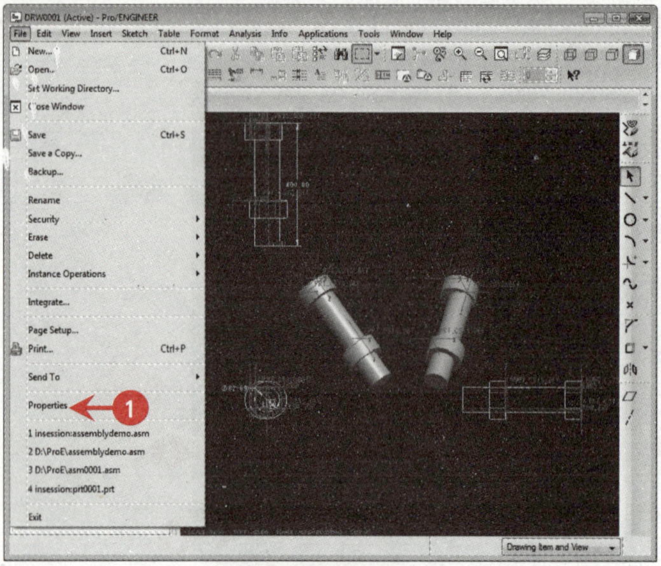

Fig.Pro/E-6.64

The **Menu Manager** menu appears and displays various options for manipulating the Pro/ENGINEER Wildfire 4.0 Drawing file properties.

2. *Select* the **Drawing Options** option from the **Menu Manager** menu, as shown in Fig.Pro/E-6.65:

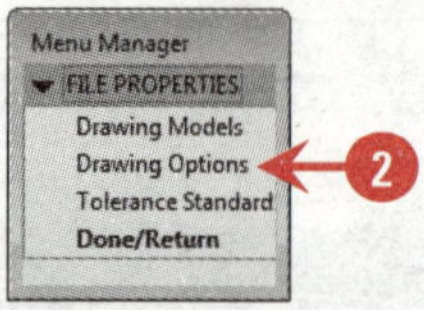

Fig.Pro/E-6.65

The **Options** dialog box appears and displays various drawing options to manipulate (Fig.Pro/E-6.66).

3. Specify **tol_display** in the **Option** edit box and specify 'yes' in the **Value** edit box, as shown in Fig.Pro/E-6.66:

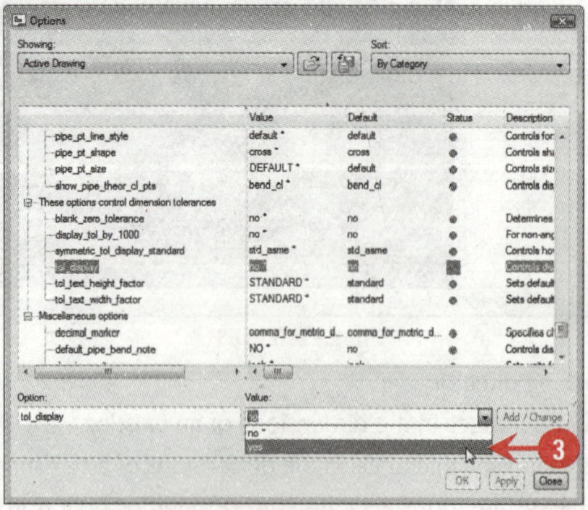

Fig.Pro/E-6.66

4. *Click* the **Add/Change** button in the **Options** dialog box to add the specified value of the **tol_display** option, as shown in Fig.Pro/E-6.67:

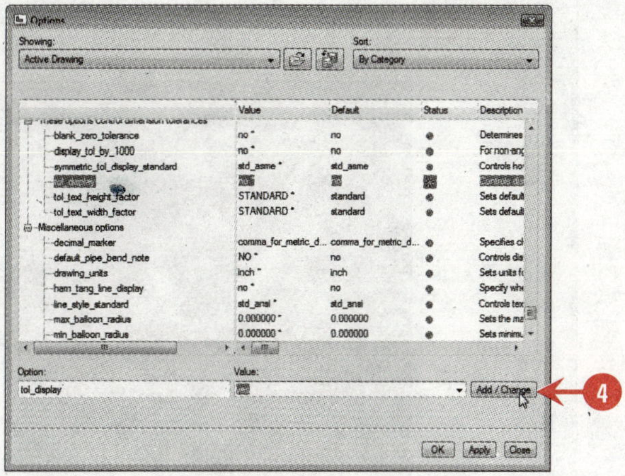

Fig.Pro/E-6.67

When the **tol_display** option is set to 'yes', and the specified value is added, the display of tolerances for the current session is enabled.

5. *Click* the **Apply** button in the **Options** dialog box to set the specified value **tol_display** option, as shown in Fig.Pro/E-6.68:

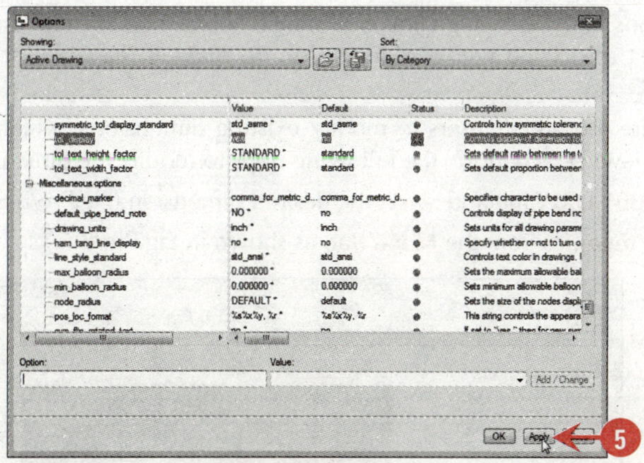

Fig.Pro/E-6.68

6. *Click* the **Close** button to exit the **Options** dialog box (Fig.Pro/E-6.68).
7. *Select* the **Done/Return** option from the **FILE PROPERTIES** sub-menu in the **Menu Manager** menu, as shown in Fig.Pro/E-6.69:

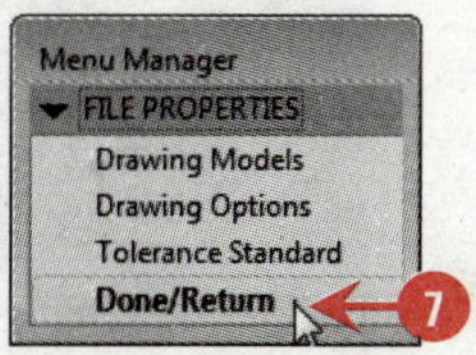

Fig.Pro/E-6.69

Geometric tolerances are induced successfully in the dimensions of the selected view, as shown in Fig.Pro/E-6.70:

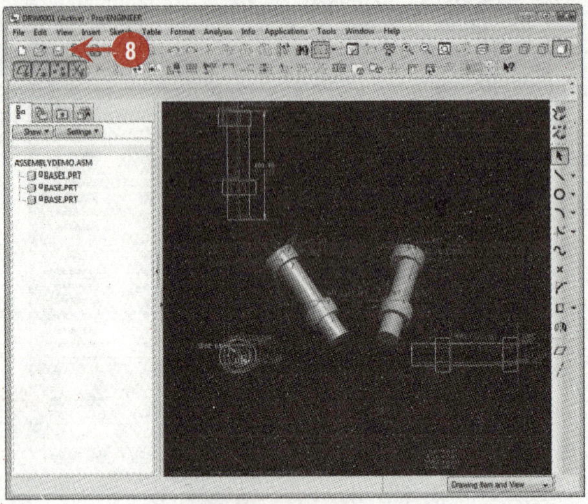

Fig.Pro/E-6.70

8. Finally, *save* the file in the working directory.

Now that the dimensions have been added to the drawing views, let's now discuss about modifying them.

Modifying Dimensions

After adding dimensions in the drawing view, you may want to modify these dimensions by either changing the position of the text associated with the dimension (to make it readable) or to change the values of the dimensions.

Pro/ENGINEER Wildfire 4.0 allows users to modify existing dimensions, which are added to specify every point in the drawing view. Let's perform the following steps to modify the dimensions of an existing view:

1. *Open* the **DRW0001.DRW** file and *select* the desired dimension (Fig.Pro/E-6.71).
2. *Select* the **Edit→Properties** from the **Menu Bar**, as shown in Fig.Pro/E-6.71:

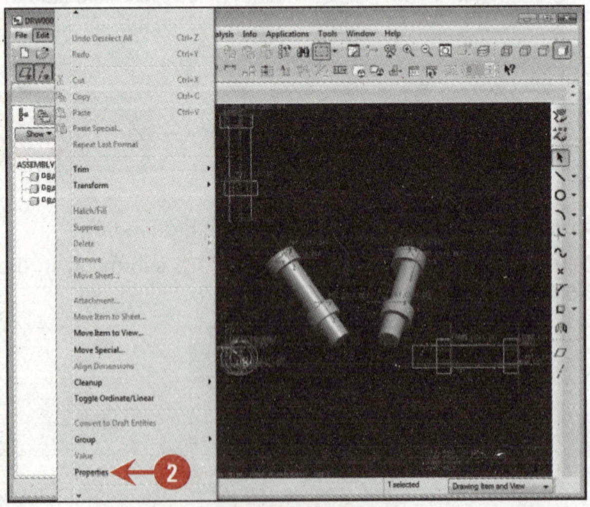

Fig.Pro/E-6.71

The **Dimension Properties** dialog box appears and the **Properties** tab is selected by default, as shown in Fig.Pro/E-6.72:

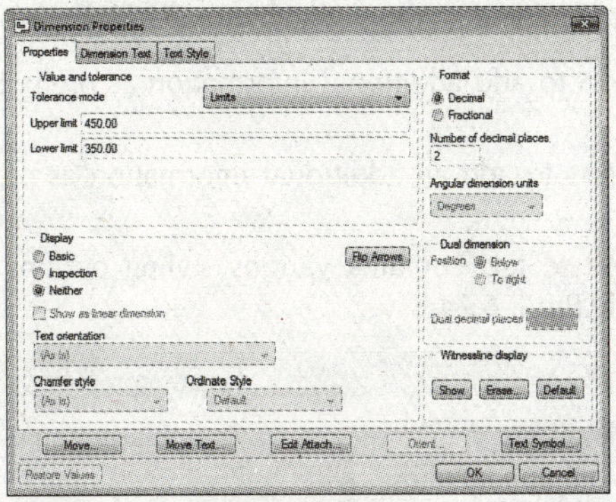

Fig.Pro/E-6.72

The **Dimension Properties** dialog box provides various options for modifying dimensions. Let's discuss the options present in **Dimension Properties** dialog box:

- **Properties tab:** The **Properties** tab contains various options for manipulating the properties of the dimensions. The **Properties** tab contains the following options:
  - **Value and tolerance Area:** Provides an option named **Tolerance** mode for manipulating tolerance in the drawing view. The **Tolerance** mode drop down list allows the user to select the type of tolerance mode such as **Limits**, **Nominal**, and so on. The other options in the **Value and tolerance** Area changes with the options selected in the **Tolerance** mode drop down.
  - **Display Area:** Provides radio buttons such as Basic, Inspection, Neither for providing display area for the dimensions to be placed. The Basic radio button allows user to display the dimension in a rectangular box. If the Inspection radio button is selected, then the dimension is displayed in an elliptical box. The Neither radio button is selected when you do not want to display the dimension neither in the Basic nor the Inspection type.
  - **Format Area:** Provides options such as Decimal and Fractional to display dimensions in decimal and fractional format.
  - **Dual dimension Area:** Provides options such as **Position** and **Dual decimal places** for specifying a position to place dual dimensions and to place number of decimal places in dual dimensions.
- **Dimension Text tab:** The **Dimension Text** tab contains various options related to text associated with the dimensions, as shown in Fig.Pro/E-6.73:

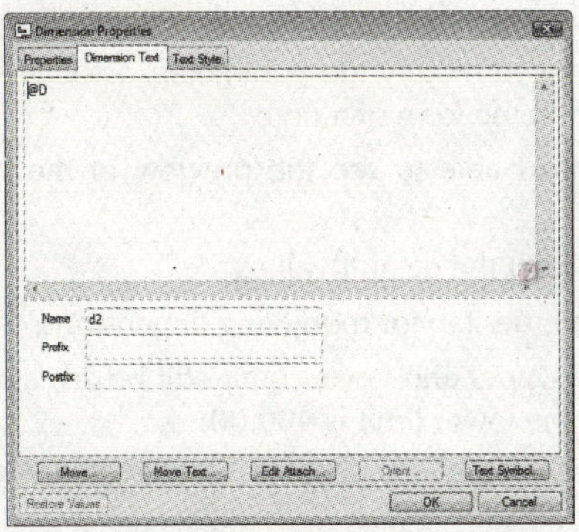

Fig.Pro/E-6.73

The **Dimension Text** tab contains:

- **Set dimension text box:** Displays the text of the selected dimension.
- **Name edit box:** Displays the symbol associated with the dimension.
- **Prefix edit box:** Allows users to add additional information as prefix to the default text associated with the dimension.
- **Postfix edit box:** Allows users to add an additional information as postfix to the default text associated with the dimension.
- **Text Style tab:** The **Text Style** tab contains various styling options for the text associated with the dimensions, as shown in Fig.Pro/E-6.74:

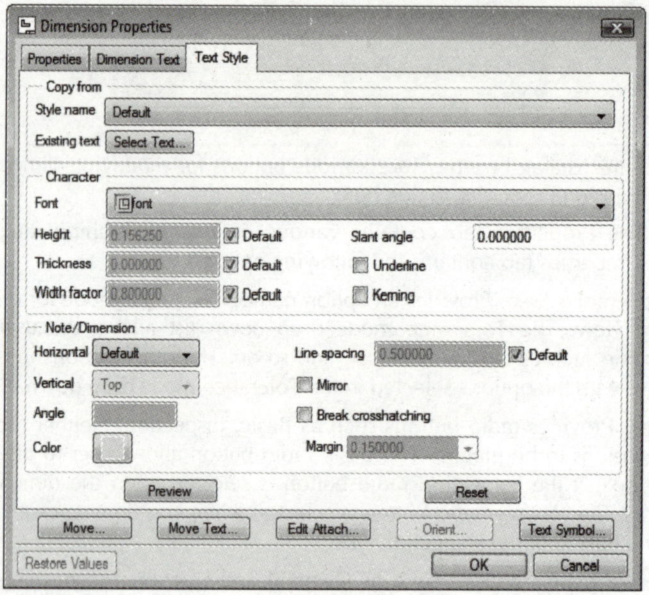

Fig.Pro/E-6.74

The **Text Style** tab contains the following options:

- **Copy from Area:** Provides options such as Style name and Existing text to modify the style of the text associated with the dimension
- **Character Area:** Provides options such as **Font**, **Height**, **Thickness**, and so on to modify the font style of the text associated with the dimension
- **Note/Dimension Area:** Provides to modify the parameters associated with the dimension such as color, line spacing, angle, and so on in the form of a note
- **Preview button:** Allows users to able to see the preview of the changes made using the Dimension Properties dialog box
- **Reset button:** Allows users to reset the default values

In our case, we will select the **Properties** tab for modifying dimensions.

3. *Specify* the new values in the **Upper** and **Lower limit** edit boxes, as shown in Fig.Pro/E-6.75.In our case, the upper limit is 500.00 and the lower limit is 400.00.
4. *Click* the **OK** button, as shown in Fig.Pro/E-6.75:

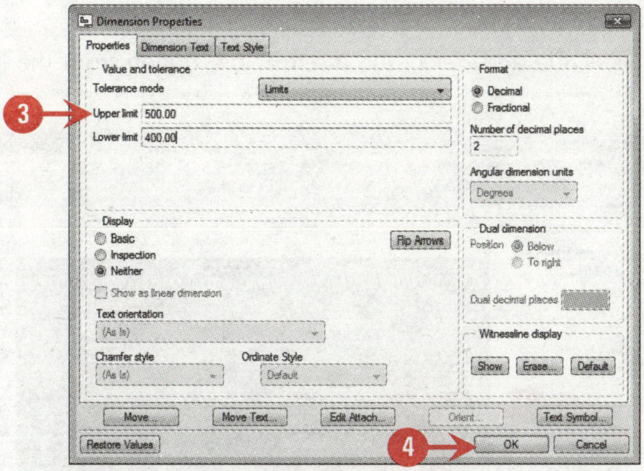

Fig.Pro/E-6.75

The Upper and Lower limits are values representing the highest and lowest size of the selected dimension. The limits associated with the selected dimension are modified, as shown in Fig.Pro/E-6.76:

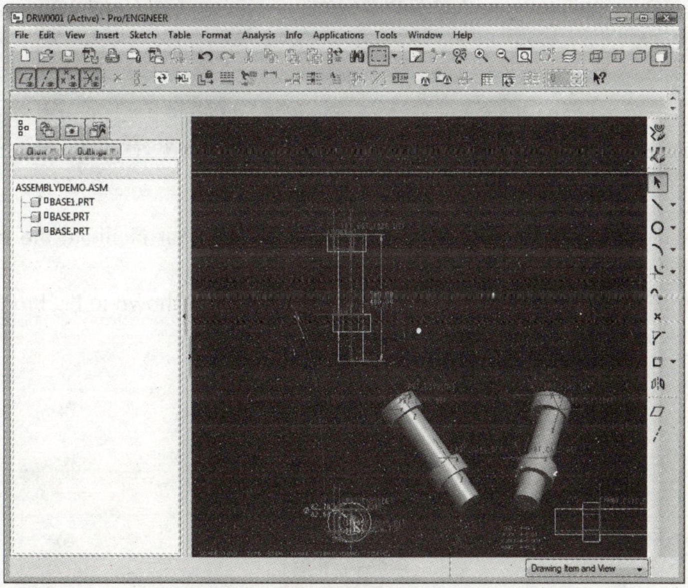

Fig.Pro/E-6.76

5. Now, save the file in the working directory.

Till now, we discussed about the addition of dimensions and tolerances in drawing views. Apart from this, we also discussed the modification of existing dimensions.

Next, let's discuss creating notes in the drawing views.

Creating Notes

Pro/ENGINEER Wildfire 4.0 allows users to provide additional information in the Drawing file by adding notes. The additional information can be specific to any Part or an Entity type. Occasionally, general information that does not apply to any particular part or feature is also specified in the note. The drawing notes contain information in textual as well as symbolic format. Let's discuss creating notes in a Drawing file.

Perform the following steps to create a note in a Drawing file:

1. *Open* the **DRW0001.DRW** file and *Select* the **Insert→Note** option from the Menu Bar, as shown in Fig.Pro/E-6.77:

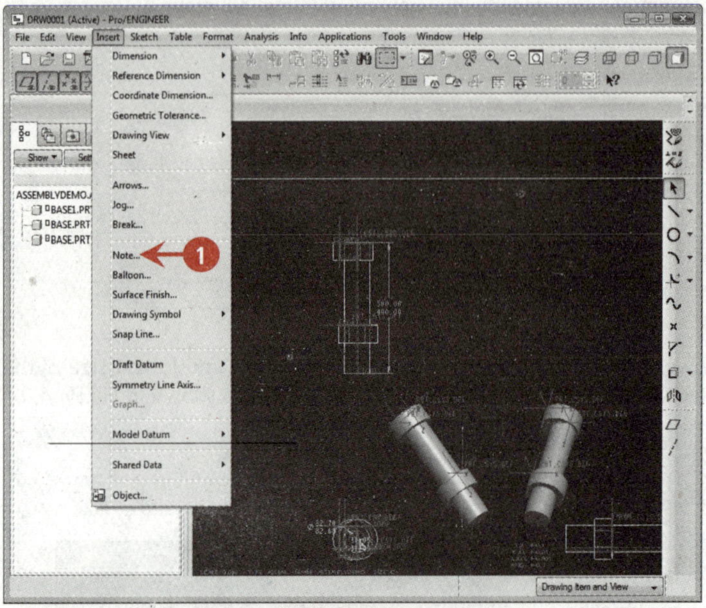

Fig.Pro/E-6.77

The **Menu Manager** menu appears and contains various options that facilitate the creation of notes in a Drawing file (Fig.Pro/E-6.78).

2. *Select* the **With Leader** option from the **Menu Manager** menu, as shown in Fig.Pro/E-6.78:

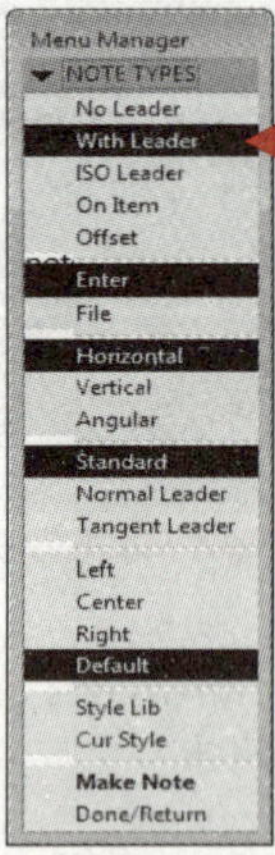

Fig.Pro/E-6.78

The **NOTE TYPES** sub menu in the **Menu Manager** menu provides various options to specify the appearance of the drawing note. Let's discuss these options before we start working with them:

- **No Leader:** The **No Leader** option allows users to avoid the leader for the note to be placed in the Drawing file.

- **With Leader:** The With Leader option allows users to specify a leader for the note to be placed in a Drawing file.
- **ISO Leader:** The ISO Leader option allows users to specify an ISO type leader. An ISO type leader contains underline text.
- **On Item:** The On Item option allows users to attach a note directly on the selected item in the Drawing file.
- **Offset:** The Offset option allows users to group the note with any selected entity such as dimension, geometric tolerance, and so on.
- **Enter:** The Enter option allows users to enter text for the note from the keyboard.
- **File:** The File option allows users to provide text for the note from a file.
- **Horizontal:** The Horizontal option allows users to create a note with the text placed horizontally.
- **Vertical:** The Vertical option allows users to create a note with the text placed vertically.
- **Angular:** The Angular option allows users to create a note with the text placed at certain angle.
- **Standard:** The Standard option allows users to use the default leader type.
- **Normal Leader:** The Normal Leader option allows users to make the leader either normal or perpendicular to any entity type.
- **Tangent Leader:** The Tangent Leader option allows users to make the leader tangent to any entity type.
- **Left:** The Left option allows users to left align the text associated with the note.
- **Center:** The Center option allows users to centrally align the text associated with the note.
- **Right:** The Right option allows users to right align the text associated with the note.
- **Default:** The Default option allows users to align the note text with the Left alignment, which is the default alignment.
- **Style Lib:** The Style Lib option allows users to create a new style for the current session in which the user is currently working.
- **Cur Style:** The Cur Style allows users to select a new style for the current session from the list of pre-defined styles available in Pro/ENGINEER Wildfire 4.0.
- **Make Note:** The Make Note option allows users to start the process of creating a note in the Drawing file.

3. *Select* the Make Note option from the Menu Manager menu, as shown in Fig.Pro/E-6.79:

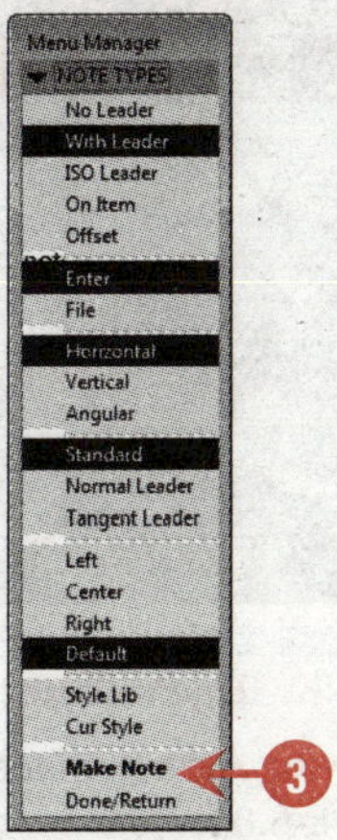

Fig.Pro/E-6.79

The **ATTACH TYPE** sub-menu appears in the **Menu Manager** menu (Fig.Pro/E-6.80).

4. *Select* the **On Surface** option from the **ATTACH TYPE** sub-menu to select a point on the surface of the front view to place a note, as shown in Fig.Pro/E-6.80:

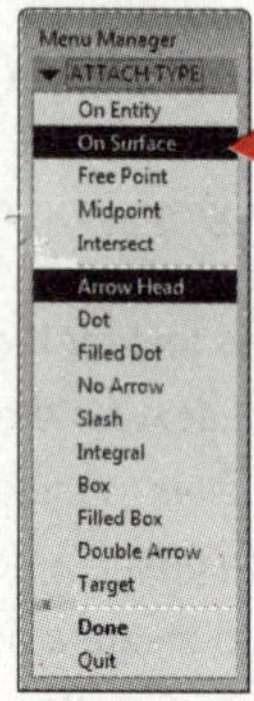

Fig.Pro/E-6.80

Apart from the **Menu Manager** menu, the **Select** dialog box also appears, as shown in Fig.Pro/E-6.81:

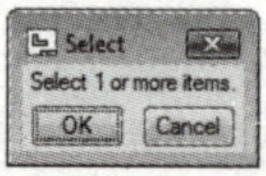

Fig.Pro/E-6.81

5. *Select* the desired surface from the front view in the Drawing file to place a note, as shown in Fig.Pro/E-6.82:

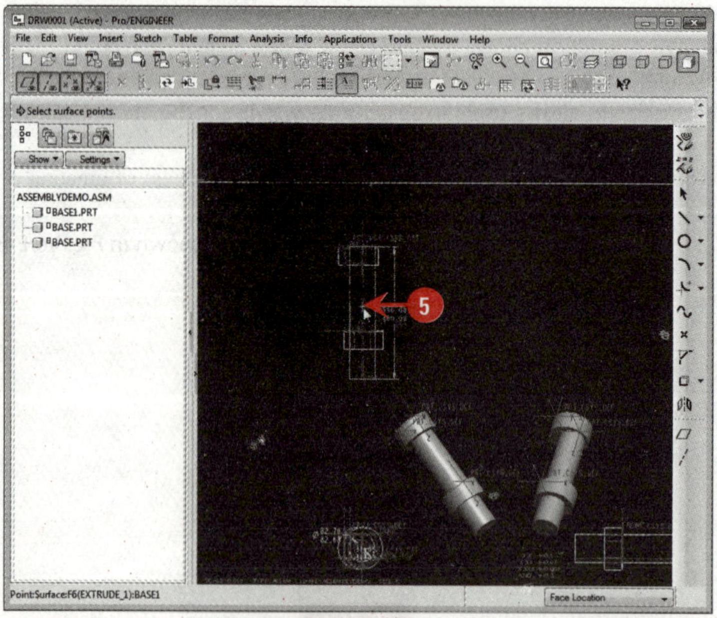

Fig.Pro/E-6.82

6. *Click* the selected surface in the **Drawing Area** to place a note. A red colored box appears in place of the selected surface, as shown in Fig.Pro/E-6.83:

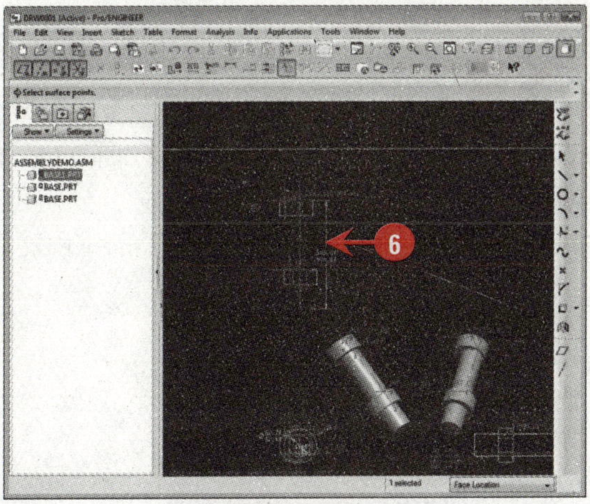

Fig.Pro/E-6.83

7. *Press* the middle mouse button in the middle of the red colored box. The **Message Input window** appears to accept a note from the user, as shown in Fig.Pro/E-6.84:

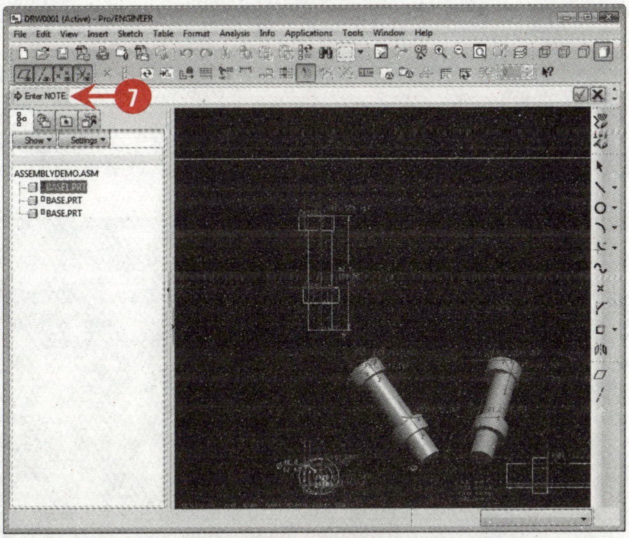

Fig.Pro/E-6.84

Apart from the **Message Input window**, the **Text Symbol** dialog box also appears, as shown in Fig.Pro/E-6.85:

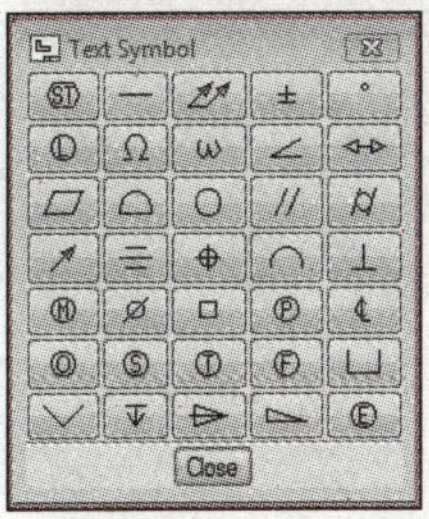

Fig.Pro/E-6.85

8. *Specify* a note in the **Message Input window**, as shown in Fig.Pro/E-6.86:

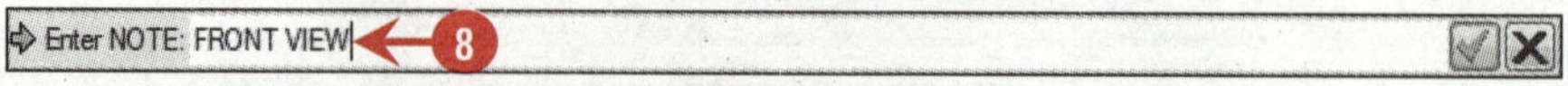

Fig.Pro/E-6.86

You can insert desired symbols in the note from the **Text Symbol** dialog box. In our case, we have selected button to insert in the note.

9. *Click* the button in the **Text Symbol** dialog box, as shown in Fig.Pro/E-6.87:

Fig.Pro/E-6.87

The selected symbol is appended to the text specified in the **Message Input** window, as shown in Fig.Pro/E-6.88:

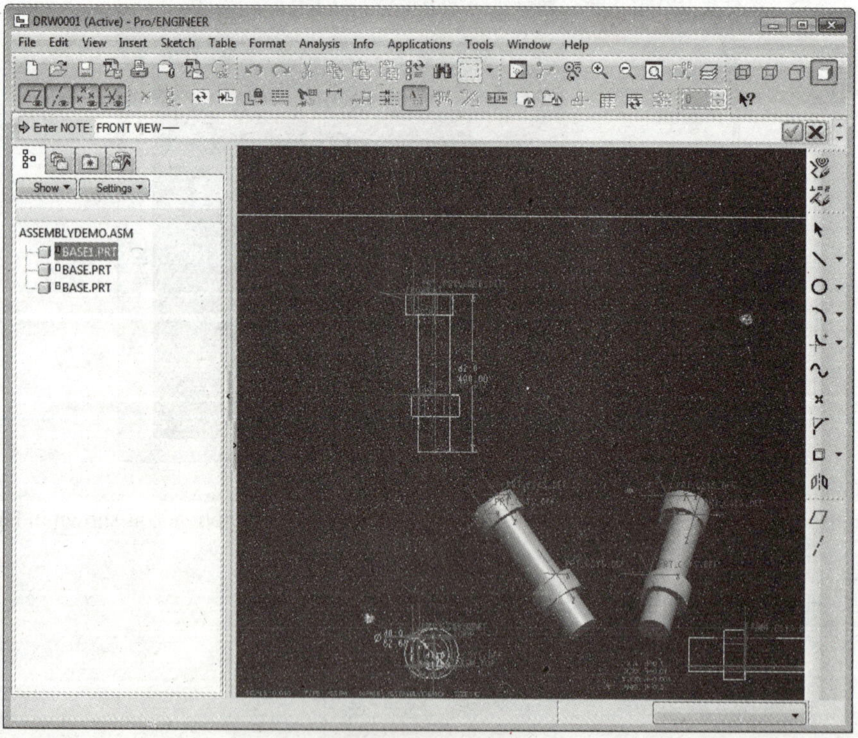

Fig.Pro/E-6.88

10. *Close* the **Text Symbol** dialog box and specify additional information in the **Message Input window**. Now, *click* the **Build** feature button ( ), as shown in Fig.Pro/E-6.89:

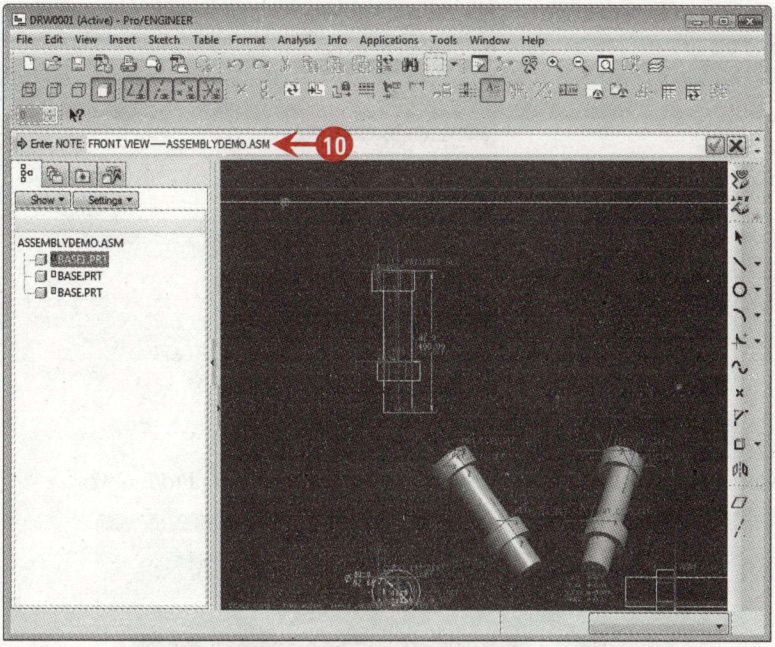

Fig.Pro/E-6.89

The Drawing Area previews the note that is created in the current session, as shown in Fig.Pro/E-6.90:

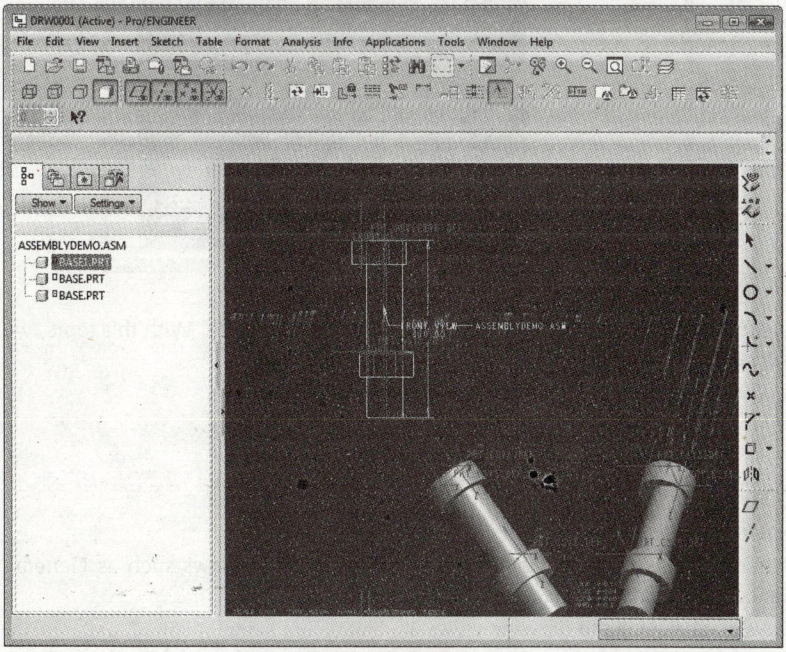

Fig.Pro/E-6.90

11. *Select* the **Done/Return** option from the **Menu Manager** menu to finish the note creation process and exit from the **Menu Manager** menu, as shown in Fig.Pro/E-6.91:

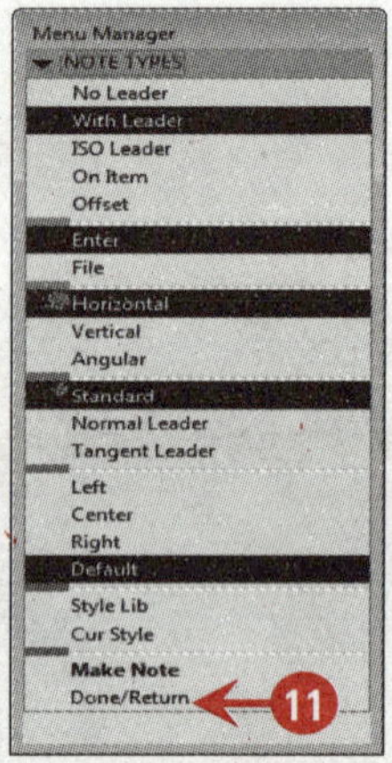

Fig.Pro/E-6.91

The note is placed successfully in the selected view, as shown in the Fig.Pro/E-6.92:

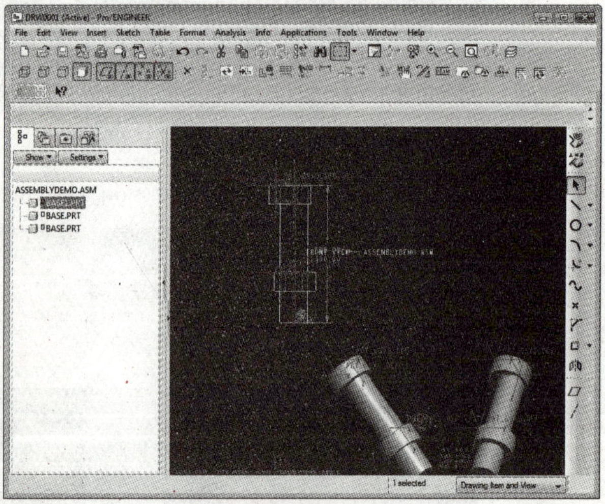

Fig.Pro/E-6.92

Now, we have completed the process to create notes in the drawing file. With this topic, we have reached to the end of this chapter.

Finally, let's now have a glance at the topics we covered in this chapter.

Summary

In this chapter, you learned about:

- The process to start the Drawing mode.
- The mechanism & procedure to generate the various drawing views such as General view, Projection view, Detailed view, Auxiliary view, and Revolved view.
- Various ways of modifying the drawing views.
- Adding dimensions and tolerances to the drawing views.
- The mechanism to modify an existing dimension of the drawing view.
- Creating notes in drawing views to provide additional information.

Chapter 7

Implementing Surface Modeling

In this Section

- Creating an Extruded Surface
- Creating the Sweep Surface
- Creating Datum Curves
- Creating Surfaces from Boundary Curves
- Merging Surfaces
- Trimming Surfaces

Surface modeling in `Pro/ENGINEER Wildfire 4.0` is used to create and manipulate surfaces. A surface is defined as the outer body of a model having neither mass nor thickness.

In this chapter, we study how to create and manage surfaces. The chapter also discusses the process to create complex surfaces using datum curves. Sometimes, in order to get the desired surface model, we may need to merge and trim surfaces. The processes to merge and trim surfaces are also discussed in the chapter.

Creating an Extruded Surface

The process to create an extruded surface is similar to the process of creating an extruded feature, which is already discussed in Chapter 4, *Exploring Pro/ENGINEER Wildfire 4.0 Part Mode*. The only difference between an extruded feature and an extruded surface is that an extruded feature has properties such as mass, thickness while an extruded surface does not have these properties. Let's follow these steps to create an extruded surface:

1. *Start* the **Part** mode.

Note

> The process to start the Part mode is already discussed in Chapter 4, Exploring Pro/ENGINEER Wildfire 4.0 Part Mode.

2. *Click* the **Extrude** button ( ) in the **Base Features** toolbar in the right tool chest. The **Extrude** dashboard with the options to create an extruded surface appears, as shown in Fig.Pro/E-7.1.
3. Now, *click* the **Extrude as surface** button ( ) in the **Extrude** dashboard to create an extruded surface (Fig.Pro/E-7.1).

Note

> By default, the Extrude as solid ( ) button is selected.

4. *Select* the **Placement** tab in the **Extrude** dashboard, as shown in Fig.Pro/E-7.1:

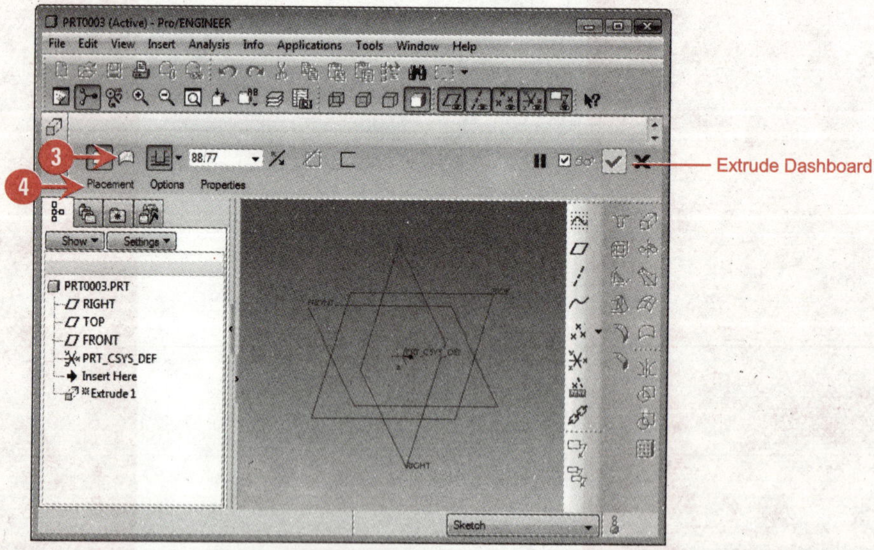

Fig.Pro/E-7.1

A slide-down panel that prompts you to select the sketching plane appears (Fig.Pro/E-7.2).

5. *Click* the **Define** button in the slide-down panel, as shown in Fig.Pro/E-7.2:

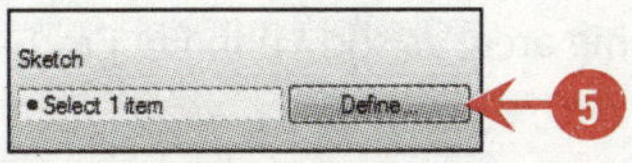

Fig.Pro/E-7.2

The **Sketch** dialog box where you have to specify the sketching plane, the reference plane, and the orientation plane appears, as shown in Fig.Pro/E-7.3:

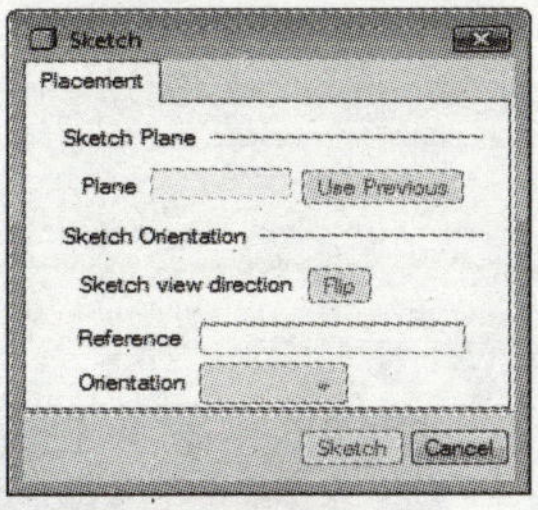

Fig.Pro/E-7.3

6. *Select* the **Front** datum plane, either in the **Model Tree** or in the drawing area (Fig.Pro/E-7.1). The selected datum plane, reference plane, and orientation plane automatically appears in the **Sketch** dialog box, as shown in Fig.Pro/E-7.4.
7. *Click* the **Sketch** button in the **Sketch** dialog box (Fig.Pro/E-7.4) to start creating the sketch of the extruded surface:

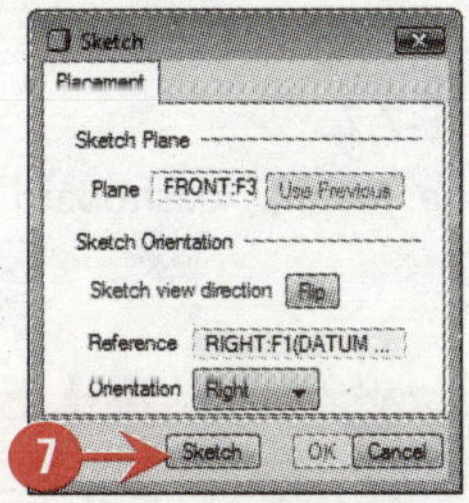

Fig.Pro/E-7.4

The **Sketcher tools toolbar** appears, as shown in Fig.Pro/E-7.5:

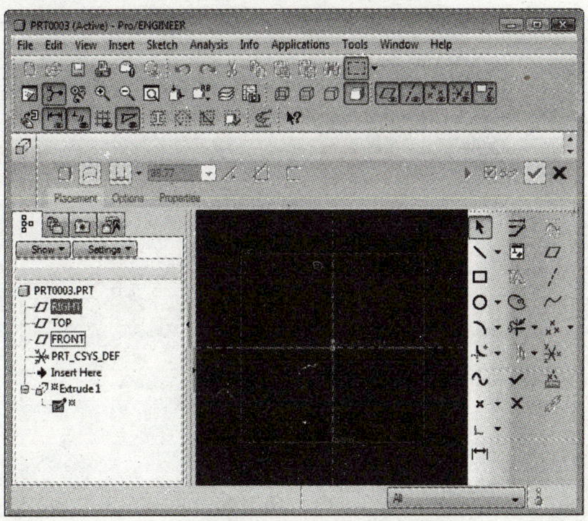

Fig.Pro/E-7.5

8. *Draw* a rectangle in the drawing area, as shown in Fig.Pro/E-7.6.

Note

> The process to draw a rectangle is already discussed in Chapter 3, Exploring Pro/ENGINEER Wildfire 4.0 Sketch Mode.

9. Now, *click* the **Done** (✓) button in the **Sketcher tools** toolbar to complete the rectangle, as shown in Fig.Pro/E-7.6:

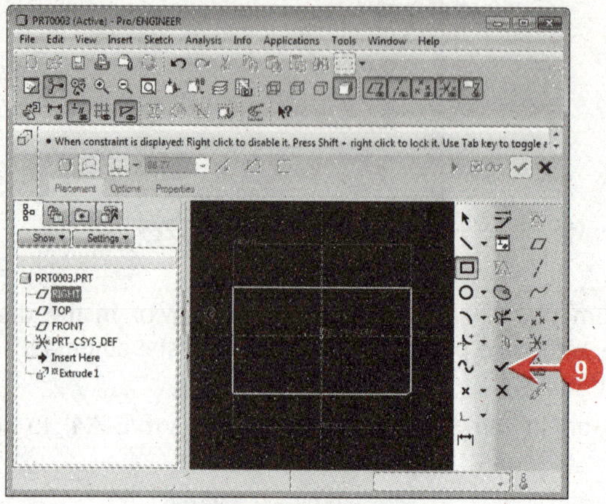

Fig.Pro/E-7.6

The rectangle appears, as shown in Fig.Pro/E-7.7.

10. *Enter* 100.00 in the **Depth** box in the **Extrude** dashboard (Fig.Pro/E-7.7) to specify the depth of the rectangle.

Note

> You can also continue with the default depth value.

11. *Press* the **ENTER** key to project the new depth value in the **Depth** box.
12. *Click* the **Build feature** (✓) button to complete the process, as shown in Fig.Pro/E-7.7:

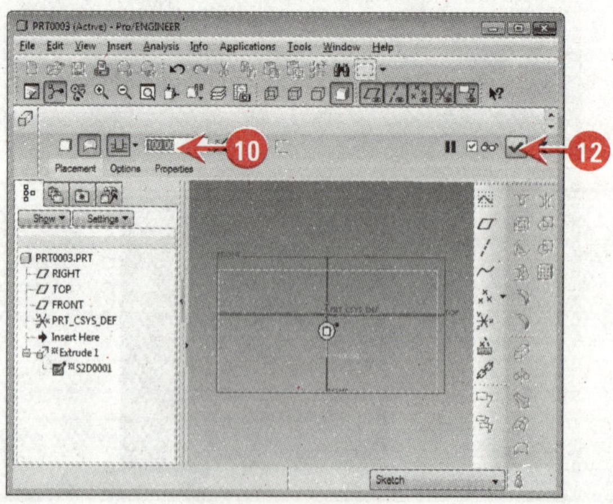

Fig.Pro/E-7.7

The extruded surface is highlighted, as shown in Fig.Pro/E-7.8:

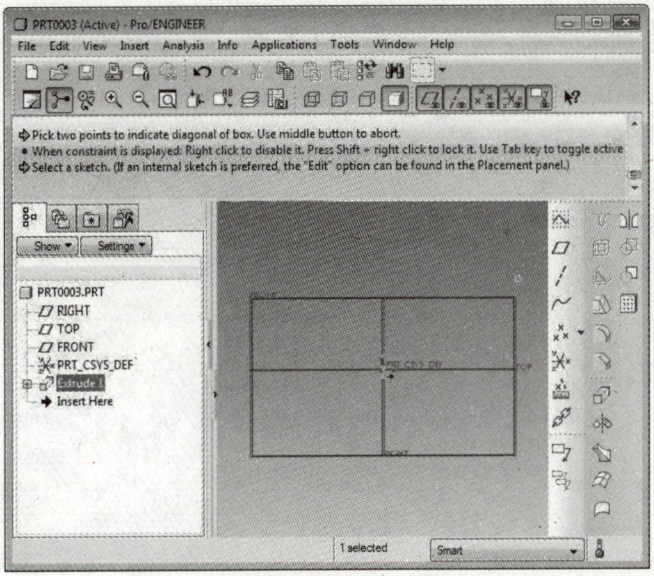

Fig.Pro/E-7.8

13. Now, press and *hold* the middle mouse button and spin the extruded surface to view it from different angles. The extruded surface appears, as shown in Fig.Pro/E-7.9:

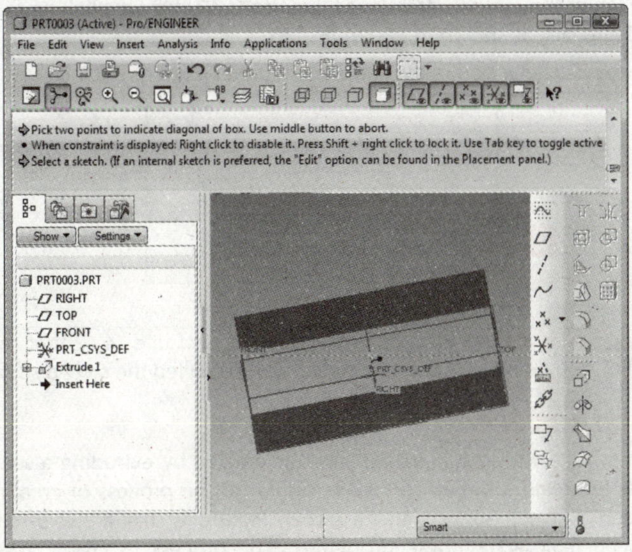

Fig.Pro/E-7.9

14. *Save* the file with the name **EXTRUDEDSURFACE.prt**.

In Pro/ENGINEER Wildfire 4.0, a surface can be open ended or capped ended. The difference between an open ended surface and a capped ended surface is that while an open ended surface is open from both the sides; both the ends are capped in the capped end surface. In capped end surface, both the ends are capped, as shown in Fig.Pro/E-7.10:

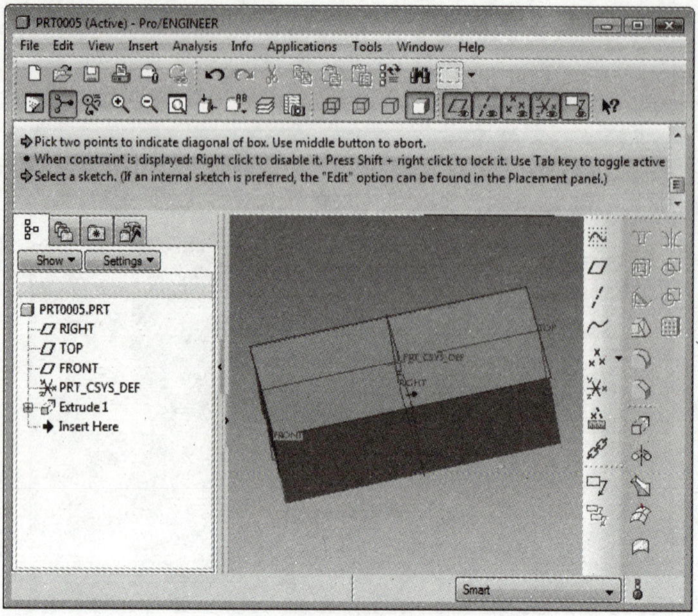

Fig.Pro/E-7.10

The process to create a capped end extruded surface is similar to the process of creating an open ended extruded surface, except the following steps:

1. *Click* the **Option** tab in the **Extrude** dashboard (Fig.Pro/E-7.1) before clicking the **Build feature** button. A slide-down panel appears, as shown in Fig.Pro/E-7.11.
2. *Select* the **Capped ends** check box in the slide-down panel (Fig.Pro/E-7.11):

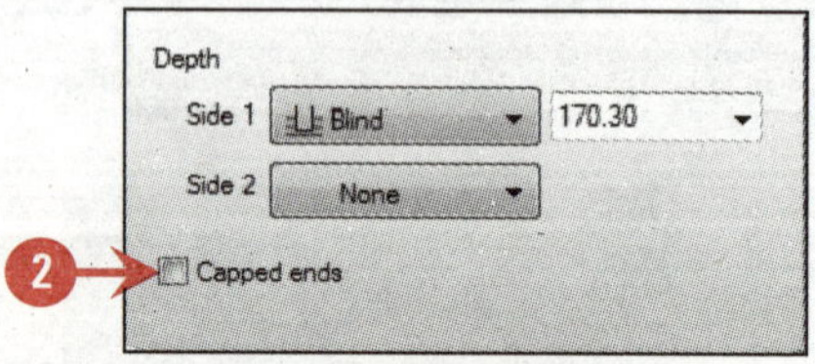

Fig.Pro/E-7.11

This enables the **Capped ends** feature in `Pro/ENGINEER Wildfire 4.0` that results in the surface with capped ends (Fig.Pro/E-7.10). Now, let's create another surface called the sweep surface.

Creating the Sweep Surface

A sweep surface in `Pro/ENGINEER Wildfire 4.0` is created by extruding a section of an entity along a trajectory. The process to create a sweep surface is similar to the process of creating a sweep feature. The only difference between a sweep surface and a sweep feature is that a sweep surface has no mass and thickness while a sweep feature has these properties. The process to create a sweep feature is already explained in Chapter 4, *Exploring Pro/ENGINEER Wildfire 4.0 Part Mode*. In this section, we study the process to create a sweep surface. Following are the steps that help you to create a sweep surface:

1. *Start* the **Part** mode.
2. Now, *select* **Insert→Sweep→Surface**, as shown in Fig.Pro/E-7.12:

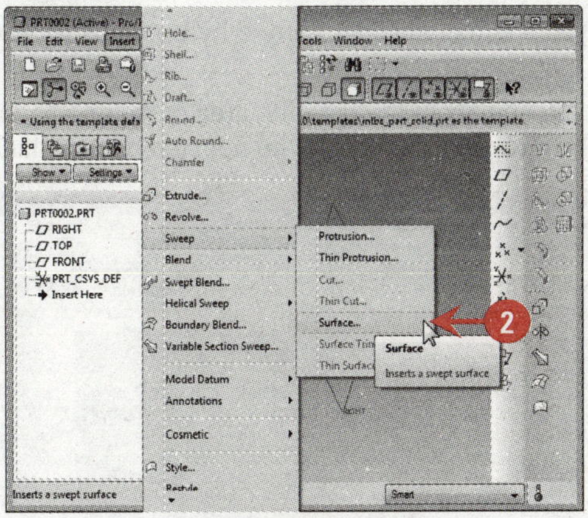

Fig.Pro/E-7.12

The **SURFACE:Sweep** dialog box (Fig.Pro/E-7.13) along with a **Menu Manager** menu (Fig.Pro/E-7.14) appears:

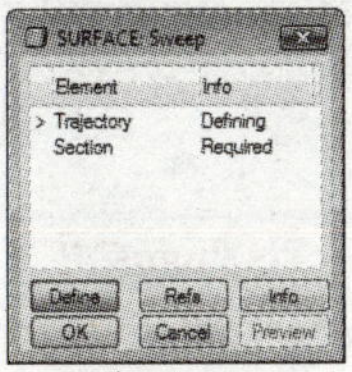

Fig.Pro/E-7.13

The **Menu Manager** menu provides two options:

- To sketch a trajectory
- To select a trajectory from the existing feature

In our case, we sketch a trajectory.

3. *Select* the **Sketch Traj** option (Fig.Pro/E-7.14) to sketch a trajectory, as shown in Fig.Pro/E-7.14:

Fig.Pro/E-7.14

The appearance of the **Menu Manager** menu changes, as shown in Fig.Pro/E-7.15:

Fig.Pro/E-7.15

A **Select** dialog box also appears along with the **Menu Manager** menu, as shown in Fig.Pro/E-7.16:

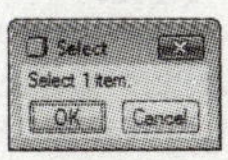

Fig.Pro/E-7.16

The **Select** dialog box prompts you to select an item, which means to select the datum plane to define a trajectory along it.

4. *Select* the **Front** datum plane either in the drawing area or in the **Model Tree** in the Part mode window. A **red** arrow appears in the direction of viewing the selected datum plane, as shown in Fig.Pro/E-7.17:

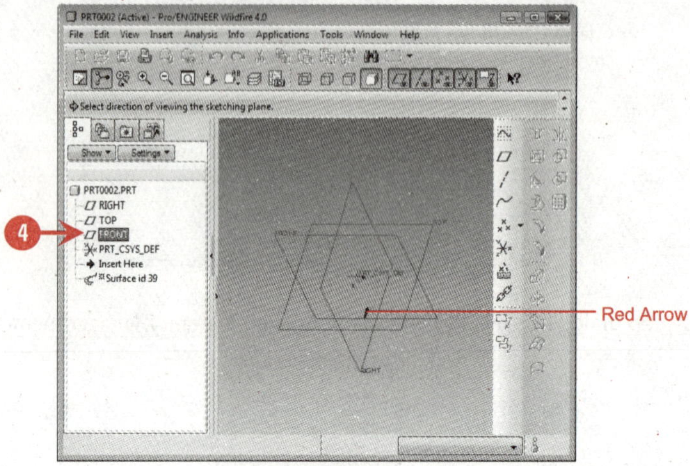

Fig.Pro/E-7.17

Appearance of the **Menu Manager** menu changes once again, as shown in Fig.Pro/E-7.18.

5. *Select* the **Okay** option to continue creating a sweep feature, as shown in Fig.Pro/E-7.18:

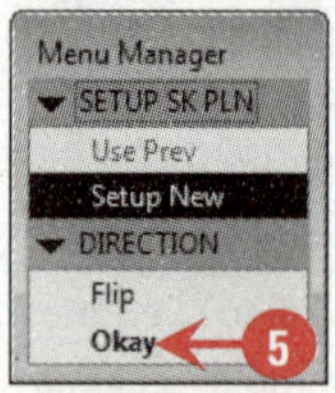

Fig.Pro/E-7.18

The appearance of the **Menu Manager** menu changes again, as shown in Fig.Pro/E-7.19.

Now, you have to select the orientation view of the sweep feature from the options given in the **SKET VIEW** submenu of the **Menu Manager** menu (Fig.Pro/E-7.19).

6. *Select* the **Top** option in the **SKET VIEW** submenu, as shown in Fig.Pro/E-7.19:

Fig.Pro/E-7.19

The appearance of the **Menu Manager** menu changes again, as shown in Fig.Pro/E-7.20:

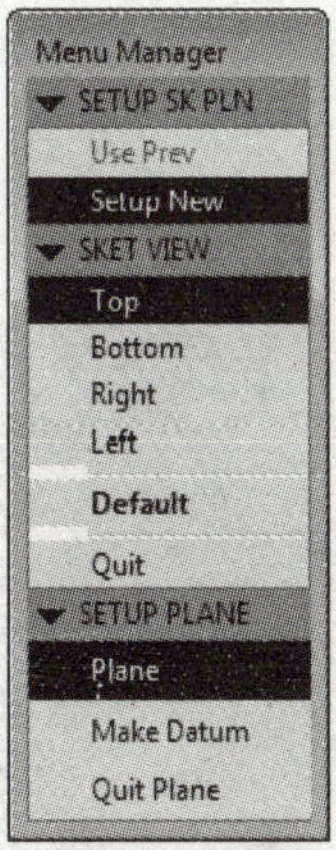

Fig.Pro/E-7.20

The Select dialog box (Fig.Pro/E-7.16) appears again. Both, the Menu Manager (Fig.Pro/E-7.20) and the Select dialog box prompt you to select the reference plane of the Front datum plane in the Part mode window.

7. *Select* the Top datum plane in the Model Tree. The Sketcher tools toolbar appears to let you sketch a trajectory (Fig.Pro/E-7.21).

8. *Select* the 3-Point/Tangent End ( ) button in the Sketcher tools toolbar to draw an arc, as shown in Fig.Pro/E-7.21:

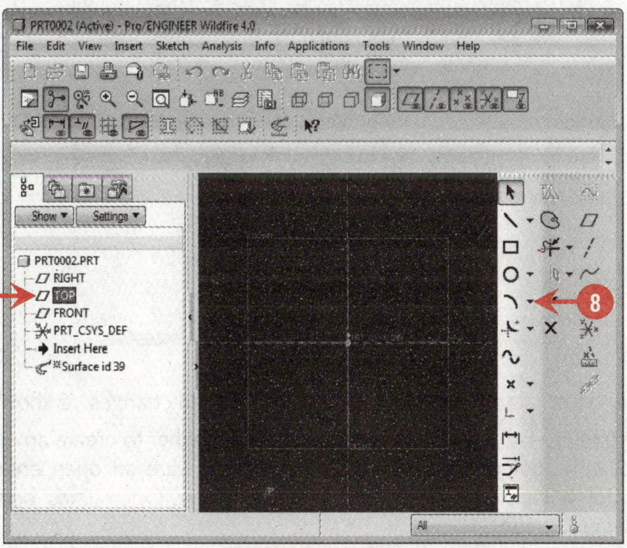

Fig.Pro/E-7.21

A yellow arrow is attached to the arc. This yellow arrow represents the start point of the trajectory and points in the direction in which the sweep surface will be created.

Note

The process to draw an arc has already been explained in Chapter 3, Exploring Pro/ENGINEER Wildfire 4.0 Sketch Mode.

9. *Click* the Done ( ) button to complete the sketch of the trajectory, as shown in Fig.Pro/E-7.22:

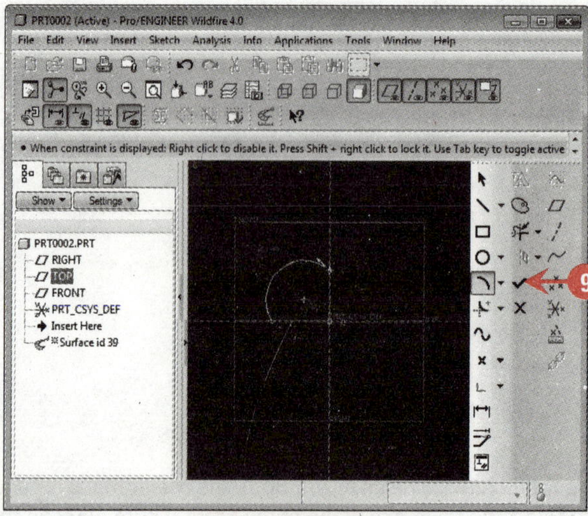

Fig.Pro/E-7.22

The view of the drawing area is changed to let you create a section for sweep surface, as shown in Fig.Pro/E-7.23:

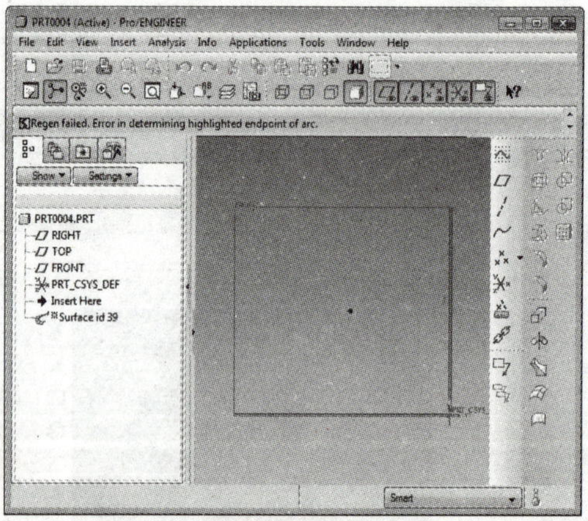

Fig/Pro/E-7.23

The appearance of the Menu Manager menu (Fig.Pro/E-7.20) again changes, as shown in Fig.Pro/E-7.24.

Now, Menu Manager (Fig.Pro/E-7.24) lets you select the option either to create an open ended sweep feature or a capped sweep feature. Select the Open Ends option to create an open ended sweep feature or the Capped Ends option for a capped end feature. By default, the Open Ends option is selected. We are continuing with the default option.

10. *Select* the Done option in the Menu Manager menu (Fig.Pro/E-7.24) to create an open ended sweep feature:

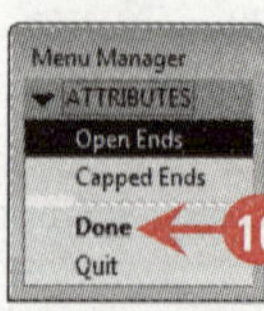

Fig.Pro/E-7.24

The view of the drawing area changes once again (Fig.Pro/E-7.25).

11. *Click* the **Center and Point** (○ ▾) button in the **Sketcher tools toolbar** (Fig.Pro/E-7.25):

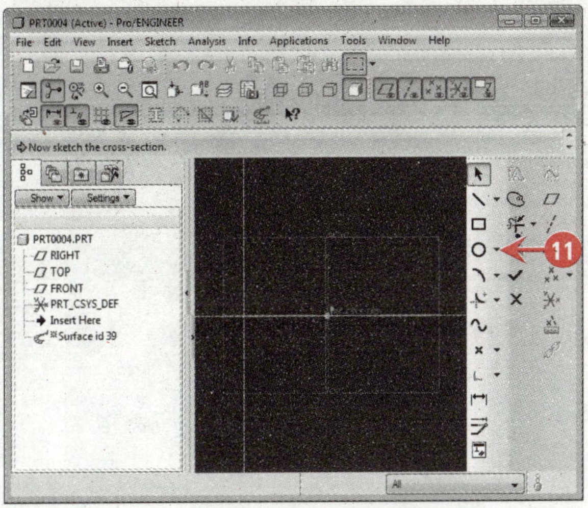

Fig.Pro/E-7.25

In Fig.Pro/E-7.25, the intersection point of two infinite yellow lines is representing the starting point of the trajectory. The start point of the sweep section should lie at the start point of the trajectory.

A circle is drawn as a sweep section, as shown in Fig.Pro/E-7.26.

Note

The process to draw a circle has already been explained in Chapter 3, Exploring Pro/ENGINEER Wildfire 4.0 Sketcher Mode.

12. *Click* the **Done** (✓) button in the **Sketcher tools toolbar** (Fig.Pro/E-7.26):

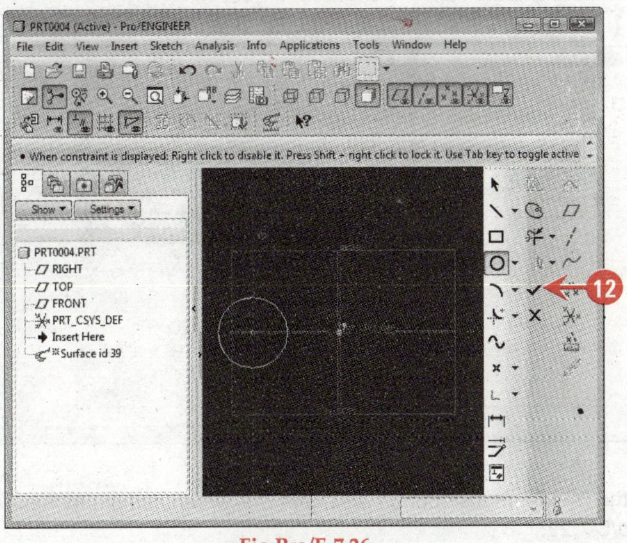

Fig.Pro/E-7.26

The sweep surface is created, as shown in Fig.Pro/E-7.27:

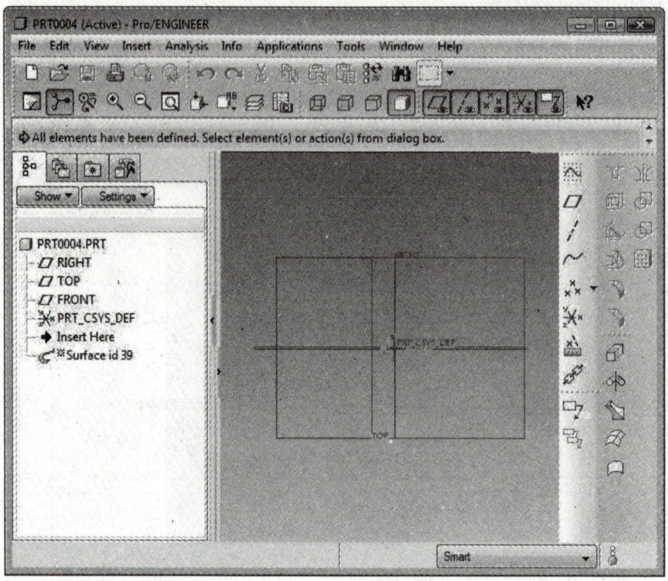

Fig.Pro/E-7.27

13. *Click* the **Preview** button in the **SURFACE: Sweep** dialog box (Fig.Pro/E-7.13). The sweep surface appears, as shown in Fig.Pro/E-7.28.
14. Now, *select* the sweep surface, as shown in Fig.Pro/E-7.28:

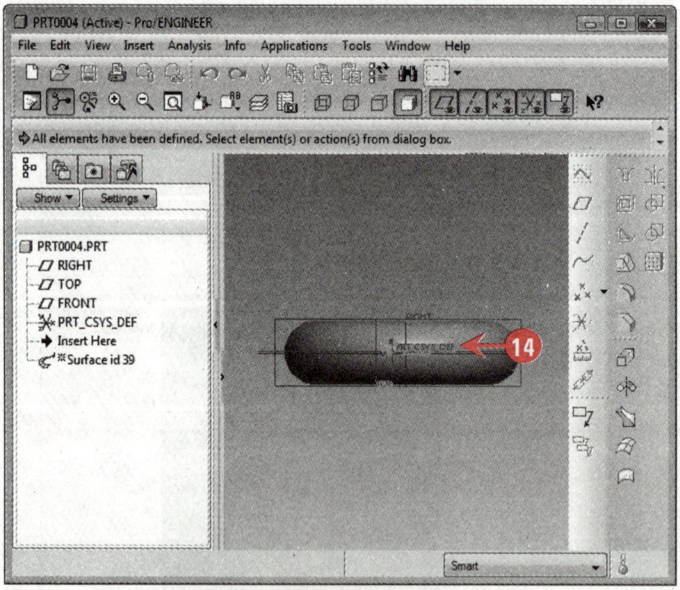

Fig.Pro/E-7.28

15. *Press* and *hold* the middle mouse button and *spin* the sweep surface to view it from different angles, as shown in Fig.Pro/E-7.29:

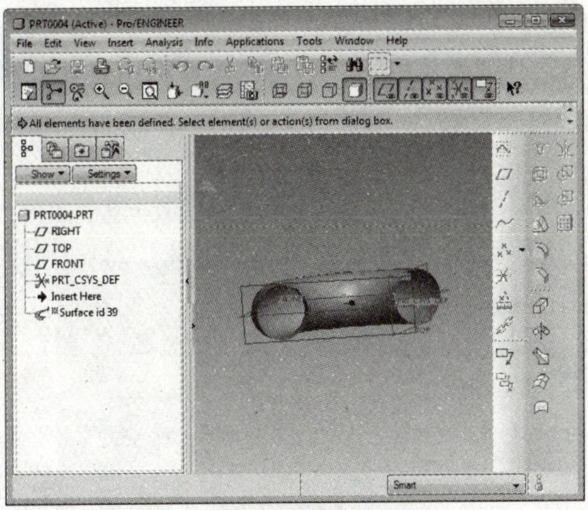

Fig.Pro/E-7.29

16. *Click* the OK button in the SURFACE: Sweep dialog box (Fig.Pro/E-7.13) to finish the process. The sweep surface is highlighted in red, as shown in Fig.Pro/E-7.30:

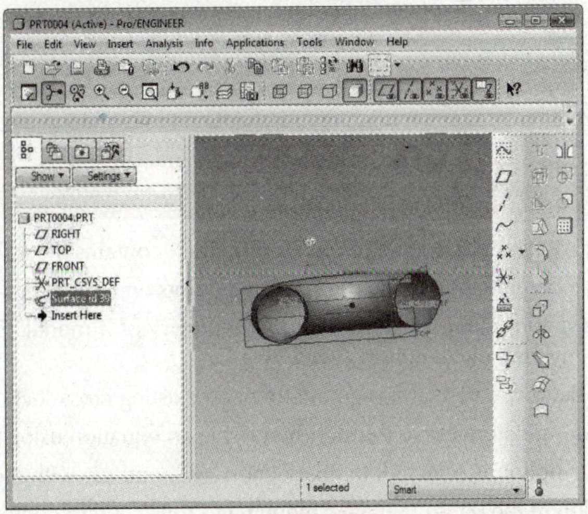

Fig.Pro/E-7.30

17. Now, *save* the file. Here, we have saved the file with name SWEEPSURFACE.prt.

Till now, we have studied how to create a sweep surface. Now, let's discuss how to create datum curves.

Creating Datum Curves

Datum curves are defined as the curves having no mass, volume, and thickness. Datum curves are used to create complex surfaces that are not easy to create with the datum planes. Once a datum curve is created, it appears in the Model Tree. A datum curve in `Pro/ENGINEER Wildfire 4.0` is created using the Curve (~) button in the Datum toolbar. A datum curve is created along with a surface; therefore, we use the extruded surface created in the `Creating an Extruded Surface` section to create a datum curve. Let's follow these steps to create a datum curve:

1. *Start* the Part mode.

2. *Open* the **EXTRUDEDSURFACE.prt** file. The **EXTRUDEDSURFACE.prt** file opens (Fig.Pro/E-7.31).
3. Now, *click* the **Curve** ( ) button in the **Datum** toolbar, as shown in Fig.Pro/E-7.31:

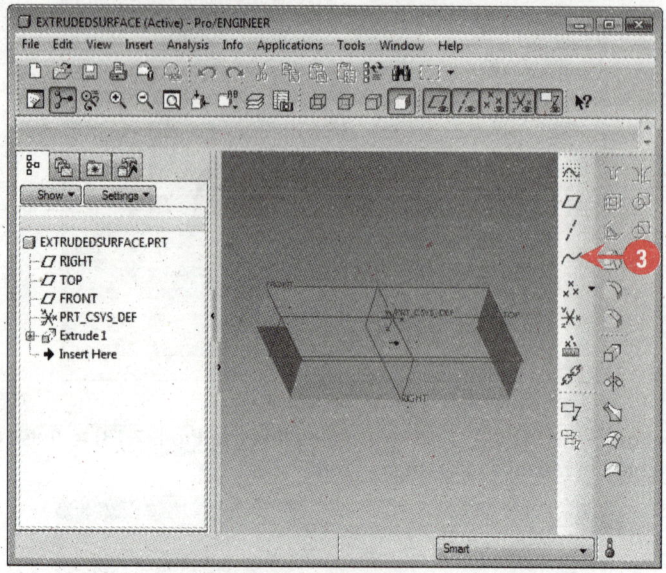

Fig.Pro/E-7.31

Note

The process to open a .prt file is already explained in Chapter 4, Exploring Pro/ENGINEER Wildfire 4.0 Part Mode.

The **Menu Manager** menu appears, as shown in Fig.Pro/E-7.32. It contains the following options:

- **Thru Points:** Creates a datum curve along the specified points on a surface.
- **From File:** Creates a datum curve by importing the geometry of a model. The imported geometry then automatically gets converted into a datum curve.
- **Use Xsec:** Creates a datum curve that has geometry of an existing cross section.
- **From Equation:** Creates a datum curve through specifying an equation using the co-ordinate systems.

By default, the **Thru Points** option is selected. In this section, we continue with the default option.

4. *Select* the **Done** option in the **Menu Manager** menu (Fig.Pro/E-7.32) to continue creating a datum curve:

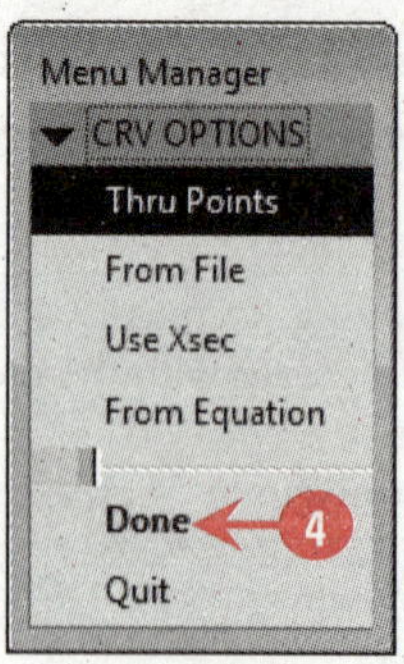

Fig.Pro/E-7.32

The CURVE: Through Points dialog box appears, as shown in Fig.Pro/E-7.33:

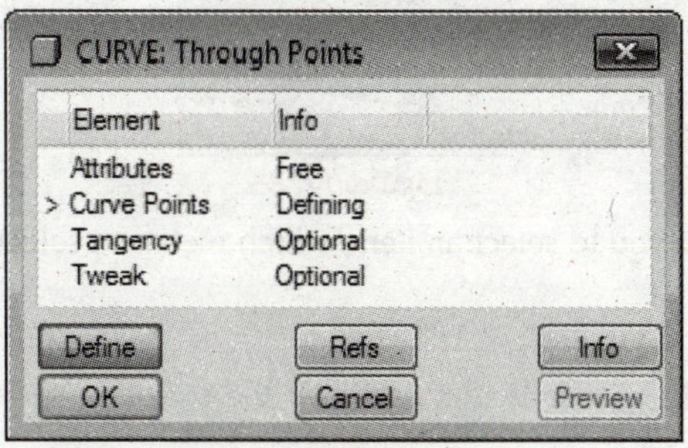

Fig.Pro/E-7.33

This dialog box allows you to select the points on a surface. Apart from this, the appearance of Menu Manager menu changes, as shown in Fig.Pro/E-7.34:

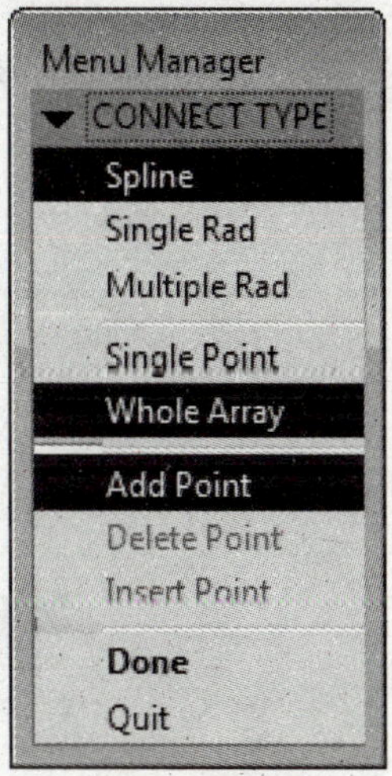

Fig.Pro/E-7.34

The changed Menu Manager menu (Fig.Pro/E-7.34) contains the following options according to which the shape of the curve is defined:

- **Spline:** Creates a datum curve in the form of a spline. This option is selected by default.
- **Single Rad:** Creates a datum curve having constant radius between two points. If there are more than two points, then the datum curve between all the points has the same radius.
- **Multiple Rad:** Creates a datum curve by specifying different radii between the points.
- **Single Point:** Selects a single point on a surface individually. By selecting the points individually, multiple datum curves can be created between the selected points.
- **Whole Array:** Selects all the points to create a single datum curve. This option is selected by default.

In this section, we continue with the default options.

Apart from the CURVE: Through Points dialog box, the *Select* dialog box also appears, as shown in Fig.Pro/E-7.35:

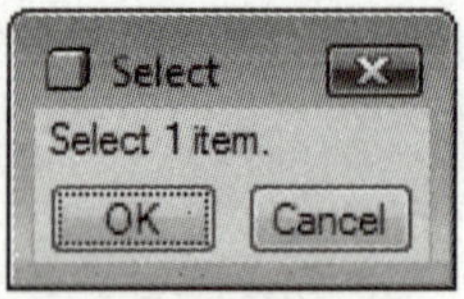

Fig.Pro/E-7.35

The *Select* dialog box prompts you to select an item, which means to select points on the surface to create a datum curve.

5. *Select* the extruded surface (Fig.Pro/E-7.31).
6. *Press* and *hold* the middle mouse button and *spin* the extruded surface in the way that it appears in Fig.Pro/E-7.36.
7. Now, *select* all the four vertices of the extruded surface, as shown in Fig.Pro/E-7.36:

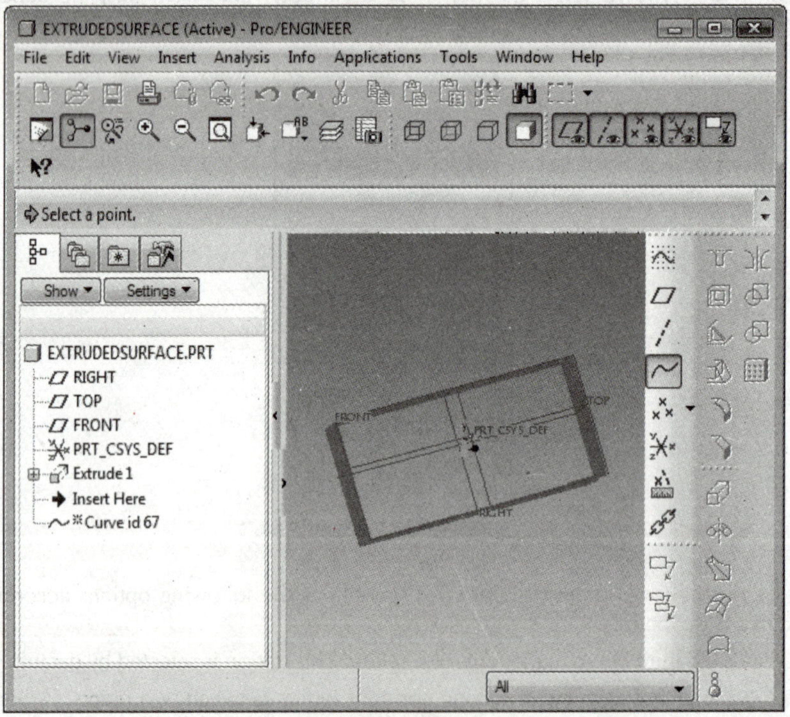

Fig.Pro/E-7.36

A datum curve is outlined, as shown in Fig.Pro/E-7.37:

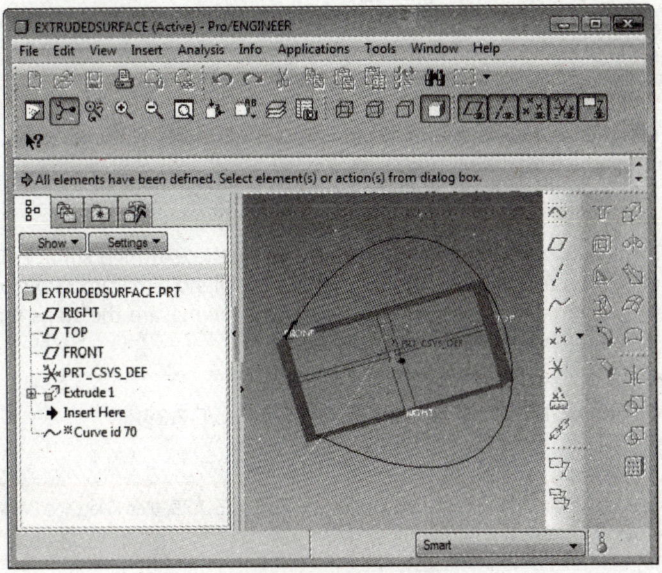

Fig.Pro/E-7.37

In Fig.Pro/E-7.37, the datum curve is outlined in blue color. A blue colored arrow attached with the datum curve indicates the start point of the datum curve.

8. *Select* the Done option in the Menu Manager menu (Fig.Pro/E-7.34).
9. Now, *click* the OK button in the CURVE: Through Points dialog box (Fig.Pro/E-7.33) to complete the process of creating the datum curve. The datum curve is highlighted in red color, as shown in Fig.Pro/E-7.38:

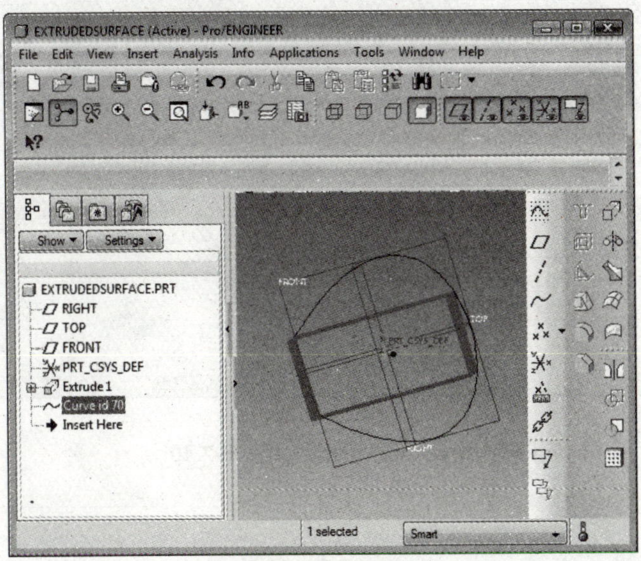

Fig.Pro/E-7.38

10. *Save* the file with the name CURVE.prt.

Note

> You can also create a datum curve along with a feature to create a complex feature.

With this, we have completed the process of creating a datum curve. Now, let's see how to create a surface by blending the datum curve with an existing surface.

Creating Surfaces from Boundary Curves

In the previous section, we have created a datum curve along the extruded surface. In this section, we study the process of blending the datum curve with a surface. To blend a datum curve with a surface to create another surface, the **Boundary Blend** dashboard is used. Following are the steps that help you to create a surface by blending a datum curve with another surface:

1. *Start* the **Part** mode.
2. *Open* the **CURVE**.prt file. The **CURVE.prt** file opens (Fig.Pro/E-7.39).

Note

> If the orientation of the file is not similar to the orientation, as shown in Fig.Pro/E-7.39, then select the BACK orientation available in the Named View List option in the View toolbar.

3. Now, *click* the **Boundary Blend** ( ) button in the **Base features** toolbar, as shown in Fig.Pro/E-7.39:

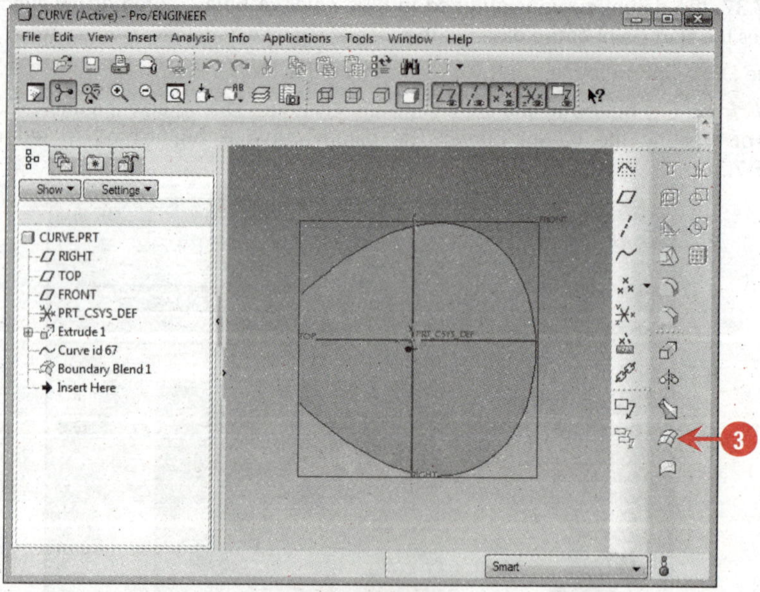

Fig.Pro/E-7.39

The **Boundary Blend** dashboard appears, as shown in Fig.Pro/E-7.40:

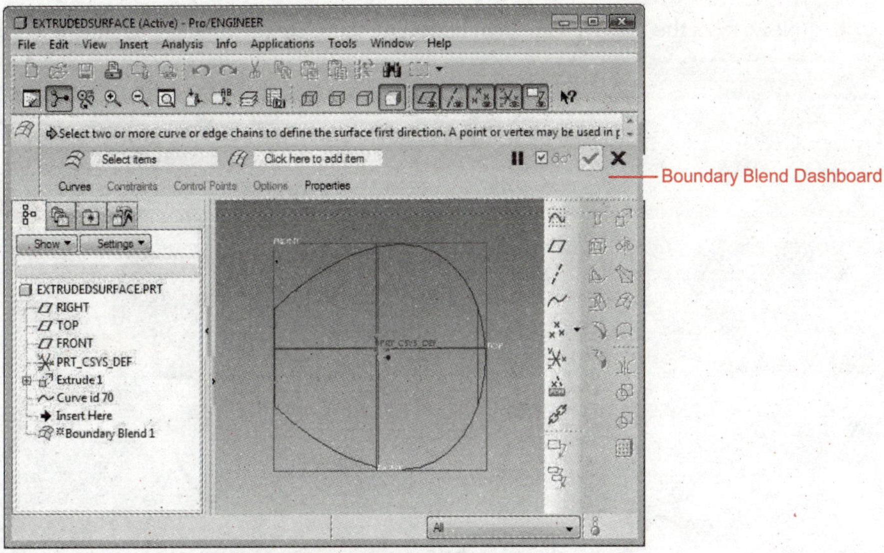

Fig.Pro/E-7.40

Now, you have to select the curve and the edges of the extruded surface to blend them.

4. *Select* an edge of the extruded surface, as shown in Fig.Pro/E-7.41:

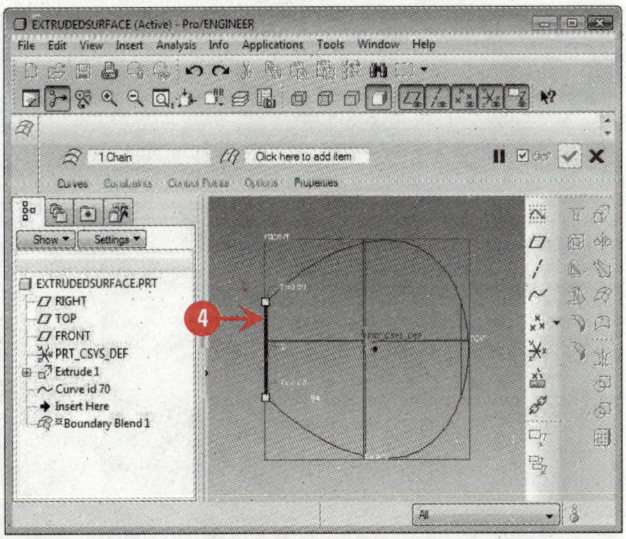

Fig.Pro/E-7.41

The selected edge is highlighted in red color. The selected edge also appears in the Boundary Blend dashboard in the left edit box. In the Boundary Blend dashboard, the selected edge is numbered as 1 Chain. Now, if you select more than one edge, then it will be successively numbered.

5. Now, *select* the datum curve while pressing the CTRL key. The surface that is created after blending is shown in Fig.Pro/E-7.42.

As soon as the curve is selected, it is highlighted in red color (Fig.Pro/E-7.42). The total number of selected items now becomes 2 Chains in the Boundary Blend dashboard.

6. Now, *click* the **Build feature** (☑) button to complete the process, as shown in Fig.Pro/E-7.42:

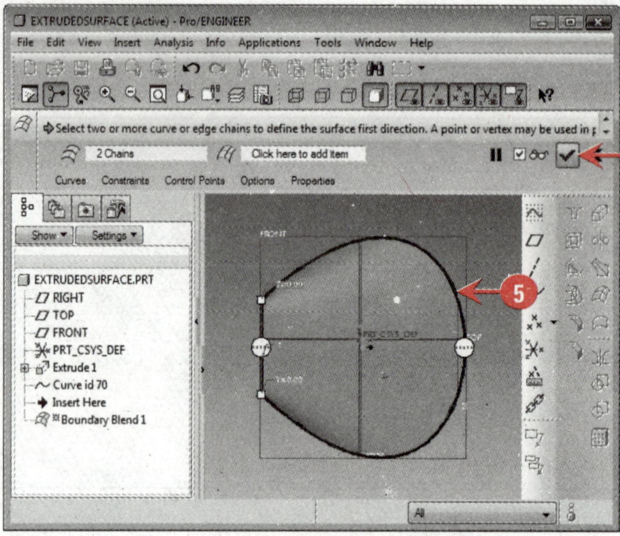

Fig.Pro/E-7.42

The new surface appears, as shown in Fig.Pro/E-7.43:

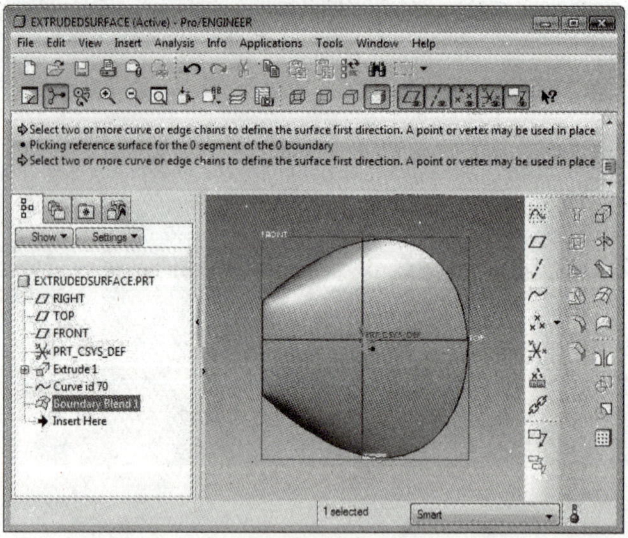

Fig.Pro/E-7.43

Note

Use the options available in the Named View List option in the View toolbar to view the surface in different orientations.

7. *Save* the file with the name **BLENDEDCURVE.prt**.

Till now, we have studied the process to create a surface by blending curves. Now, let's study how to merge surface.

Merging Surfaces

In Pro/ENGINEER Wildfire 4.0, merging is defined as the process to join or intersect two surfaces to create a single surface. The surfaces that have to be merged should be attached at their edges. In Pro/ENGINEER Wildfire 4.0, the **Merge** dashboard is used to merge two surfaces. The **Merge** dashboard is invoked by clicking the **Merge** () button in the **Edit Features** toolbar. Along with merging, the surfaces are also automatically trimmed. Let's follow these steps to merge two surfaces:

1. *Start* the **Part** mode.
2. *Open* the **EXTRUDEDSURFACE.prt** file (Fig.Pro/E-7.9). The **EXTRUDEDSURFACE.prt** file works as the first surface.
3. Now, *create* the second surface along with the extruded surface, as shown in Fig.Pro/E-7.44:

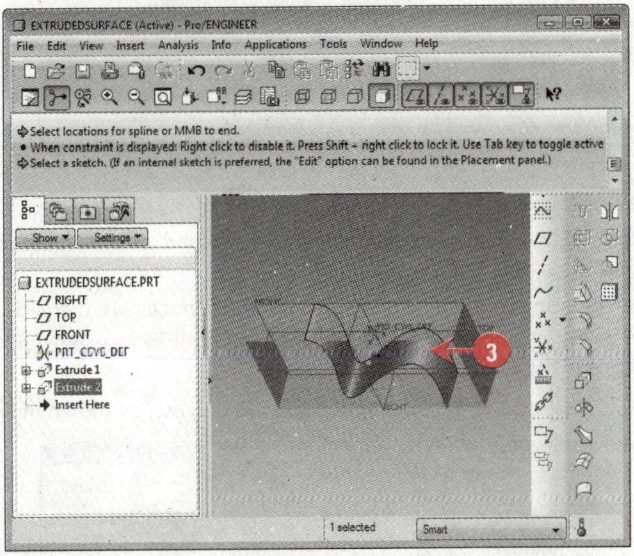

Fig.Pro/E-7.44

In Fig.Pro/E-7.44, the sketch of the second surface is created using the **Spline** button in the **Sketcher** tools toolbar.

Note

> The process to draw a sketch using the Spline button has already been discussed in Chapter 3, Exploring Pro/ENGINEER Wildfire 4.0 Sketch Mode.

4. Now, *hold* the **CTRL** key and *select* both the surfaces that are to be merged. Selected surfaces are highlighted in red. The **Merge** button also gets enabled.

Note

> By default, the Merge button is disabled. The Merge button is enabled only when the surfaces to be merged are selected.

5. *Click* the **Merge** button, as shown in Fig.Pro/E-7.45:

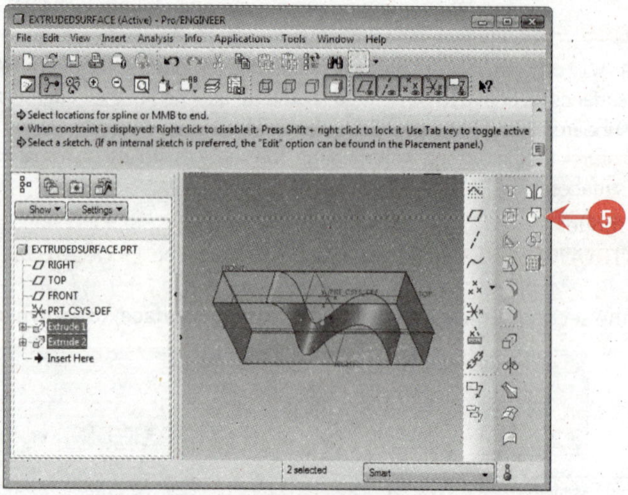

Fig.Pro/E-7.45

The **Merge** dashboard appears (Fig.Pro/E-7.46).

The yellow dots in Fig.Pro/E-7.46 represent the area that is retained after merging the surfaces. The area that is not covered with the yellow dots is trimmed.

The yellow arrows are pointing in the direction in which the surfaces are merged. You can also change the direction of these arrows by clicking the **Change side of first quilt to keep** ( ) and **Change side of second quilt to keep** ( ) buttons in the **Merge** dashboard.

6. Now, *click* the **Build feature** ( ) button to complete the process, as shown in Fig.Pro/E-7.46:

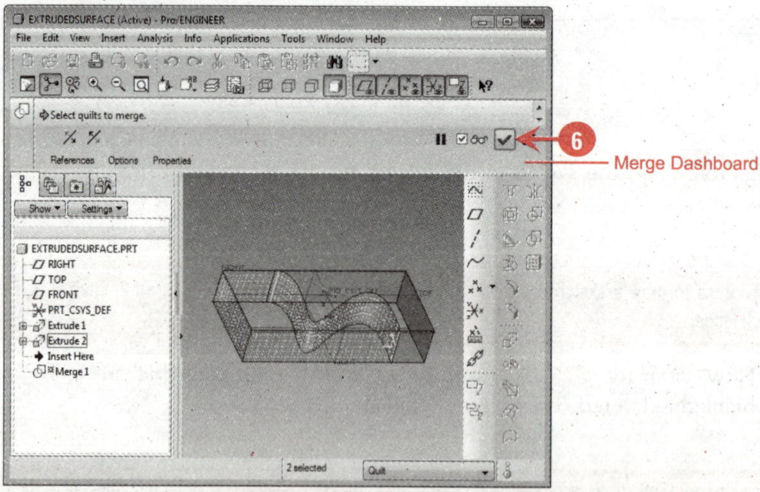

Fig.Pro/E-7.46

Note

> In Pro/ENGINEER Wildfire 4.0, you can merge surfaces, either by intersecting them or by joining them. Both the Intersect and Join options appear when the Options tab in the Merge dashboard is selected. By default, the Intersect option is selected.

In our case, we are continuing with the default option. The output is shown in Fig.Pro/E-7.47:

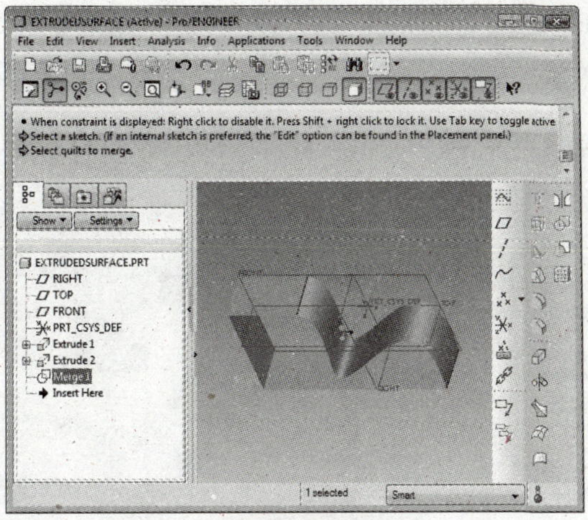

Fig.Pro/E-7.47

7. *Save* the file with the name MERGEDSURFACE.prt.

After discussing the merging process, let's now study how to trim surfaces.

Trimming Surfaces

Trimming in `Pro/ENGINEER Wildfire 4.0` is used to crop the surfaces. Trimming is done by using the Trim dashboard. However, the Trim dashboard is invoked by clicking the Trim button ( ) in the Edit Feature toolbar. Here, we use the sweep surface created in the section, `Creating a Sweep Surface` to trim. Let's now follow these steps to understand the process of trimming surfaces:

1. *Start* the Part mode.
2. *Open* the `SWEEPSURFACE.prt` (Fig.Pro/E-7.30) file to access the sweep surface created in the section, `Creating a Sweep Surface`.
3. Now, *select* the sweep surface that you want to trim (Fig.Pro/E-7.48).

Note

By default, the Trim button is disabled. The Trim button is enabled only when the surface to be trimmed is selected.

4. *Click* the Trim ( ) button in the Edit Feature toolbar, as shown in Fig.Pro/E-7.48:

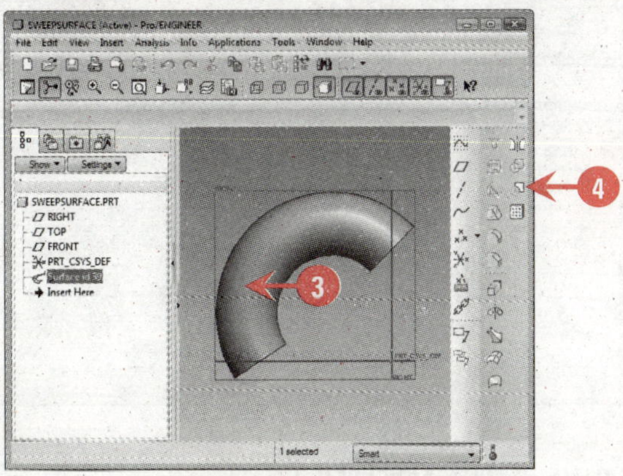

Fig.Pro/E-7.48

The Trim dashboard appears, as shown in Fig.Pro/E-7.49:

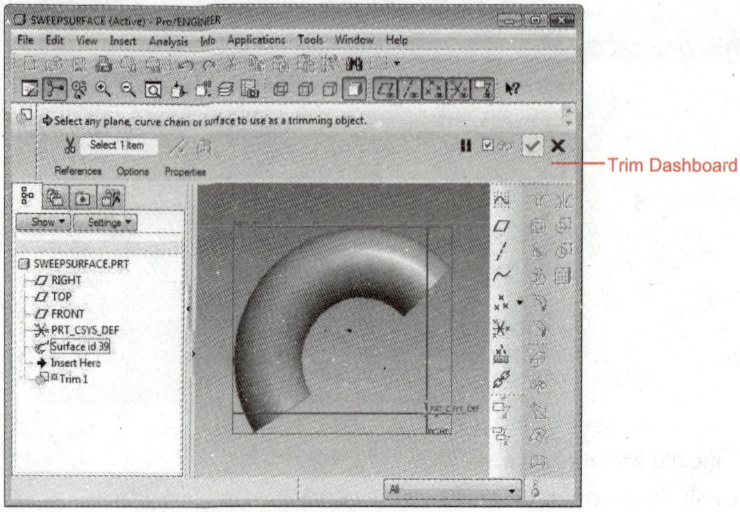

Fig.Pro/E-7.49

In Fig.Pro/E-7.49, you can see a statement written in the message area. This statement prompts you to select datum plane, curve chain, or surface to use as a trimming object.

5. Now, *select* the Top datum plane in the Model Tree (Fig.Pro/E-7.50).

The selected portion gets highlighted with yellow dots. These dots represent the area that will be retained after trimming. Besides the highlighted area, there is a yellow arrow that represents the direction in which the surface is trimmed. You can change the direction by clicking the Flip ( ) button in the Trim dashboard.

6. *Click* the Build feature ( ) button to complete the process, as shown in Fig.Pro/E-7.50:

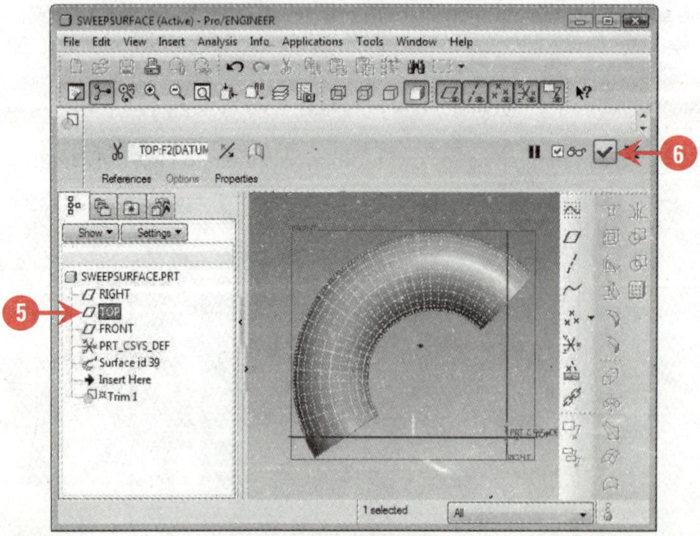

Fig.Pro/E-7.50

The trimmed surface appears, as shown in Fig.Pro/E-7.51:

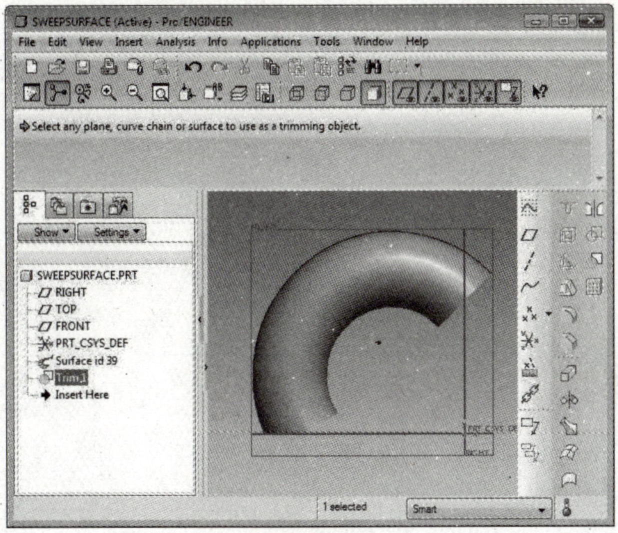

Fig.Pro/E-7.51

7. *Save* the file with the name TRIMMEDSURFACE.prt.

In this way, surfaces are trimmed in `Pro/ENGINEER Wildfire 4.0`. Let's now summarize the topics covered in this chapter.

Summary

In this chapter, you have learned about:

- Creating an extruded surface
- Creating a sweep surface
- Designing complex surfaces using datum curves
- Blending datum curves and surfaces
- Merging two surfaces to create a single complex surface
- Trimming a surface